John Francis Twisden

Elementary Introduction to Practical Mechanics

John Francis Twisden

Elementary Introduction to Practical Mechanics

ISBN/EAN: 9783337176860

Printed in Europe, USA, Canada, Australia, Japan

Cover: Foto ©ninafisch / pixelio.de

More available books at **www.hansebooks.com**

ELEMENTARY INTRODUCTION

TO

PRACTICAL MECHANICS

ILLUSTRATED BY NUMEROUS EXAMPLES

BEING THE EIGHTH EDITION OF

ELEMENTARY EXAMPLES IN PRACTICAL MECHANICS'

BY THE

REV. JOHN F. TWISDEN, M.A.

Professor of Mathematics in the Staff College; formerly Scholar of Trinity College, Cambridge; and Author of 'First Lessons in Theoretical Mechanics'

LONDON
LONGMANS, GREEN, AND CO.
1884

PREFACE
TO
THE SIXTH EDITION.

IN ISSUING the Sixth Edition of the present work, it may not be improper to mention that the Second Edition, published in 1863, was altered in several respects from the First Edition, published in 1860, and that each subsequent Edition has undergone a careful revision. In the present Edition some changes have been made in the technical terms employed, and two chapters have been rewritten: viz. Chapter IX., Part I., on the 'Deflection and Rupture of Beams;' and Chapter III., Part II., on 'Force and Motion.' At the end of the latter will be found the three laws of motion, as stated by Newton, together with his illustrations of them, translated from the Introduction to the 'Principia.' All the Examples have been worked through several times, and it may be presumed that the Answers are correct, with few exceptions. The unit of force in which they are expressed—with such exceptions as are apparent from the context—is the *gravitation unit*, the force of one pound, as defined on p. 48. What is meant by the absolute unit of force is explained in Chapter III. of Part II., and the subject is illustrated by some Examples. On the whole, it is hoped that the later Editions have been considerably improved.

J. F. T.

October 1, 1880.

PREFACE
TO
THE FIRST EDITION.

THE FOLLOWING TREATISE is designed to be an Introduction to the science of *Applied Mechanics*: in this it differs from all the elementary works commonly in use, which are introductory to *Rational Mechanics*. How great a difference is caused by this circumstance will appear from an inspection of the Contents; it may, however, be mentioned that, at the least, one-half of the present work has no counterpart in any *Elementary* Treatise that has fallen under the author's notice. That so great a divergence from the usual type should be possible seems sufficient reason for believing that something is wanting in the ordinary works; but how far the present will supply that want is, of course, another question. It was originally intended to be a book of Examples, and a supplement to others already in existence: it was, however, found that by a few additions it could be made independent, and it was thought that what was gained in point of convenience by completeness, would more than compensate a small increase of size and cost.

The work is intended to comprise two courses: the first is contained in Chapter I., the first section of Chapter II., and Chapter III. of Part I., and in Chapter I. of Part II.; the second forms the remainder of the book. The first course may be read by any one who understands arithmetic, a little algebra, practical geometry, and the rules of mensuration; in many of the Examples it is intended that a geometrical construction should take the place of calculation: instances of the use of construction are given in Examples 178, 216, &c. In this course the principles of the science are merely stated, their formal demonstration being reserved to the second course; in other words, the order most convenient for teaching and learning has been followed at some sacrifice of the systematic development of the subject. The second course presupposes that the reader is acquainted with Euclid, algebra, and trigonometry, as commonly taught in schools; a very few Examples are inserted which require some acquaintance with co-ordinate geometry and the differential calculus;* the reason for their insertion will generally be obvious from the context in which they occur. Frequent use has been made of simple geometrical limits; they will probably present but little difficulty to the reader: some remarks on the subject of limits will be found in the Appendix.

Very many Examples require numerical answers; it is hoped that but few of the arithmetical operations will prove laborious to any one who possesses a proper facility in manipulating numbers, and it must be remembered

---

* Most of these Examples are contained in Chap. IX., Part I.; the others are distinguished by an asterisk.

that few things are more important to a learner in the earlier stages of his progress than that he should be continually referred to the numerical results that follow from the formulæ he investigates. Hints and explanations have been freely given in connection with the more difficult Examples, and it is hoped they will be found sufficient to enable the reader to complete the solutions, though many of them are important mechanical theorems, and some of them but rarely to be met with (e.g. Examples 134, 149, 393, 429, 522, 553, 566, &c.).

A list is subjoined of the principal works referred to in drawing up the present Treatise; particular instances of obligation are acknowledged in the footnotes in the course of the work. A more explicit recognition of assistance is due to the Rev. H. Moseley, Canon of Bristol; about two hundred of the Examples were given by him to his classes at King's College, London, in the years 1840, 1, 2, 3; these he very kindly placed at the author's disposal, and also gave him permission to use freely his excellent Treatise on the 'Mechanical Principles of Engineering'—a permission of which great use has been made.

STAFF COLLEGE: *August* 1860.

WORKS REFERRED TO.

M. POISSON, Traité de Mécanique.
M. PONCELET, Introduction à la Mécanique industrielle.
M. PONCELET, Traité de Mécanique appliquée aux Machines.
M. MORIN, Aide-Mémoire de Mécanique pratique.
M. MORIN, Notions fondamentales de Mécanique pratique.
DR. T. YOUNG, Lectures on Natural Philosophy.
REV. W. WHEWELL, D.D., History of the Inductive Sciences.
REV. H. MOSELEY, Mechanical Principles of Engineering.
REV. R. WILLIS, Principles of Mechanism.
DR. RANKINE, Applied Mechanics.
SIR W. THOMSON and MR. P. G. TAIT, Natural Philosophy, Vol. I.

CONTENTS.

PART I.

×CHAPTER I.
                                                                     PAGE
Properties of Materials—Weight—Expansion by Heat—Elongation by Tension—Resistance to Rupture by Tearing and Crushing .   1

ᵧCHAPTER II.
×Sect. I. Efficiency of Agents—Modulus of Steam Engines—Waterwheels—Work of Animate Agents—Cost of Labour . . .   18
Sect. II. Work done by a Variable Force—Steam Indicator—Work of Raising Weights   36

ₓ CHAPTER III.
Statement of the Fundamental Principles of Statics—Parallelogram of Forces—Principle of Moments—Applications, viz.: the Lever —the Steel-yard—Equilibrium of Walls—Thrusts of Rafters .   47

CHAPTER IV.
The Fundamental Theorems of Statics   77

CHAPTER V.
Centre of Gravity—Geometrical Applications   112

CHAPTER VI.
Limiting Angle of Resistance—Laws of Friction—Inclined Plane— Wedge—Screw—Friction on a Pivot—Endless Screw . .   131

CONTENTS.

CHAPTER VII.
Equilibrium of Body resting on an Axle—Rigidity of Ropes—Pulley
—Capstan—Carriage-wheel 157

CHAPTER VIII.
The Stability of Walls—The Line of Resistance—Pressure of Water
—Pressure of Earth 173

CHAPTER IX.
Bending Moment—Deflection of Beams—Equation of three Moments
—Strength of Beams 184

CHAPTER X.
Virtual Velocities—Work done by a Force—Machines in a State of
Uniform Motion—Modulus of Machine—Toothed Wheels—
Epicycloid—Form of Teeth—Equilibrium of Toothed Wheels . 202

PART II.

CHAPTER I.
Velocity—Motion of Falling Bodies—Motion produced by a given
Force—Accumulated Work—Motion on a Curve—Centrifugal
Force—Oscillation of Simple Pendulum—Centre of Oscillation . 231

CHAPTER II.
Uniformly accelerated Motion—Composition of Velocities—Motion
of Projectiles 252

CHAPTER III.
Mass, Momentum—Absolute Unit of Force and of Work—Motion on
an Inclined Plane—Newton's Laws of Motion 265

CHAPTER IV.
Constrained Motion of a Particle—Centrifugal Force—Small Circular
Oscillations—Longitudinal Vibrations of a Rod. . . . 279

CONTENTS.

CHAPTER V.

	PAGE
Moment of Inertia—Radius of Gyration	289

CHAPTER VI.

Angular Velocity—Effective Force—D'Alembert's Principle—Pressure on a Fixed Axis of Rotation 297

CHAPTER VII.

Kinetic Energy of a Rotating Body—Smeaton's Machine—Atwood's Machine—Fly-wheel—M. Morin's Experiments on Friction—Compound Pendulum 306

CHAPTER VIII.

Impulsive Action—Impact of Smooth Balls—Impulse on Axis of Rotation—Centre of Percussion—Axis of Spontaneous Rotation—Ballistic Pendulum 321

APPENDIX.

On Limits 335

TABLES.

		Page
Table of Specific Gravity		2
„	Expansion produced by Heat	9
„	Moduli of Elasticity	11
„	Tenacity	12
„	Crushing Pressures	14
„	Moduli of Steam Engines	23
„	Moduli of Water-wheels	26
„	Work done by Living Agents	28
„	Horizontal Transport of Burdens	30
„	Cost of Labour	34
„	Coefficients of Friction	139
„	Friction of Axles	159
„	Rigidity of Ropes	164
„	Moduli of Rupture	200
„	Acceleration due to Gravity	250

PRACTICAL MECHANICS.

CHAPTER I.

ON SOME OF THE PHYSICAL PROPERTIES OF MATERIALS.

1. *Properties of Materials.*—The present chapter is intended to serve as an introduction to those that follow. It contains examples illustrative of the more obvious physical properties of the materials commonly used in construction and machinery. These physical properties are (1) Weight; (2) Expansion or Contraction, produced by change of temperature; (3) Elongation and Compression, produced by Tension or Pressure; (4) Resistance offered to Rupture by Tension; (5) Resistance offered to Rupture by Compression.

2. *Weight.*—For estimating the weight of bodies with sufficient accuracy it may be assumed that the weight of a cubic foot of water is 1000 oz. This number is easily remembered, and is within a very little of the truth. In every example contained in the following pages wherein the weight of bodies is concerned, it will be assumed that the weight of a cubic foot of water is 1000 oz., unless the contrary is specified. As a matter of fact, a cubic foot of pure water at 39° F. (when its density is greatest) weighs 998·8 oz. It may also be convenient for the reader to remember that a gallon contains 277·274 cubic inches,

and that a gallon of water at the standard temperature (62° F.) weighs 10 lbs.

Ex. 1.—A reservoir is internally 12 ft. long, 5 ft. wide, and 3 ft. deep: determine the weight of the water it contains when full, and the least error produced by considering that each cubic foot weighs 1000 oz.
Ans. Weight, 5 tons, 0 cwt. 50 lbs.
Error, 13½ lbs.

Ex. 2.—A cylindrical boiler terminated by plane ends, is internally 15 ft. long and 4 ft. in diameter; through the lower half pass lengthwise 50 fire-tubes, 3 in. in external diameter: determine the volume and weight of the water contained in it when the surface of the water passes through the centres of the ends. *Ans.* Vol. 57·43 cubic ft.
Weight, 1 ton, 12 cwts. 0 qr. 5·5 lbs.

Ex. 3.—The surface of a pond measures 10 acres; in the course of a period of dry weather the surface falls 1½ in. by evaporation: what is the weight of the water that has been withdrawn? *Ans.* 1520 tons, nearly.

3. *Specific Gravity.*—The specific gravity or specific density of a solid or liquid substance is the proportion which the weight of a certain volume of that substance bears to the weight of an equal volume of water; thus when it is stated that the specific gravity of cast iron is 7·2070, it means that a cubic foot, or a cubic inch, &c., of cast iron weighs 7·2070 times as much as a cubic foot, cubic inch, &c., of water; consequently a cubic foot of cast iron will weigh 7207 oz., and in general, if s is the specific gravity of a substance, a cubic foot of it will weigh 1000 s oz., at least with sufficient accuracy in almost all cases. The following table gives the specific gravities of some common materials:—

TABLE I.

SPECIFIC GRAVITIES.

Metals.

Platinum (laminated) .	22·0690	Brass (cast) . . .	8·3958
Pure Gold (hammered) .	19·3617	Steel (hard) . . .	7·8163
Gold 22 carat (do.) . .	17·5894	Iron (cast) . . .	7·2070
Mercury	13·5681	,, (wrought) . .	7·7880
Lead (cast) . . .	11·3523	Tin (cast) . . .	7·2914
Pure Silver (hammered) .	10·5107	Zinc (cast) . . .	7·1908
Copper (cast) . . .	8·7880		

SPECIFIC GRAVITY.

Stones and Earth.

Marble (white Italian)	2·638	Portland Stone	2·145
Slate (Westmoreland)	2·791	Coal (Newcastle)	1·2700
Granite (Aberdeen)	2·625	Brick (Red)	2·168
Paving Stone	2·4158	Clay	1·919
Mill Stone	2·4835	Sand (River)	1·886
Grindstone	2·1429	Chalk (mean)	2·315

Woods (Dry).

Elm	0·588	Oak (English)	0·934
Fir (Riga)	0·753	Teak (Indian)	0·657
Larch	0·522	Cork	0·2400
Mahogany (Spanish)	0·800		

1 foot length of Hempen rope weighs in lbs. $0·045 \times (circ.\ in\ inches)^2$.
1 ,,   ,,   Cable weighs in lbs. $0·027 \times (circ.\ in\ inches)^2$.
1 cubic foot of Brickwork weighs 112 lbs.

NOTE.—The above numbers, where printed to four places of decimals, are taken from Dr. Young's *Lectures on Natural Philosophy*, vol. ii. p. 503 ; where printed to three places of decimals, from Mr. Moseley's *Mechanics of Engineering*, 1st ed. p. 622. A definite specific gravity is assigned to each substance to prevent ambiguity in working the following examples. It will be remarked, however, that different specimens of the same substance have different specific gravities : thus of 16 specimens of cast iron the specific gravities have been found to vary from 7·295 to 6·963. The reader must, therefore, bear in mind that the numbers in the text give mean values from which the specific gravity of any specimen of a given substance will not largely vary—though the limits of variation are greater with some substances than with others. A similar remark applies to all quantities determined by experiment.

Ex. 4.—What is the weight of a rectangular block of marble 63 ft. long, and in section 12 ft. square ? *Ans.* Weight, 667 tons, 14 cwts. 3 qrs.

Ex. 5.—The girth of a tree is 3 ft. at top, 3 ft. 9 in. at bottom, it is 14 ft. long. Determine its weight according as it is larch, oak, or mahogany. Also, its value at the following prices: larch, 2*s.* 6*d.*; oak, 7*s.*; mahogany, 19*s.* per cubic foot rough.
Ans. Vol. 12·74 cubic ft.
Weight: Larch, 416 lbs.   Oak, 744 lbs.   Mah. 637 lbs.
Price: ,,   1*l.* 11*s.* 10*d.*   ,, 4*l.* 9*s.* 2*d.*   ,, 12*l.* 2*s.* 1*d.*
[The volume to be determined as that of the frustum of a cone.]

Ex. 6.—Find the weight of a rectangular mass of oak, 12 ft. long, 4 ft. broad, and 2½ feet thick. What would be the weight of a mass of granite of the same dimensions? *Ans.* Oak,   62 cwts. 2 qrs. 5 lbs,
Granite, 175 cwts. 3 qrs. 3½ lbs.

Ex. 7.—Find the separate weights of a cast-iron ball, 4 in. in radius, and of a copper cylinder 3 ft. long, the diameter of whose base is 1 in. Determine also the diminution in the weight of the ball if a hole were cut through it which the cylinder would exactly fit, the axis of the cylinder passing through the centre of the sphere. Also, find the error that results from considering the part cut away a perfect cylinder.

Ans. Weight of sphere, 1118·09 oz.
,,   cylinder, 143·8 oz.
,,   part cut from sphere, 26·204 oz.
Error, 0·102 oz.

Ex. 8.—If a 10-in. shell were of cast iron, and were 2 in. thick, what would be its weight supposing it complete? If the weight of a 10-in. shell were 86 lbs., what would be its thickness supposing it complete?

Ans. (1) 107 lbs.   (2) 1·41 in.

Ex. 9.—A hammer consists of a rectangular mass of wrought iron, 6 in. long, and 3 in. by 2 in. in section; its handle is of oak, and is a cylinder 3 ft. 6 in. long, on a base of 1 in. in radius. Determine its weight.

Ans. 12·83 lbs.

Ex. 10.—A pendulum consists of a cylindrical rod of steel 40 in. long, on a base whose diameter measures $\frac{1}{4}$ in.; to the end of this is screwed a steel cylinder $\frac{1}{2}$ in. thick, and $1\frac{1}{2}$ in. in radius, which fits accurately a hollow cylinder of glass, containing mercury 6 in. deep, the glass vessel weighing 3 oz. Determine the weight of the pendulum. *Ans.* 360·8 oz.

Ex. 11.—Determine the weight of a leaden cone whose height is 1 ft. and radius of base 6 in.; determine also the external radius of that hollow cast iron sphere which is 1 in. thick, and equals the cone in weight.

Ans. (1) 185·74 lbs.   (2) 8·02 in.

Ex. 12.—A rectangular mass of cast iron 6 ft. long, 6 in. wide, and 3 in. deep, has fitted square to its end a cube of the same materials whose edge is $1\frac{1}{4}$ ft. long; find its weight. *Ans.* 1858 lbs.

Ex. 13.—It is reckoned that a foot length of iron pipe weighs 64·4 lbs. when the diameter of the bore is 4 in. and the thickness of the metal $1\frac{1}{4}$ in.: what does this assume to be the specific gravity of iron? *Ans.* 7·197.

Ex. 14.—A cast-iron column 10 ft. high and 6 in. in diameter will safely support a weight of $17\frac{1}{2}$ tons, whether it be solid, or hollow and 1 in. thick; determine:—(1) the weight of a solid column; (2) the number of equally strong hollow columns that can be made out of 500 solid columns; (3) the price of 500 solid columns at 10s. per cwt. and of 500 hollow columns at 11s. 3d. per cwt.; (4) the cost of sending the 500 solid and the 500 hollow columns to a given place at the rate of 10s. 6d. per ton.

Ans. (1) 884·4 lbs.   (2) 900.   (3) 1974*l*. 3s. solid.   1236*l*. 16s. hollow.
(4) 103*l*. 13s. solid.   57*l*. 12s. hollow.

Ex. 15.—Determine the weight of a hollow leaden cylinder whose length is 3 in., internal radius 1½ in., and thickness 1½ in. *Ans.* 26·121 lbs.

Ex. 16.—Determine the weight of a grindstone 4 ft. in diameter and 8 in. thick, fitted with a wrought-iron axis of which the part within the stone is 2 in. square, and the projecting parts each 4 in. long with a section 2 in. in diameter. *Ans.* 1135 lbs.

Ex. 17.—Determine the weight of an oak door 7 ft. high, 3 ft. wide, and 1½ in. thick. *Ans.* 153¼ lbs.

Ex. 18.—There is a fly wheel of cast iron the external radius of whose rim is 5 ft. and internal radius 4 ft. 6 in.; it is 4 in. thick, and is connected with the centre by 8 spokes 4 in. wide and 1 in. thick, strengthened by a flange on each side 1 in. square (so that their section is a cross 4 in. long and 3 in. wide), each spoke is 4 ft. long; the centre to which they join the rim has the same thickness as the rim, is solid, and (of course) 6 in. in radius : determine the weight of the whole. *Ans.* 2959 lbs.

Ex. 19.—There are two rooms each 100 ft. long and 30 ft. wide; the one is floored with oak planking 1¼ in. thick; the other with deal planking (Riga fir) 1¼ in. thick. Determine the weights of the floors and their cost, the price of deal being 3s. and oak 7s. per cubic foot.
  Ans. Deal floor weighs 17,648 lbs. costs 56l. 5s.
    Oak   „   18,242 lbs. „ 109l. 7s. 6d.

Ex. 20.—A cubic foot of copper is drawn into wire $\frac{1}{16}$ of an inch in diameter; what length of wire is made? *Ans.* 46,936 ft.

Ex. 21.—It is said that gold can be drawn into wire one millionth part of an inch thick; what will be the length of such a wire that can be made from an ounce of pure gold? *Ans.* 1,793,448 miles.

Ex. 22.—It is said that silver leaf can be made $\frac{1}{150000}$ of an inch thick; how many ounces of silver would be required to make an acre of such silver leaf? *Ans.* 254·3 oz.

4. *Brickwork*.—The measurement and determination of the weight of brickwork depend upon the following data :—

(1) A rod of brickwork has a surface of 1 square rod (or 30¼ square yards) and a thickness of a brick and a half, i.e. of 1 ft. 1½ in., or it contains 306 cubic feet.

(2) A rod of brickwork contains about 4500 bricks in mortar, or 5000 bricks laid dry.

(3) A rod of brickwork requires 3½ loads (i.e. 3½ cubic yards) of sand and 18 bushels of stone lime.

(4) A brick measures $8\frac{3}{4} \times 4\frac{1}{4} \times 2\frac{3}{4}$ inches, i.e. a quarter of an inch each way less than $9 \times 4\frac{1}{2} \times 3$ inches.

(5) A bricklayer's hod measures $16 \times 9 \times 9$ inches, and can contain 20 bricks. Labourers, however, commonly put 10 or 12 bricks into it.*

Ex. 23.—How many rods of brickwork are there in a square tower 117 ft. high and 28 ft. by 7 ft. at its base, externally, and 3 bricks thick? Determine the number of bricks required to build the tower and their price at $1l.$ 10s. per thousand.

Ans. (1) 52·43 rods. (2) 236,000 bricks. (3) 354l.

Ex. 24.—A tower the base of which measures externally 9 ft. square is 50 ft. high and 2 bricks thick; how many bricks are required to build it, and how many loads of sand and bushels of lime? Determine also the cost of the materials if the bricks cost 1l. 10s. per thousand, sand 5s. 4d. per load, and lime 1s. 8d. per bushel.

Ans. (1) 7·35 rods. (2) 33,000 bricks, 25$\frac{3}{4}$ loads of sand, 132$\frac{1}{2}$ bushels of lime. (3) Cost 67l. 8s. 2d.

Ex. 25.—How many rods of brickwork are there in a reservoir of a rectangular form, the internal measurements of which are 20 ft. long, 6 ft. wide, and 12 ft. deep; the work being 2 bricks thick, viz. both walls and floor; and the reservoir being open at the top?   *Ans.* 4·43.

Ex. 26.—Find how many rods of brickwork there are in a wall 360 ft. long, 17 ft. high, and 2 bricks thick; and determine the cost of the material from the data in *Ex.* 24.   *Ans.* (1) 30 rods. (2) 275l. 10s.

Ex. 27.—If the wall in the last example had an additional 2 ft. of foundation 3 bricks thick, and were supported by 20 square buttresses reaching to the top of the wall 2 bricks thick, on foundations 3 bricks thick, and measuring 2$\frac{1}{2}$ ft. in a direction perpendicular to the face of the wall; determine the number of rods of brickwork in the foundations and buttresses.

Ans. 10·2 rods.

Ex. 28.—What would be the cost of the carriage of the bricks in the wall described in the last two examples at 5s. 6d. per thousand?

Ans. 49l. 15s.

Ex. 29.—The following are the actual dimensions of the brickwork of the outer shell of the chimney of St Rollox, Glasgow. Commencing from the top, there are five divisions; the tops of these divisions are respectively 435$\frac{1}{2}$, 350$\frac{1}{2}$, 210$\frac{1}{2}$, 114$\frac{1}{2}$, 54$\frac{1}{2}$ ft. above the ground; the external diameters at the *tops* of the divisions are respectively 13 ft. 6 in., 16 ft. 9 in., 24 ft., 30 ft. 6 in., 35 ft. The diameter on the ground is 40 ft.; the thicknesses of

---

* Weale's *Contractor's Price Book* for 1859, p. 280.

EXPANSION AND CONTRACTION BY HEAT. 7

the divisions are respectively 1½, 2, 2½, 3, and 3½ bricks; below ground the brickwork reaches 14 ft., with a uniform external diameter of 40 ft.; the first 8 feet are 3 ft. thick; in the remaining 6 feet the thickness gradually increases to 12 ft. Determine the number of rods of brickwork contained in the chimney; the number of bricks employed, their cost at 1*l*. 11*s*. 3*d*. per thousand; also, if the mortar were of sand and stone lime, determine the number of loads of sand and bushels of stone lime required, and their cost at 5*s*. 4*d*. per load, and 1*s*. 8*d*. per bushel respectively.

[The surface of each division of the chimney may be considered as that of a conic frustum; the real volume of each division will be the difference between the volumes of two conic frustums. A sufficiently close approximation may be obtained by multiplying the mean surface by the thickness and considering the slant side equal to the height; the volume of the part below ground is to be determined accurately.]

Ans. (1) 218 rods, or 981,000 bricks. (2) Cost of bricks, 1532*l*. 16*s*. 3*d*. (3) 763 loads of sand, costing 203*l*. 9*s*. 4*d*. (4) 3924 bushels of lime, costing 327*l*.

5. *Expansion and contraction by heat.*—It is found that all bodies experience a small change of volume on the application of heat. In general, the change is one of increase,* and with sufficient accuracy may be considered to obey the following law within moderate ranges of temperature. If a volume v be increased by k v when its temperature is raised one degree, it will be increased by $n \times k$ v when the temperature is raised n degrees, i.e. the increase of volume is proportional to the increase of temperature. The same rule holds for the expansions in *length*, which a body experiences from an increase of temperature. In order to fix the conception of a degree of temperature (with sufficient accuracy for our present purpose), it will be proper to mention that when heat is applied to ice the water produced by melting retains a constant temperature until the whole of the ice is melted. This temperature serves as one fixed point, and is called the freezing point. Moreover, boiling water in free contact with the air also keeps at a constant temperature (at least when the baro-

* Water, near freezing point, is a conspicuous exception.

meter stands at a given height). This fact, therefore, supplies a second fixed point, and is called the boiling point, viz., when the barometer stands at thirty inches. These two points being fixed, the graduation is arbitrary. The scale of Fahrenheit's thermometer (which is commonly used in England) is constructed by dividing the space between the freezing and boiling points into 180 equal parts, termed degrees, and by commencing the graduation 32° below freezing point, so that the freezing point is marked 32°, and the boiling point 212°. In the centigrade thermometer (now commonly used in scientific investigation) the graduation begins at the freezing point and the interval between the freezing and boiling points is divided into 100 equal parts called degrees.* It is easy to see that if at any temperature Fahrenheit's thermometer stood at F° and the centigrade at C°, we should have

$$\frac{F°-32}{180} = \frac{C°}{100}$$

Ex. 30.—The density of water is greatest at 3°·9 on the centigrade scale; what is the same temperature called on Fahrenheit's scale? *Ans.* 39°·02 F.

Ex. 31.—The standard temperature not unfrequently referred to in English experiments is 60°F.; what would the same temperature be called on the centigrade scale? *Ans.* 15°·55 C.

Ex. 32.—If the centigrade thermometer stood at 5° below zero, or at −5° C, what would the same temperature be marked on Fahrenheit's scale? *Ans.* 23° F.

Ex. 33.—What degree on the centigrade scale would be the equivalent to −4° on Fahrenheit's scale? *Ans.* −20° C.

The following table gives the fractional part of the whole by which substances expand when heated: †—

---

* In Réaumur's thermometer the freezing point is marked zero, and the boiling point 80°: consequently $\frac{F°-32}{180} = \frac{R°}{80}$.

† From Dr. Young's *Natural Philosophy*, vol. ii. p. 390.

EXPANSION PRODUCED BY HEAT.

TABLE II.

EXPANSION PRODUCED BY HEAT.

	Temperature raised from 32° to 212° F.	Temperature raised 1° F.	Authority
In length : Glass Tube	0·00077615	0·00000431	Roy
,, Platinum	0·000856	0·00000476	Borda
,, Cast Iron	0·0011094	0·00000617	Roy
,, Wrought Iron	0·001156	0·00000642	Borda
,, Steel rods	0·0011447	0·00000636	Roy
,, Brass rods	0·0018928	0·00001052	Roy
,, Lead	0·002867	0·00001592	Smeaton
,, Copper	0·001700	0·00000944	Smeaton
In volume : Mercury	. .	0·00010415	Roy
,, ,, in glass (apparent)	. .	0·00008696	Committee of Royal Society

Ex. 34.—The length of the base line of the Ordnance Survey on Hounslow Heath was found to be 27,404 ft.; this was measured first by glass tubes, and then by steel chains; if, in correcting the glass tubes for temperature, a uniform error of 1° in excess had been committed, and in correcting the steel chain an error of 1° in defect had been committed, what would have been the difference between the apparent measurements?
Ans. 3·51 in.

Ex. 35.—If the wrought-iron rails on a railway are 10 miles long when at a temperature of 32° below freezing, by how much will they lengthen if their temperature is raised to 88° F.? *Ans.* 29·83 ft.

Ex. 36.—Ramsden's brass yard exceeded Shuckburgh's by 0·002505 of an inch; what would be the difference of their temperatures when accurately the same length? *Ans.* 6°·6 F.

Ex. 37.—Two rods, respectively of iron and brass, A B and C D, are fastened together in the middle; they are accurately the same length, at 62° F.; to their ends are fastened by pivots tongues C A E and D B F which are perpendicular to the bars, at 62° F.; in consequence of the unequal expansion or contraction of the bars the tongues will assume different positions, as shown by the dotted lines; it is required to determine the length of C E, that the point E may remain unmoved by the expansion or contraction of the bar. The length of A B is 10 ft. and the distance A C is 1·725 in. *Ans.* C E = 4·426 in.

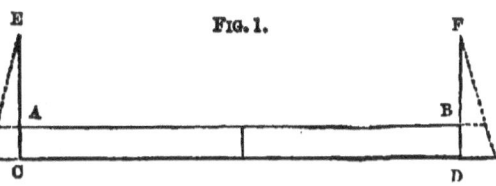

FIG. 1.

Ex. 38.—If the expansion in length of a substance is *e* times the length at a given temperature, show that the expansion in volume will be very nearly 3 *e* times the volume at that temperature.

Ex. 39.—The volume of a mass of lead being a cubic foot at 60° F., what will be its volume at 0° F.? and what at 88° F.?

Ans. At 0° F. 0·997134 cubic ft.
At 88° F. 1·00133728 cubic ft.

Ex. 40.—There is half a cubic inch of mercury in a thermometer at 32° F.; when the temperature is raised to 92° F. the mercury ascends 4 in.; what is the diameter of the bore of the glass tube? *Ans.* 0·0288 in.

6. *Elongation produced by tension.*—The principle on which this determination is made is the following:— Suppose the length of a beam or bar to be L feet, the area of its section to be K square inches, then if by the application of a tension of P lbs. its length becomes L + *l*, it appears from experiment that

$$l : L :: \frac{P}{K} : E$$

where E is a constant number depending on the nature of the material, and called the Modulus of Elasticity.

It is found that all substances obey this law when the degree of extension does not exceed certain limits; the limits are different in different substances, and in many are very narrow. It appears also that within these limits (i.e. the limits of elasticity) a tension producing a certain degree of extension will, if applied in the opposite direction so as to become a pressure, produce an equal degree of compression.

It will be observed that $\frac{P}{K}$ is the tension or pressure per square inch of the section of the beam or bar. It is also plain that if $\frac{P}{K}$ were equal to E then would *l* be equal to L, so that the modulus of elasticity is that tension per square inch of the section of a bar which would double its length if its elasticity continued perfect. It is, perhaps, unnecessary to remark that no solid substance has limits of elasticity any way approaching this in extent.

ELONGATION PRODUCED BY TENSION

TABLE III.
MODULI OF ELASTICITY.*

Material	Modulus	Material	Modulus
Wrought Iron bars	29,000,000	Oak (English)	1,450,000
Cast Iron	17,000,000	Larch	1,050,000
„ Brass	8,930,000	Fir (Riga)	1,330,000
Steel (hard)	29,000,000	Elm	700,000
Copper wire	17,000,000		

Ex. 41.—By how much would a bar of wrought iron $\frac{1}{4}$ of an inch square and 100 ft. long lengthen under a tension of 2 tons (neglecting the weight of the bar)? *Ans.* 0·247 ft.

Ex. 42.—Determine the elongation of a steel bar 2 in. square and 40 ft. long when subjected to a tension of 40 tons. What would have been its elongation had it been of cast brass? *Ans.* Steel 0·03 ft. Brass 0·1 ft.

Ex. 43.—A bar of wrought iron 2 in. square has its ends fixed between two immovable blocks when the temperature is 20° F.; what pressure will it exert against them if the temperature becomes 96° F.? *Ans.* 25¼ tons.

Ex. 44.—A wall of brickwork 2 ft. thick and 12 ft. high is supported by columns of oak 6 in. in radius, 18 ft. high and 14 ft. apart from centre to centre; determine the pressure per square inch of the section of the columns, and the amount of their compression.
Ans. (1) 332·7 lbs. (2) $\frac{1}{20}$ in. nearly.

Ex. 45.—In the last example if the wall had been of Portland stone and 1¼ ft. thick, what would have been the pressure per square inch and the degree of compression? *Ans.* (1) 248·9 lbs. (2) $\frac{3}{80}$ in.

Ex. 46.—In the last example if the oak columns were replaced by wrought-iron bars 2 in. square what would be the degree of compression? and at what temperature would one of the iron bars have the same length as it has when unpressed at 32° F.? *Ans.* (1) $\frac{21}{400}$ in. (2) 69·8° F.

Ex. 47.—A bar of wrought iron a square inch in section is fixed firmly between two *immovable* blocks which are 50 ft. apart; if the temperature is raised 50° F. above that which the bar had when fixed, find the pressure produced against these blocks. *Ans.* 9309 lbs.

Ex. 48.—In the last example, if only one of the blocks were immovable and the other were capable of revolving round a joint 12 ft. below the point at which it is met by the rod, determine the angle through which it will be turned by the expansion of the rod. *Ans.* 0° 4' 36".

---

* Based on Mr. Moseley's *Mech. Eng.* p. 622, compared with Mr. Rankine's *Applied Mechanics*, p. 631.

Ex. 49.—It is observed that two opposite walls of an ancient building are each 3° out of the vertical, the inclination being outward; to bring them into the perpendicular, the following means are employed; at certain intervals iron bars are placed across the building, their ends passing through the walls and projecting on the outside, on these ends strong plates or washers are screwed; the rods are then heated and expand; in this state the washers are screwed tightly against the outside of the walls and the rods allowed to cool, when they contract and draw the walls together; the process being continued until the walls become vertical.* If we suppose the rods to be 50 ft. long and 3 square inches in section, and to be fastened 15 ft. above the joint of the masonry, round which walls will be made to turn; and if the range of temperature is from 60° F. to 240° F.; determine the number of times the bars must be heated before the operation is complete, and the tension which would tend to draw the walls together if they were entirely immovable. *Ans.* (1) 27 times. (2) 100,572 lbs.

7. Resistance to rupture by tearing or tenacity.—

When the tension which elongates a bar attains a certain magnitude, the bar will break. If we determine by experiment this tension in lbs. per square inch, we obtain the *tenacity* of the substance. It is manifest that the tension which will tear a bar whose section is n square inches will be n times the tenacity.

TABLE IV.
TENACITIES.

Material	Tenacity	Material	Tenacity
Wrought Iron (bars) . . .	67,200 lbs.	Oak (English) .	17,300 lbs.
		Larch . . .	10,000 „
Cast Iron (average)	16,500 „	Fir (Riga) .	12,000 „
Iron wire ropes	90,000 „	Elm . . .	13,500 „
Cast Brass . .	18,000 „	Hempen ropes .	5,600 „
Copper wire .	60,000 „		

Ex. 50.—How great a tension will a cylindrical bar of wrought iron bear which is ¼ of an inch in diameter? and by what fraction of its length would it lengthen under this tension if the elasticity continued perfect?
Ans. (1) 3298 lbs. (2) 0·0023.

Ex. 51.—How many iron wires $\frac{1}{10}$ of an inch in diameter must be put together to sustain a weight of 3 tons? *Ans.* 13.

---

* The walls of Armagh Cathedral were restored to a vertical position by this process. Daniell's *Chemistry*, p. 103.

RUPTURE BY PRESSURE. 13

Ex. 52.—What is the length of a bar of wrought iron which, being suspended vertically, would break by its own weight?   *Ans.* 19,880 ft.

Ex. 53.—What tension will a bar of oak 1½ in. square sustain?
Ans. 38,925 lbs.

Ex. 54.—What tension will a cylindrical bar of larch 1½ in. in diameter sustain?   *Ans.* 17,671 lbs.

Ex. 55.—If a rope be made of wires whose diameter is d, show that the number of wires in each square inch of the section of the rope is very nearly given by the formula $\dfrac{2}{\sqrt{3}\,d^2}$ or $\dfrac{8}{7\,d^2}$.

Ex. 56.—How many wires $\frac{1}{10}$ of an inch in diameter must be put together to form a rope a square inch in section ?   *Ans.* 115.

Ex. 57.—If the number of wires $\frac{1}{20}$ of an inch in diameter which must be put together to form a rope one square inch in section be determined by each of the formulæ in *Ex.* 55, what is the difference between the results?
Ans. 4·8.

Ex. 58.—Show that the number of lbs. weight in a foot length of a rope made of iron wire is given by the formula (circ. in inches)² × 0·244 very nearly; the specific gravity of iron wire being assumed to be the same as that of wrought iron.

Ex. 59.—Show that if a rope of hemp and a rope of iron wire have the same strength, the circumference of the latter is about ¼ of the circumference, and its weight about ⅓ of the weight of the former.

8. *Resistance to rupture by pressure.*—There are as many as five forms which the results of crushing assume in different bodies. They are enumerated as follows by Mr. Rankine : *—

(1) *Crushing by splitting*, when the substance divides in a direction nearly parallel to the direction of the pressure. This occurs in the case of hard homogeneous substances of a glassy texture.

(2) *Crushing by shearing*, when the substance divides along a plane inclined at a certain angle to the direction of the force, the upper part of the substance sliding upon the lower. This fact was ascertained, and its conditions investigated, by Mr. Hodgkinson. It takes place in the case of substances of a granular texture, such as cast iron,

* *Applied Mechanics*, p. 303. See also Mr. Moseley's *Mechanics of Engineering*, pp. 549, 579.

and most kinds of stone and brick. To exhibit its effects the height of the block to be crushed must be at the least one and a half times its thickness. In the above cases the resistance to crushing is considerably greater than the tenacity. In the case of cast iron the resistance is more than *six* times the tenacity.

(3) *Crushing by bulging*, when the material spreads like compressed dough. This takes place with ductile substances, such as wrought iron in short blocks. In this case the resistance is somewhat less than the tenacity, being in wrought iron about $\frac{2}{3}$ of the tenacity.

(4) *Crushing by crippling*, which is characteristic of fibrous substances, and takes place when the thrust acts along the fibres in timbers and in bars of wrought iron that are too long to yield by bulging. It consists in a lateral yielding, and sometimes separation of the fibres. In the case of dry timber the resistance is about $\frac{1}{2}$ of the tenacity, in the case of moist timber about $\frac{1}{4}$ of the tenacity; consequently moist timber is only half as strong as dry when subjected to a crushing force.

(5) *Crushing by crossbreaking*, which is the mode of fracture in columns and struts where the length greatly exceeds the diameter. Under the breaking load they yield sideways, and are broken across like beams under a transverse pressure.

TABLE V.

CRUSHING PRESSURE IN LBS. PER SQUARE INCH.

Material	Pressure	Material	Pressure
Wrought Iron	36,000	Granite (average)	8,000
Cast Iron (average)	112,000	Oak (English) dry	9,500
„ Brass	10,300	Larch dry	5,500
Brick	800	Fir (Riga) dry	6,000
Sandstone	4,000	Elm	10,300
Limestone (granular)	4,000		

WORKING STRESS. 15

Ex. 60.—What must be the height of a column of cast iron producing that pressure per square inch which would crush a small column of the same material? *Ans.* 35,805 ft.

Ex. 61.—Compare the heights of columns of cast iron, wrought iron, cast brass, and larch fir, which would produce the pressure per square inch requisite for crushing short columns of their respective materials?
Ans. 1·475 : 0·439 : 0·116 : 1.

9. *Ultimate and proof strength and working stress.*— It must be borne in mind that no material is in practice subjected to the stress which it is capable of supporting. This will appear very clearly from the following definitions : *—

(1) *The ultimate strength* of a solid is the stress required to produce fracture in some specified way.

(2) *The proof strength* is the stress required to produce the greatest strain in some specified way consistent with safety. A stress exceeding the proof strength, though it does not produce immediate fracture, will produce it by long application or frequent repetition.

(3) *The working stress* is always made less than the *proof strength* in a certain ratio determined by experience.

In cases of wrought-iron boilers, timber, brick, and stone, the *ultimate strength* is from 2 to 3 times the *proof strength*, and from 8 to 10 times the *working stress*. In the following examples the *working stress* is assumed to be $\frac{1}{10}$th of the ultimate strength :—

Ex. 62.—A wall of brickwork 3 ft. thick is supported at intervals of 10 ft. by sandstone columns 9 in. in diameter; to what height can the wall be carried? *Ans.* 7·6 ft.

Ex. 63.—If in the last example the columns had been of brickwork 2 ft. thick, to what height could the work then be carried? *Ans.* 10·8 ft.

Ex. 64.—To what height could the wall in *Ex.* 44 be carried with safety so far as the strength of the columns is concerned? *Ans.* 34·26 ft.

Ex. 65.—Make the same determination with regard to *Ex.* 45.
Ans. 45·8 ft.

---

* Rankine, *Applied Mechanics*, p. 273.

Ex. 66.—What would have been the heights in each of the last examples if the columns had been of brickwork? What if of limestone? What if of granite? *Ans.* Brickwork, 2·9 ft. 3·9 ft.
Limestone, 14·4 ft. 19·3 ft.
Granite, 28·9 ft. 38·6 ft.

Ex. 67.—A wall of brickwork, 50 ft. high and 3 ft. thick, is to be carried by columns of brickwork 20 ft. apart, from centre to centre; determine the least diameter consistent with safety. Make the same determination if the columns were of granite. *Ans.* 73⅛ in. brickwork. 23⅛ in. granite.

10. *Strength of cast-iron columns.*—The columns in the preceding examples are supposed to follow the law of the crushing of short columns. It may be instructive to add the following particulars, which have reference to the crushing of cast-iron columns exceeding that length. The greatest part of our knowledge of this subject is due to experiments conducted by Mr. Hodgkinson, who thus states his conclusions with regard to the form of the ends of iron columns:—' 1st. A long circular pillar, with its ends flat, is about three times as strong as a pillar of the same length and diameter with its ends rounded in such a manner that the pressure would pass through the axis. . . . . 2nd. If a pillar of the same length and diameter as the preceding has one end rounded and one flat, the strength will be twice as great as that of one with both ends rounded. 3rd. If, therefore, three pillars be taken, differing only in the forms of their ends, the first having both ends rounded, the second having one end rounded and one flat, and the third both ends flat, the strength of these pillars will be as 1—2—3 nearly.' Mr. Hodgkinson further considers that the breaking weight w of a hollow column is given in tons by the formula

$$w = M \times \frac{D^{3.5} - d^{3.5}}{l^{1.63}}$$

and that of a solid column by the formula

$$w = m \times \frac{D^{3.5}}{l^{1.63}}$$

where M and m are constants depending on the nature of the iron, D the external and d the internal diameters of the column in inches, and l the length in feet. The values of M and m vary considerably with different kinds of iron, but may be taken at 42 tons. The limits of variation in the values of m are 49·94 and 39·60.*

Ex. 68.—Determine the breaking weight of a solid cast-iron column 20 ft. high and 6 in. in diameter. *Ans.* 168·3 tons.

Ex. 69.—Determine the breaking weight of the column in the last example if it were hollow and 1 in. thick. *Ans.* 127·6 tons.

Ex. 70.—Determine the thickness of a column 20 ft. high and 7 in. in external diameter, which is as strong as that in *Ex.* 68. *Ans.* 0·774 in.

* *Proceedings of the Royal Society,* vol. viii. p. 318.

CHAPTER II.

ON WORK; OR, THE EFFICIENCY OF AGENTS.

11. *Definition of work*.—An agent is said to do work when it causes the point of application of the force it exerts to move through a certain distance; thus a carpenter employed in planing wood *works*, since he causes the point of application of the force he exerts to move through a certain distance, and the same is true of any agent that works in the sense here intended. For the sake of distinctness it may be observed that the union of *force* and *motion* is essential to the conception of *work*; thus when the expansive force of steam lifts the piston of a steam engine it does work. In the boiler, though it produces an enormous pressure, it does no work, since the pressure is unaccompanied by motion. The unit by which the work of different agents is expressed numerically according to the practice of English writers is called a foot-pound; it is defined as follows:—

Def.—The work done when the force of 1 lb. is exerted through a distance of 1 ft. *in the direction of the force* is a foot-pound.

The following important principle is closely connected with this definition. When a force of P lbs. is exerted through a distance of S ft., it does P S foot-pounds, the force being exerted along the line in which its point of application is made to move. For since a foot-pound is done when a force of 1 lb. is exerted through 1 ft., there must be 2 foot-pounds done when a force of 2 lbs. is exerted through 1 ft., 3 foot-pounds when a force of 3 lbs. is exerted through 1 ft., and generally P foot-pounds when a force of P lbs. is

COMPARISON OF THE EFFICIENCY OF AGENTS. 19

exerted through 1 ft. Again, since P foot-pounds are done when a force of P lbs. is exerted through 1 ft., there must be 2 P done when it is exerted through 2 ft., 3 P when it is exerted through 3 ft., and generally P S foot-pounds must be done when the force of P lbs. is exerted through S ft.

Ex. 71.—How many foot-pounds are expended in raising 2 cwts. through 30 fathoms? *Ans.* 40,320.

Ex. 72.—The mean pressure on the piston of a steam engine is 15 lbs. per sq. in., the length of the stroke is 6 ft.; if the area of the piston is 448 sq. in., how many foot-pounds are done per stroke? *Ans.* 40,320.

12. *Comparison of the efficiency of agents.*—If the above examples are compared, it will be seen that the work done during each stroke by the steam on the piston of the engine is equivalent to the work expended in raising 2 cwts. through a height of 30 fathoms; and whatever agent raises this weight must do as much work as that done by the steam. In these examples we have not considered the *time* in which the work is done; let us then suppose that the engine in Ex. 72 makes 10 strokes per minute; the expansive force of the steam will then do 403,200 foot-pounds per minute. Now, if we suppose an agent, or a number of agents, to raise a weight of 1 ton through 30 fathoms in one minute, they will do exactly 2240×180 or 403,200 foot-pounds per minute. It is plain that under these circumstances the comparison is complete between the efficiency of the expansive force of the steam and the efficiency of the other agents, and that they are reciprocally equivalent. Hence we infer the general principle—

The number of foot-pounds of work yielded by any agent in a given time is a true measure of its efficiency or working power, i.e. of its rate of doing work.

Of course it follows from this principle that the working powers of two agents or their rates of doing work are in the ratio of the number of foot-pounds done by them in the same time.

20   PRACTICAL MECHANICS.

The most familiar instance of this mode of measuring the power of an agent is furnished by the steam engine, whose efficiency is estimated in horse-power, as when we speak of an engine of 'twenty horse-power.' From some experiments, Mr. Watt concluded that a horse is capable of yielding 33,000 foot-pounds per minute. The conclusion, as far as regards the efficiency of the animal, is not very correct; it has, however, fixed the meaning of the term horse-power when applied to a steam engine. Hence

Def.—*A steam engine works with one horse-power when it yields* 33,000 *foot-pounds per minute.*

Of course an engine of n horse-power yields n times 33,000 foot-pounds per minute.

Ex. 73.—The piston of a steam engine is 15 in. in diameter, its stroke is $2\frac{1}{2}$ ft. long; it makes 40 strokes per minute; the mean pressure of the steam on it is 15 lbs. per square inch; what number of foot-pounds is done by the steam per minute, and what is the horse-power of the engine?
Ans. 265,072 ft.-pds.   8·03 H.-P.

Ex. 74.—A weight of $1\frac{1}{2}$ tons is to be raised from a depth of 50 fathoms in 1 minute; determine the horse-power of the engine capable of doing the work.   *Ans.* 30 $\frac{8}{11}$ H.-P.

Ex. 75.—The resistance to the motion of a certain body is 440 lbs; how many foot-pounds must be expended in making this body move over 30 miles in one hour? What must be the horse-power of an engine that does the same number of foot-pounds in the same time?
Ans. 69,696,000 ft.-pds. $35\frac{1}{5}$ H.-P.

13. *Application of the foregoing principles.*—A considerable number of practical questions can be answered by means of the principles already laid down, viz. such questions as the horse-power of the engine required to do a certain amount of work, the time in which an engine of a certain power will do a certain amount of work, &c. . . . . They are all done by following the same method, viz. First, from a consideration of the work to be done, obtain the number of foot-pounds that must be expended in a certain time. Next from a consideration of the power of the agent

EXAMPLES OF THE WORK OF STEAM. 21

obtain the number of foot-pounds yielded in the same time. One of these expressions will contain an unknown quantity, but, since by the terms of the question they are equal, they will form an equation from which the unknown quantity can be readily determined.

Ex. 76.—An engine is required to raise a weight of 13 cwts. from a depth of 140 fathoms in 3 minutes; determine its horse-power.

Let x be the required horse-power; then the number of foot-pounds yielded in 3 minutes will equal $33,000 \times x \times 3$; also the number of foot-pounds required to raise 13 cwts. from a depth of 140 fath. equals $13 \times 112 \times 140 \times 6$. And since these two numbers are equal we have

$$33,000 \times 3 \times x = 13 \times 112 \times 140 \times 6.$$
$$\therefore x = 12\cdot35 \text{ H.-P.}$$

Ex. 77.—In how many minutes would an engine working at 25 horse-power raise a load of 12 cwts. from a depth of 160 fathoms?
Ans. 1·564 min.

Ex. 78.—A locomotive engine draws a gross load of 60 tons at the rate of 20 miles an hour; the resistances are at the rate of 8 lbs. per ton; what must be the horse-power of the engine?

[The reader must bear in mind that the work to be done is to overcome a resistance of 480 lbs. through 20 miles in one hour.] *Ans.* 25·6 H.-P.

Ex. 79.—What must be the horse-power of an engine that raises 20 cubic feet of water per minute from a depth of 200 fathoms? *Ans.* $45 \tfrac{5}{11}$ H.-P.

Ex. 80.—How many cubic feet of water would an engine working at 100 horse-power raise per minute from a depth of 25 fathoms? *Ans.* 352.

Ex. 81.—How many cubic feet of water will an engine of 250 horse-power raise per minute from a depth of 200 fathoms? *Ans.* 110 cub. ft.

Ex. 82.—It being required to raise 100 cubic feet of water per minute from a depth of 495 ft., what must be the horse-power of the engine?
Ans. $93\tfrac{3}{4}$ H.-P.

Ex. 83.—There is a mine with three shafts which are respectively 300, 450, and 500 ft. deep: it is required to raise from the first 80, from the second 60, from the third 40 cubic feet of water per minute; what must be the horse-power of the engine? *Ans.* $134\tfrac{31}{88}$ H.-P.

Ex. 84.—At what rate per hour will a locomotive engine of 30 horse-power draw a train weighing 90 tons gross, the resistances being 8 lbs. per ton? *Ans.* 15·625 miles.

Ex. 85.—What is the gross weight of a train which an engine of 25 horse-power will draw at the rate of 25 miles an hour, resistances being 8 lbs. per ton? *Ans.* 46·875 tons.

Ex. 86.—A train whose gross weight is 80 tons travels at the rate of 20 miles an hour; if the resistance is 8 lbs. per ton, what is the horse-power of the engine? *Ans.* $34\tfrac{2}{15}$ H.-P.

Ex. 87.—An engine working with the same power as that in the last example draws a train at the rate of 30 miles an hour; the resistances being 7 lbs. per ton, what is the gross weight of the train?
Ans. $60\frac{20}{21}$ tons.

Ex. 88.—What must be the length of the stroke of the piston of an engine, the surface of which is 1500 square inches, which makes 20 strokes per minute, so that with a mean pressure of 12 lbs. on each square inch of the piston, the engine may be of 80 horse-power? *Ans.* $7\frac{1}{2}$ ft.

Ex. 89.—The diameter of the piston of an engine is 80 in., the length of the stroke is 10 ft., it makes 11 strokes per minute, and the mean pressure of the steam on the piston is 12 lbs. per square inch: what is the horse-power? *Ans.* 201·06 H.-P.

Ex. 90.—Find the horse-power of an engine that will raise in one minute 100 cubic feet of water from a depth of 600 feet. *Ans.* $113\frac{7}{11}$ H.-P.

Ex. 91.—A train weighing 50 tons is drawn along a railway at the rate of 20 miles an hour; the resistances being 8 lbs. per ton, find the horse-power of the engine. *Ans.* $21\frac{1}{3}$ H.-P.

Ex. 92.—The cylinder of a steam engine has an internal diameter of 3 ft.; the length of the stroke is 6 ft.; it makes 6 strokes per minute; under what effective pressure per square inch would it have to work in order that 75 horse-power may be done on the piston? *Ans.* 67·54 lbs.

Ex. 93.—What must be the horse-power of a stationary engine that draws a weight of 150 tons along a horizontal road at the rate of 30 miles per hour, friction being 8 lbs. per ton? *Ans.* 96 H.-P.

14. *Modulus of a machine.*—An agent rarely, if ever, does a considerable amount of useful work *directly*, but nearly always through the intervention of a machine, by which the motive power of the agent is so applied as to overcome the resistance in the most convenient manner. For instance, when a steam engine raises water out of a shaft, the motive power is the pressure of the steam on the piston, the resistance to be overcome is the weight of the water, the beam, crank, &c., of the engine are the means by which the motive power is applied so as to overcome the resistance. Now it will be remarked that each part of the machine offers more or less resistance to the motion, so that a certain part of the work done by the motive power must be expended in overcoming these resistances, i.e. in reference to the purpose of the

MODULI OF STEAM ENGINES.

machine, must be expended *uselessly*. The remainder of the work done by the motive power will be expended *usefully* in accomplishing that purpose.

If the number of foot-pounds done by the agent is represented by U, the number expended in overcoming prejudicial resistances by U_0, and the number expended usefully by U_1, all in the same given time, then it admits of proof in the case of a machine moving uniformly, that

$$U = U_0 + U_1.$$

It also appears that in most machines U_1 bears to U a constant ratio, so that

$$U_1 = K\, U$$

where the letter K denotes some proper fraction, depending on the nature of the machine; this fraction is called the modulus of the machine; the following table, taken from General Morin's *Aide-Mémoire de Mécanique Pratique*, gives the value of K for different classes of steam engines:—

TABLE VI.

MODULI OF STEAM ENGINES.

Description of Machine	Horse-power	Value of K	
		Best working	Ordinary do.
Watt's low-pressure engine	4 to 8	0·50	0·42
	10 „ 20	0·56	0·47
	30 „ 100	0·60	0·54
Cornish engines, working by expansion and condensation	up to 30	0·44	0·35
	30 „ 40	0·49	0·39
	40 „ 50	0·57	0·46
	50 „ 60	0·62	0·50
	60 „ 70	0·66	0·53
	70 „ 80	0·82	0·66
	80 „ 100	0·70	0·59
High-pressure engines, working without expansion or condensation	up to 10	0·50	0·40
	10 „ 20	0·55	0·44
	20 „ 30	0·60	0·48
	30 „ 40	0·65	0·52
	above 40	0·70	0·56

24   PRACTICAL MECHANICS.

Ex. 94.—The diameter of the piston of a steam engine is 60 in.; it makes 11 strokes per minute; the length of each stroke is 8 ft.; the mean pressure per square in. 15 lbs. The modulus of the engine being 0·65, determine the number of cubic feet of water it will raise per hour from a depth of 50 fathoms.

[The number of foot-pounds done by steam on piston in one hour equals $\pi \times 30^2 \times 8 \times 15 \times 11 \times 60$; this number multiplied by 0·65 will give the number of foot-pounds usefully spent in raising water; hence the number of cubic feet of water is found.]   *Ans.* 7763 cub. ft.

Ex. 95.—The diameter of the piston of an engine is 80 in., the mean pressure of the steam is 12 lbs. per square inch, the length of the stroke is 10 ft., the number of strokes made per minute is 11. How many cubic feet of water will it raise per minute from a depth of 250 fathoms, its modulus being 0·6?   *Ans.* 42·46 cub. ft.

Ex. 96.—If the engine in the last example had raised 55 cubic feet of water per minute from a depth of 250 fathoms, what would have been its modulus?   *Ans.* 0·7771.

Ex. 97.—How many strokes per minute must the engine in *Ex.* 95 make in order to raise 15 cubic feet of water per minute from the given depth?
      Ans. 4.

Ex. 98.—What must be the length of the stroke of an engine whose modulus is 0·65, and whose other dimensions and conditions of working are the same as in *Ex.* 95, if they both do the same quantity of useful work?
      Ans. 9·23 ft.

Ex. 99.—The diameter of the cylinder of an engine is 80 inches, the piston makes per minute 8 strokes of $10\frac{1}{4}$ ft. under a mean pressure of 15 lbs. per square inch; the modulus of the engine is 0·55. How many cubic feet of water will it raise from a depth of 112 ft. in one minute?
      Ans. 485·78 cub. ft.

Ex. 100.—If in the last example the engine raised a weight of 66,433 lbs. through 90 ft. in one minute, what must be the mean pressure per square inch on the piston?   *Ans.* 26·37 lbs.

Ex. 101.—If the diameter of the piston of the engine in *Ex.* 99 had been 85 in., what addition in horse-power would that make to the *useful* power of the engine?   *Ans.* 13·28 H.-P.

15. Work of water-wheels.

—Hitherto we have considered only one kind of motive power, viz. the pressure of steam. The same principles are applicable to machines worked by any other motive power, as by the muscular force of animal agents, the pressure of moving air, or of falling water. The last of these, viz. the power of falling water, is, next to steam, the most conspicuous example of

WORK OF WATER-WHEELS

work done on a large scale by an inanimate agent. We shall therefore consider somewhat particularly the application of this power by means of water-wheels.

It is plain that 1 lb. of water, in descending through 1 foot, must accumulate as much work as would be required to raise it through 1 foot, and hence if P lbs. of water descend through h feet, they will accumulate P h foot-pound of work; and if, moreover, we suppose this water to descend against an obstacle, such as the float boards of a water-wheel, the amount of work so accumulated will be done upon the wheel, and this work may then be applied to any useful purpose after a certain deduction has been made on account of prejudicial resistances.

It must be borne in mind that the height of the fall is the difference between the levels of the surface of the water

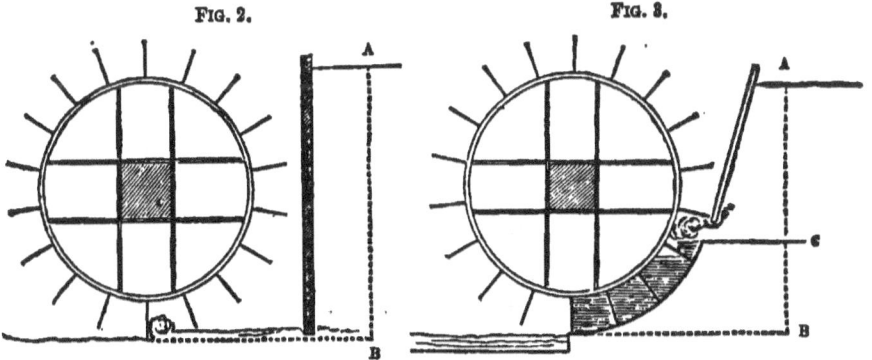

FIG. 2.   FIG. 3.

in the reservoir and in the exit canal or tail-race; in the case of overshot wheels it is supposed that the extreme circumference of the wheel is just in contact with the surface of the water in the tail-race. The height is represented by A B in the accompanying figures; of which fig. 2 represents the ordinary undershot wheel with plane float boards; fig. 3 the breast wheel, in which the water acts upon the float boards considerably above the level of the tail-race. Fig. 4 represents the overshot wheel.

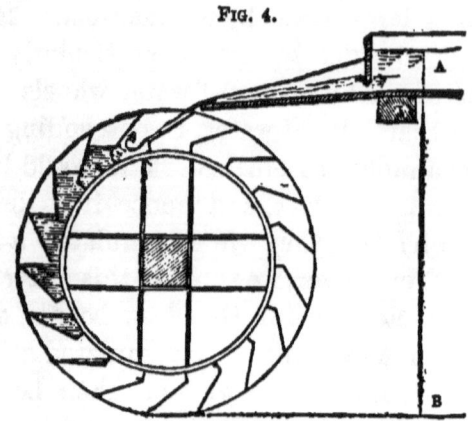

Fig. 4.

The following table exhibits the moduli of various kinds of water-wheels. It is founded on results given in General Morin's *Aide-Mémoire*. In the table H denotes the length of the line A B in figs. 2, 3, 4, and h denotes the length of B C in fig. 3 :—

TABLE VII.
MODULI OF WATER-WHEELS.

Description	Modulus
(1) Undershot wheels, with flat float boards . .	0·25 to 0·30
(2) Breast wheels with flat float boards	
(a) when $\frac{h}{H} = \frac{1}{4}$	0·40 to 0·45
(b) „ $\frac{h}{H} = \frac{2}{5}$	0·42 „ 0·49
(c) „ $\frac{h}{H} = \frac{2}{3}$	0·47
(d) „ $\frac{h}{H} = \frac{3}{4}$	0·55
(e) „ $\frac{h}{H} = 1$	0·65 „ 0·70
(3) Breast wheels with curved float boards (Poncelet's construction) . . . . for H greater than 6½ feet	0·60 to 0·65
(4) Overshot wheels, when the velocity is small and the buckets half filled	0·70 to 0·75

WORK OF WATER-WHEELS. 27

Ex. 102.—The mean section of a stream is 5 ft. by 2 ft.; its mean velocity is 35 ft. per minute; there is a fall of 13 ft. on this stream, at which is erected a water-wheel whose modulus is 0·65; determine the horse-power of the wheel. *Ans.* 5·6 H.-P.

Ex. 103.—In how many hours would the wheel in the last example grind 1000 quarters of wheat, it being assumed that each horse-power will grind 1 bushel per hour? *Ans.* 1428 hours.

Ex. 104.—How many quarters of wheat will the same wheel grind in 72 hours? *Ans.* 50·41 quarters.

Ex. 105.—Suppose the wheel in *Ex.* 102 to have replaced an undershot wheel with flat float boards, whose modulus was 0·25, determine the number of quarters of wheat each wheel will grind in 24 hours.
Ans. (1) 6·5. (2) 16·8.

Ex. 106.—How many cubic feet of water must be made to descend the fall per minute in *Ex.* 102, 3, that the wheel may grind at the rate of $3\frac{1}{2}$ quarters per hour? *Ans.* 1749·5.

Ex. 107.—Given the stream in *Ex.* 102, 3, what must be the height of the fall to grind $1\frac{1}{4}$ quarters per hour; first, if the modulus of the wheel is 0·40, next, if it is 0·47, and lastly, if it is 0·65?
Ans. (1) 37·7 ft. (2) 32 ft. (3) 23·2 ft.

Ex. 108.—The mean section of a stream is 8 ft. by 1 ft.; its mean velocity is 40 ft. per minute; it has a fall of $17\frac{1}{2}$ ft.; it is required to raise water to a height of 300 ft. by means of a water-wheel whose modulus is 0·7; how many cubic feet will it raise per minute? *Ans.* 13·07 cub. ft.

Ex. 109.—To what height would the wheel in the last example raise $2\frac{1}{4}$ cubic feet of water per minute? *Ans.* $1742\frac{2}{9}$ ft.

Ex. 110.—The mean section of a stream is $1\frac{1}{2}$ ft. by 11 ft.; its mean velocity is $2\frac{1}{2}$ miles per hour; there is on it a fall of 6 ft. on which is erected a wheel whose modulus is 0·7; this wheel is employed to raise the hammers of a forge, each of which weighs 2 tons, and has a lift of $1\frac{1}{4}$ ft.; how many lifts of a hammer will the wheel yield per minute? *Ans.* 142 nearly..

Ex. 111.—In the last example determine the mean depth of the stream if the wheel yields 135 lifts per minute. *Ans.* 1·43 ft.

Ex. 112.—In *Ex.* 110 how many cubic feet of water must descend the fall per minute to yield 97 lifts of the hammer per minute? *Ans.* 2483 cub. ft.

Ex. 113.—Determine how many quarters of corn the mill in *Ex.* 110 might be made to grind in six days if it were to work for 13 hours daily.
Ans. 281·5 quarters.

Ex. 114.—Down a 14-ft. fall 200 cub. ft. of water descend every minute, and turn a wheel whose modulus is 0·6. The wheel lifts water from the bottom of the fall to a height of 54 ft.; how many cubic feet will be thus raised per minute? If the water were raised from the top of the fall to the same point, what would the number of cubic feet then be?
Ans. (1) 31·1 cub. ft. (2) 34·7 cub. ft.

[Of course in the second case the number of cubic feet of water taken from the top of the fall being x, the number of feet that turn the wheel will be $200-x$.]

Ex. 115.—Water has to be raised from a mine 120 ft. deep, the whole of the water raised forms a stream with a fall of 30 ft., the machinery by which the water is raised is worked by a steam engine of 50 horse-power, and an overshot wheel whose modulus is 0·715 turned by the steam; determine the whole number of cubic feet raised per minute.   *Ans.* 267·8 cub. ft.

Ex. 116.—In the last example, if the ground allowed an exit to be made for the water 30 ft. below the mouth of the shaft (by which of course the fall is entirely lost), what must be the horse-power of the engine to raise per minute the same amount of water as before?   *Ans.* 45·6 H.-P.

16. *The work of living agents.*—The efficiency of men and animals is estimated in the same manner as that of the inanimate agents already considered, viz., by the number of foot-pounds of work they are capable of yielding in a given time. The number yielded under given circumstances by any particular agent must of course be determined by experiment. The results of experiment on this matter are registered in the tables that follow; they are based on similar tables given in General Morin's *Aide-Mémoire*. It must be borne in mind that these tables give mean results when the agent works in the best manner. It would be very possible for the agents to work with greater velocities than those assigned, but were this done they would yield a much smaller daily amount of work—compare the work done by a horse walking with that done by a horse trotting.

TABLE VIII.
WORK DONE BY MEN AND ANIMALS.

Nature of Labour	Daily Duration of Work in Hours	Foot-pounds per Day	Foot-pounds per Min.	Pounds raised or Mean Force exerted	Velocity	
					Feet per Min.	Miles per Hour
(1) *Raising weights vertically.* A man mounting a gentle incline or ladder without burden, i.e. raising his own weight.	8·0	2,032,000	4230	145	29	0·33
Labourer raising weights with rope and pulley, the rope returning without load	6·0	563,000	1560	40	39	0·44

WORK OF LIVING AGENTS.

TABLE VIII. (*continued*).

Nature of Labour	Daily Duration of Work in Hours	Foot-pounds per Day	Foot-pounds per Min.	Pounds raised or Mean Force exerted	Velocity Feet per Min.	Velocity Miles per Hour
Labourer lifting weights by hand	6·0	531,000	1480	44	34	0·38
Labourer carrying weights on his back up a gentle incline or up a ladder and returning unladen	6·0	406,000	1130	145	8	0·09
Labourer wheeling materials in a barrow up an incline of 1 in 12 and returning with the empty barrow	10·0	313,000	520	130	4	0·045
Labourer lifting earth with a spade to a mean height of 5¼ feet	10·0	281,000	470	6	78	0·9
(2) *Action on Machines.* Labourer walking and pushing or pulling horizontally	8·0	1,500,000	3130	27	116	1·32
Labourer turning a winch	8·0	1,250,000	2600	18	144	1·64
Labourer pulling and pushing alternately in a vertical direction	8·0	1,146,000	2390	11	216	2·70
Horse yoked to a cart and walking	10·0	15,688,000	26,150	150	175	2·00
Do. to a whim gin	8·0	8,440,000	17,600	100	175	2·00
Do. do. trotting	4·5	7,036,000	26,060	66⅔	391	4·44
Ox yoked to a whim gin and walking	8·0	8,127,000	16,930	145	117	1·33
Mule do. do.	8·0	5,627,000	11,720	66⅔	176	2·00
Ass do. do.	8·0	2,417,000	5030	30	168	1·95

The following table gives the useful effect of men and animals employed in the horizontal transport of burdens. The second and third columns give the useful effect, viz. the product of the weight in lbs. and the distance in feet. The reader must not mistake this for the foot-pounds done by the agent, the agent being employed *not* in raising the weight, but in overcoming the passive resistances, friction, &c., which depend on the weight indeed, but are only a fraction of it.

TABLE IX.

USEFUL EFFECT OF AGENTS EMPLOYED IN THE HORIZONTAL TRANSPORT OF BURDENS.

Agent	Duration of Daily Work	Useful Effect Daily	Useful Effect per Minute	Pounds Transported.*	Velocity	
					Feet per Min.	Miles per Hr.
Man walking on a horizontal road without burden, i.e. transporting his own weight	10·0	25,398,000	42,330	145	292	3·32
Labourer transporting materials in a truck on two wheels, returning with it empty for a new load	10·0	13,025,000	21,710	220	99	1·12
Do. do. in a wheelbarrow	10·0	7,815,000	13,030	130	100	1·14
Labourer walking with a weight on his back	7·0	5,470,000	13,030	90	145	1·64
Labourer transporting materials on his back and returning unburdened for a new load	6·0	5,087,000	14,110	145	97	1·10

* Exclusive of the weight of the barrow, truck, cart, &c. (Poncelet, *Méc. Ind.* p. 247.)

WORK OF LIVING AGENTS.

TABLE IX. (continued).

Agent	Duration of Daily Work	Useful Effect Daily	Useful Effect per Minute	Pounds Transported	Velocity Feet per Min.	Velocity Miles per Hr.
Do. do. on a hand-barrow	10·0	4,298,000	7160	110	65	0·74
Horse transporting materials in a cart, walking, always laden	10·0	200,582,000	334,300	1500	223	2·53
Do. do. trotting	4·5	90,262,000	334,300	750	446	5·06
Do. transporting materials in a cart, returning with the cart empty for a new load	10·0	109,408,000	182,350	1500	121	1·38
Horse walking with a weight on his back	10·0	34,385,000	57,310	270	212	2·41
Do. do. trotting	7·0	32,092,000	76,410	180	424	4·82

Ex. 117.—How many men would be required to raise by means of a capstan an anchor weighing 1 ton from a depth of 30 fathoms, in 15 minutes? *Ans.* 9 nearly.

Ex. 118.—In what time would 20 men raise the anchor in the last example? *Ans.* 6·4 min.

Ex. 119.—Through how great a distance would 30 men raise the anchor in *Ex.* 117 in each minute? *Ans.* 42 ft. nearly.

Ex. 120.—There is a well 150 ft. deep, a labourer raises water from it by a rope and pulley; how many cubic feet of water will he raise in a day?
Ans. 60 cub. ft.

Ex. 121.—How many cubic feet of water would a steam engine of 10 horse-power raise from this well in 24 hours? How many labourers would be required to do the same amount of work if they raised the water by wheel-and-axles, and how many if they raised it by means of capstans? How many horses would do the same amount of work walking in whim gins? *Ans.* (1) 50,688 cubic feet. (2) 380 labourers.
(3) 317 labourers. (4) 56 horses.

Ex. 122.—In how many minutes could 20 men working on a capstan raise an anchor weighing 2 tons from a depth of 200 fathoms?
Ans. 85·88 min.

Ex. 123.—How many men would in 40 minutes raise the anchor in the last example? *Ans.* 43 men.

Ex. 124.—Through how many fathoms could 15 men raise the anchor of *Ex.* 122 in 10 minutes? *Ans.* $17\frac{1}{2}$ nearly.

Ex. 125.—If 13 men are required to raise an anchor through 180 fathoms in 20 minutes, what must be the weight of that anchor? *Ans.* $753\frac{1}{2}$ lbs.

Ex. 126.—A town is situated 25 miles from the mouth of a coal pit, from which coal is taken to the town by a level railway on which the resistance is 10 lbs. per ton; the engine employed is of 15 horse-power and weighs with its tender 10 tons; each truck weighs 3 tons and contains 7 tons of coal; on each journey the engine takes 5 full trucks and returns with 5 empty trucks; supposing no time to be lost at the ends of the journey, how many tons of coals will be taken to the town in 48 hours? How many horses would be required to convey the same quantity of coals in the same time? *Ans.* (1) 445 tons. (2) 665 horses.

17. *Remarks on the work yielded by different agents.*—The following remarks upon the preceding tables and examples are worthy of the attention of the reader:—

(1) Every agent must be allowed to move at a certain rate in order to do the greatest amount of work it is capable of yielding; thus, a horse walking does considerably more work than a horse trotting, as an inspection of the tables will show. And this is true not of animate agents only, but also of inanimate; thus the work yielded by the consumption of a given quantity of coal will be larger in the case of a slow than of a fast engine.

(2) Also, in order that an animate agent may do its greatest amount of work, it must not be required to exert more than a certain force. This is also plain from an inspection of the table.

(3) It follows from the above considerations that though two agents may be capable of doing the same work in the same time, it may be in practice impossible or disadvantageous to substitute the one for the other. Thus an ox and a horse walking in a whim gin do very nearly the same amount of work; but since the ox moves more slowly, and exerts a greater force than the horse, it would generally be disadvantageous to substitute a horse

for an ox in a machine requiring a slow heavy pressure. Again, in cases where great speed is a *desideratum*, it would generally be impracticable by any machinery to make the slow agent perform the labour of the rapid agent; as, for instance, in the case of locomotion.

18. *On the cost of labour.*—The chief elements in the cost of labour may be enumerated as follows :—

(1) In the case of human labour, the whole cost is the wages paid.

(2) In the case of a horse, the elements of expense are attendance, keep, and the original cost; the last is but a small portion of the expense. Thus, if we suppose a horse to cost 20*l.* and to continue in working order for ten years, and reckon the value of money at four per cent. per annum, the element of cost would be 2·465*l.* yearly, or not quite 1*s.* per week.

(3) In the case of a steam engine, the chief elements are the original cost and subsequent repairs, attendance, and fuel. Of these elements the most important is that of fuel; and accordingly there is a special definition of the power of an engine with reference to the consumption of fuel. The definition is as follows :—

Def.—The number of foot-pounds of work yielded by an engine in consequence of the consumption of 1 bushel (i.e. 84 lbs.) of coal, is called the *duty* of that engine.

The extent to which the economy of fuel may be carried is very remarkably illustrated by the engines employed to drain the mines in Cornwall. In 1815, the average duty of these engines was 20 millions; in 1843, by reason of successive improvements, the average duty had become 60 millions, effecting a saving of 85,000*l.* per annum; *

---

* *Bourne on the Steam Engine*, p. 171. It may be remarked that this result depends largely on the construction of the boiler; 1 lb. of coal in the Cornish boiler evaporates 11¼ lbs. of water, while in the waggon-shaped boiler 8·7 is the maximum.—FAIRBAIRN, *Useful Information*, p. 177.

it is stated also, that, in the case of one engine, the duty was raised to 125 millions.

The actual cost of 1,000,000 foot-pounds of work, when done by different agents, cannot be specified with great precision; but a sufficiently accurate notion of the relative cost of different agents may perhaps be obtained from the annexed table, which has been calculated upon the following suppositions:—

(1) The wages of a labourer, 3s. a day.

(2) Keep of a horse, 2s. a day; attendance of 6 horses, 3s. a day; cost of each horse, 2d. a day.

(3) Steam engine of 50 horse-power, at an annual cost of 5l. per horse-power; attendance, 12s. a day; coal, 6d. a bushel.*

TABLE X.

COST OF LABOUR.

Character of Agent	Cost per Million Foot-pounds
(1) Labourer carrying weights up a ladder	88·67 pence
(2) Labourer raising weights by rope and pulley	63·94 ,,
(3) Labourer turning a winch	28·80 ,,
(4) Labourer turning a capstan	24·00 ,,
(5) Horse in a whim gin trotting	4·548 ,,
(6) Horse in a whim gin walking	3·791 ,,
(7) Horse walking in a cart	2·040 ,,
(8) Steam engine, duty 20 millions	0·429 ,,
(9) Steam engine, duty 90 millions	0·196 ,,

* In *Weale's Contractor's Price Book* for 1859 the prices of various steam engines are estimated to be from 25l. to 35l. per horse-power, boilers and fittings included; as the nominal horse-power (which is determined by measurement) is considerably less than the working horse-power, the estimate in the text is *very* ample; that estimate assumes 50l. to be the cost of a horse-power, and that 10 per cent. will represent interest on capital, repairs, and restitution. It may interest the reader to consider the following statement taken from Mr. R. Stephenson's paper on Railway Economy which forms an appendix to Mr. Smiles's *Life of George Stephenson*. In 1854 there were in the United Kingdom 5,000 locomotive engines costing from 2000l. to 2500l. apiece, and consuming annually 13 million tons of coke, made from 20 million tons of coal. It appears moreover that if a railway

ON THE COST OF LABOUR.

Ex. 127.—How many bushels of coal must be expended in a day of 24 hours in raising 150 cubic feet of water per minute from a depth of 100 fathoms; the duty of the engine being 60 millions? *Ans.* 135 bushels.

Ex. 128.—Determine the number of horses working in whim gins required to do the work of the last example. Determine also the weekly saving effected by employing steam power, supposing the total weekly expense of the engine to be double the price of coals consumed; the coals costing 10s. a ton; and each horse 20s. a week.

Ans. (1) 960 horses. (2) 924*l*. 11s. 0d. weekly saving.

Ex. 129.—There are three distinct levels to be pumped in a mine, the first 100 fathoms deep, the second 120, the third 150; 30 cubic feet of water are to come from the first, 40 from the second, and 60 from the third per minute; the duty of the engine is 70 millions. Determine its working horse-power and the consumption of coal per hour.

Ans. (1) 191 H.-P. (2) 5·4 bushels.

Ex. 130.—In the last example suppose there is another level of 160 fathoms to be pumped, that the engine does as much work as before for the other levels, and that the utmost power of the engine is 275 H.-P. Find the greatest number of cubic feet of water that can be raised from the fourth level. *Ans.* $46\frac{1}{4}$ cub. ft.

Ex. 131.—An engine raises every minute A cubic feet of water from a depth of a fathoms, B cubic feet of water from a depth of b fathoms, and C cubic feet of water from a depth of c fathoms. The diameter of the piston of the steam engine is d in., the length of the stroke l ft., it makes n strokes per minute; also it consumes o bushels of coal in twenty-four hours, and has a modulus m. Determine (1) the pressure per square inch upon the piston; (2) the horse-power of the engine (as measured by pressure of steam on piston); (3) its duty.

Ans. (1) $\dfrac{1500(\text{A}a + \text{B}b + \text{C}c)}{\pi d^2 lmn}$ (2) $\dfrac{\text{A}a + \text{B}b + \text{C}c}{88m}$

(3) $\dfrac{540,000(\text{A}a + \text{B}b + \text{C}c)}{om}$

Ex. 132.—Water is to be raised from three levels of 20, 30, and 40 fathoms respectively; 10 cubic feet of water are to be taken per minute from the first, 20 from the second, and 40 from the third. The engine consumes 15 bushels of coal in a day. The diameter of the piston is 4 ft., it makes 10 strokes of 6 ft. each per minute. The modulus of the engine is 0·65. Find the pressure per square inch on the piston, the horse-power (as measured by pressure of steam) and the duty of the engine.

Ans. (1) 12·75 lbs. (2) (nearly) 42 H.-P. (3) 133,000,000 duty.

---

company start with 100 new engines, about 20 or 25 will need repair at the end of four years, and after that there will always be about 25 in the workshop.

Ex. 133.—In *Ex.* 126 suppose the engine and trucks on the one hand, and the horses and carts on the other, to want renewal every ten years ; suppose also that each horse and cart costs 40*l.*, that one man attends to every six horses and is paid 3*s.* a day, that each horse's keep is 1*s.* 6*d.* a day, that there are two turnpikes on the road at each of which there is a toll of 6*d.* ; determine the cost of transporting 445 tons of coals. Next suppose the engine and tender to cost 1000*l.*, each truck 120*l.* (15 trucks are required to prevent loss of time); that there are three drivers and three stokers each at 6*s.* a day ; that money is worth 5 per cent., and that each mile of road cost 10,000*l.* to make, and 365*l.* a year to keep in repair ; determine in this case the cost of transporting 445 tons of coals. Also if coal cost 3*s.* a ton at the pit mouth, what will it cost in the town according to each method of transport, neglecting profit ?

Ans. (1) 214*l.* (2) 123*l.* (3) 12*s.* 6*d.* a ton by cart.
(4) 8*s.* 6*d.* a ton by rail.

[Interest on the cost price of engine, trucks, horses and carts can be neglected.]

Section II.

19. *On the work done by a variable force.*—There are two important questions in the subject of work which we shall treat in the present section: they are, (1) the work done by a variable force, when exerted through a certain distance; (2) the total amount of work done in raising a number of weights through different heights.

Fig. 5.

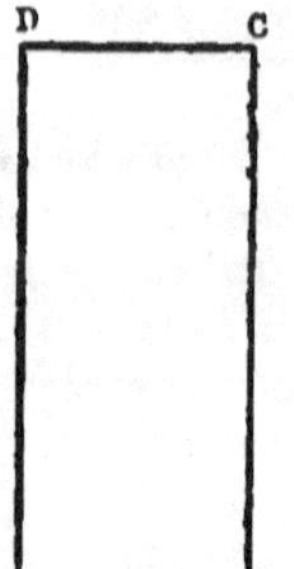

As an introduction to the theorem which follows, it may be remarked that, if a constant force of P lbs. act through a distance of s feet, and if a rectangle A B C D be drawn, of which the base A B represents the s feet on scale, and the perpendicular A D represents the P lbs. on the same scale: then, since the area of A B C D contains P S square units on the same scale, the area will correctly represent the work done by P.

WORK DONE BY A VARIABLE FORCE. 37

Proposition 1.

If a variable force act through a certain distance, and if a curve be drawn in such a manner that the abscissa and corresponding ordinate of any point represent respectively the distance through which the force has acted and the magnitude of the force, then will the area of the curve between any two ordinates represent the work done by the force while acting through a distance represented by the difference between the extreme abscissæ.

When the force has acted through a distance represented on a certain scale by A N, suppose it to be represented on the same scale by PN; also, when it has acted through a distance A M, suppose it to be represented on the scale by Q M; let the curve P Q be drawn in such a manner that any ordinate (P_3N_3) represents the force when it has acted through a distance AN_3; we have to prove that the area P N M Q represents the work done by the force while acting through the distance N M.

FIG. 6.

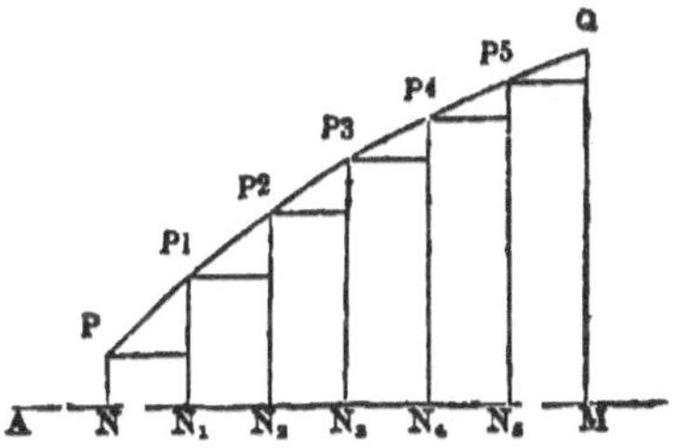

Divide N M into any number of equal parts in N_1, N_2, N_3, ... draw the ordinates P_1N_1, P_2N_2, P_3N_3 .... and complete the rectangles PN_1, P_1N_2, P_2N_3 .... Now, we shall nearly represent the actual case if we suppose the force, while acting successively through the short distances NN_1, N_1N_2, N_2N_3 .... to retain unchanged the magnitude it has at the beginning of those distances respectively; and we shall represent the case more nearly the smaller we make the distances, i.e. the greater the number of parts into which we divide N M: the actual case being the limit continually approached as the number of parts is increased.

But if the force acts uniformly through each distance,

it will do a number of units of work represented by the sum of the rectangular areas PN₁, P₁N₂, P₂N₃ . . . ., and this being true whatever be the number of the small distances, the work actually done will be properly represented by the limit of the sum of these rectangles, i.e. by the curvilinear area P N M Q.

Cor.—It must be borne in mind that the scale must be the same for lbs. and for feet; thus, if the scale be in inches, P N must be as many inches long as the force contains lbs., and N M must be as many inches long as the distance represented contains feet; this being so, the area of the curve in square inches will give the number of foot-pounds of work.

Ex. 134.—A rope l ft. long and weighing w lbs. per foot hangs by one end; determine the number of foot-pounds of work required to wind up a ft. of the length.

FIG. 7.

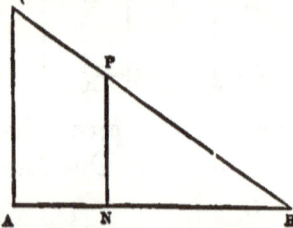

Take A B on scale equal to l, draw A C at right angles to A B and on the same scale equal to $w\,l$, join B C; in A B take any point N, draw P N parallel to A C, then

$$PN : NB :: CA : AB :: w : 1.$$

Therefore PN = w N B, i.e. the ordinate PN represents on scale the weight of the rope left hanging when the extremity has been raised through a space A N. Therefore the area A B C represents the number of foot-pounds required to wind up the whole rope, and the area C A P N the number of foot-pounds required to wind up a length A N of the rope. Hence if U is the required number of foot-pounds,

$$U = wa(l - \tfrac{a}{2}).$$

Hence also the number of foot-pounds (U₁) required to wind up the whole rope is given by the formula

$$U_1 = \tfrac{1}{2}wl^2.$$

Ex. 135.—A weight of 2 cwt. has to be raised from a depth of 100 fathoms by a rope 3 in. in circumference; determine the number of foot-pounds that must be expended in raising it, and the number of minutes in which 4 men would do the work by means of a capstan.

Ans. (1) 207,300 ft.-pds. (2) 16·5 min.

WORK DONE BY A VARIABLE FORCE.

Ex. 136.—How heavy will that anchor be which 13 men will raise by means of a capstan from a depth of 180 fathoms in 40 min., supposing the cable to weigh 1125 lbs. (neglecting the buoyancy of the water)?

Ans. 945 lbs.

Ex. 137.—A chain each foot of which weighs 8 lbs. is suspended from the top of a shaft, the depth of which is 50 fathoms; determine the number of foot-pounds required to wind up each successive 100 ft. of its length; determine also the length of the chain which will require twice as many foot-pounds to wind it up.

Ans. (1) 200,000, 120,000, 40,000 ft.-pds. respectively. (2) 424 ft.

Ex. 138.—If a chain 300 ft. long and weighing 8 lbs. per foot is wound up in 4 min., how many men working on a capstan would do it? How many horses walking in a whim gin? How many steam horses? How many of each agent would be required if the weight per foot of the chain were doubled? And how many if the length of the chain were doubled?

Ans. (1) 29 men.   (2) 5·1 horses.   (3) $2\frac{8}{11}$ horse-power.
(4) 57 men.   10·2 horses. $5\frac{5}{11}$ horse-power.
(5) 115 men.   20·4 horses. $10\frac{10}{11}$ horse-power.

Ex. 139.—A chain is a ft. long; divide it into n parts such that the winding up of each may require the same number of foot-pounds.

Ans.

$$\frac{a}{\sqrt{n}}(\sqrt{n}-\sqrt{n-1}),\ \frac{a}{\sqrt{n}}(\sqrt{n-1}-\sqrt{n-2}),\ \frac{a}{\sqrt{n}}(\sqrt{n-2}-\sqrt{n-3})\ \&c.$$

Ex. 140.—Coal is raised from the bottom to the mouth of a pit 150 ft. deep in loads of a quarter of a ton; the box containing it weighs 1 cwt., the rope by which it is raised is 3 in. in circumference; determine the number of foot-pounds spent in raising the coal, and the number spent in raising the box and rope. If the lifting engine works with 10-horse power, determine the weight of coals raised in 2 hours, supposing the ascent and descent of the box to take equal times.

Ans. (1) 84,000 ft.-pds. to raise coal.   (2) $21,356\frac{1}{4}$ ft.-pds. to raise box and rope.   (3) 47 tons.

Ex. 141.—In the last example suppose machinery to be employed by means of which the same drum winds up the rope of an ascending box and unwinds that of a descending box. Determine the number of tons raised in 2 hours.*   *Ans.* 118 tons.

[Of course the work done by the descending box and rope will nearly equal that expended on the ascending box and rope—the weight of box and rope can therefore be neglected.]

---

* The primary object of this mode of working was, probably, to save time, the saving of labour being an accidental result; though that saving is very considerable.

PRACTICAL MECHANICS.

Ex. 142.—Determine the number of tons raised under the conditions of *Ex.* 140 and 141, supposing ½ a minute is expended in filling or emptying the box. *Ans.* (1) 18¼ tons. (2) 39¾ tons.

Ex. 143.—If 4 cwt. of material are drawn from a depth of 80 fathoms by a rope 5 in. in circumference, how many foot-pounds are expended in raising them, and what horse-power is necessary to raise them in 4½ minutes?
Ans. (1) 344,640 ft.-pds. (2) 2·32 H.-P.

Ex. 144.—A rope 3 in. in circumference is strong enough to bear a working tension of 4 cwt.; how many foot-pounds are wasted in the last example by using a rope 5 in. in circumference? *Ans.* 82,944 ft.-pds.

Ex. 145.—A winding engine raises to the surface a load of 12 cwt. in 6½ minutes from a depth of 115 fathoms; the rope employed is a flat rope composed of 3 ropes each 3 in. in circumference. What is the horse-power of the engine? *Ans.* 5·67 H.-P.

Ex. 146.—If the engine in the last example have a cylinder 20 in. in diameter, and makes per minute 15 strokes of 2 ft. 10 in., under what mean pressure per square inch of steam does it work if its modulus is 0·55?
Ans. 25·5 lbs.

20. *The steam indicator.*—A very instructive application of Proposition 1 occurs in the steam indicator, which may be sufficiently described as follows: A B is a small hollow cylinder containing a powerful spring, which can be partly seen through the aperture E F; within the indicator is a small piston or plunger (marked in the figure by dotted lines) which is kept down by the spring, so that if it is forced up, the compression of the spring gives the amount of the compressing force, which can be read off on the scale C D by means of the pointer G H, which rises and falls with the plunger. The end H of the pointer carries a pencil, the point of which rests against a sheet of paper wrapped round a cylinder K L; if this cylinder be stationary, and the pencil move, a vertical straight line will be described; if the pencil be

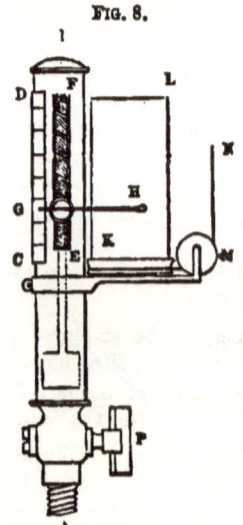
FIG. 8.

stationary, and the cylinder revolve, a horizontal straight line will be described; but if both the pencil move and the cylinder revolve, a curved line will be described. To obtain the required curve it is necessary that the cylinder K L should turn in contrary directions during the up and down strokes of the piston. This is effected by means of a clockspring placed within the cylinder K L. On the up stroke the string M N, which is fastened round the cylinder, is pulled in the direction M N, causing the cylinder to turn from left to right and winding up the spring. On the down stroke the string tends to slacken, the spring uncoils and turns K L back from right to left.

The instrument is used in the following manner:— The end A being screwed into an aperture properly constructed, the steam in the interior of the cylinder of the steam engine can be admitted into the indicator by opening the cock P; at first, however, the cock P is shut, so that the pointer remains stationary. The end of the string M N is attached to some part of the engine * in such a manner that the cylinder K L makes one revolution while the piston of the steam engine makes a stroke; this being done, and the cock kept shut, the pencil will trace on the paper a straight line called the atmospheric line: on the next stroke the cock is opened, and now the steam pressing on the plunger the pencil will rise or fall according as the pressure of the steam is greater or less than that of the atmosphere, and will describe a curve that will return into itself at the end of a double stroke (or revolution). The area of the curve thus described will give the amount of work done by the steam during a *single* stroke.

FIG. 9.

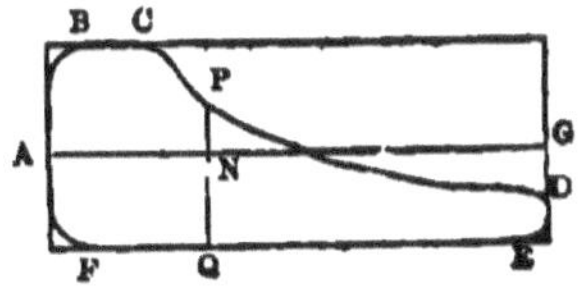

---

* Generally the radius-shaft.

To explain this, suppose A B C D E F to be the curve given by the indicator (which, it may be remarked, is described continuously in the direction A B C 1 E F A), A G the atmospheric line; draw P N Q any double ordinate, then P N represents the excess of the steam pressure above that of the atmosphere when the ascending piston is at a certain point, and N Q represents the defect of the vacuum pressure below that of the atmosphere when the descending piston is at the same point. Now the effective pressure of the steam is the excess of the steam pressure above the vacuum pressure; but

P N = steam pressure — atmospheric pressure,
N Q = atmospheric pressure — vacuum pressure,
∴ P N + N Q = steam pressure — vacuum pressure;

therefore P Q represents the effective pressure of the steam when the ascending piston is at the point corresponding to N, i.e. assuming the vacuum pressure at any point of one stroke to be the same at the same point of the next stroke. If, then, for the sake of distinctness,* we suppose each inch of the ordinate to denote a pressure of 1 lb. and each inch of the abscissa (i.e. of the atmospheric line) to denote a foot of the stroke, the area of the curve will give the number of foot-pounds of work done during a *single* stroke by the steam on an area equal to that of the plunger, and if the area of the piston of the steam engine be n times that of the plunger, the work done by the steam during a *single* stroke will be n times that given by the curve.

The area of the curve may be found by Simpson's rule, viz.—Divide A G into any even number of equal parts, and draw the corresponding ordinates; take the sum of the extreme ordinates, four times the sum of the even ordinates, and twice the sum of the odd ordinates (i.e. ex-

---

* In practice the scale would be considerably less than this.

WORK OF LENGTHENING BARS.

cepting the first and last), add them together, and multiply the sum by one of the parts of the abscissa; the product will be three times the area of the curve.*

Ex. 147.—Let the curve shown in the figure be that given by a stroke of 5 ft.; let A B be divided into 10 equal parts, and let the ordinates 1, 2, 3, 4, ... be drawn; suppose them to represent respectively 19, 22, 22, 17·5, 13, 11, 9, 7·5, 6, 5·5, 4 lbs. pressures per square inch. The radius of the piston being 20 in., determine the number of foot-pounds of work done per stroke, and the *mean* effective pressure per square inch on the piston—i.e. the constant pressure that would do the same work. *Ans.* (1) 79,000 ft.-pds. (2) 12·6 lbs.

FIG. 10.

Ex. 148.—Determine the number of foot-pounds of work and the mean pressure per square inch on a piston 3½ feet in diameter having a stroke of 5 feet, if the ordinates measured at intervals corresponding to three inches of the stroke give the following pressures—5·03, 12·57, 18·04, 20·73, 21·03, 21·11, 21·25, 20·72, 20·14, 18·63, 15·45, 13·24, 10·83, 8·53. 6·49, 4·87, 3·99, 3·74, 3·52, 3·25, 2·75. *Ans.* (1) 87,600 ft.-pds. (2) 12·65 lbs. per sq. in.

21. *Work expended on the elongation of bars.*—It is plain that if a rod be lengthened by a gradually increasing force, the force exerted at any degree of elongation will be proportional to that elongation; so that if the abscissæ represent the degree of elongation, and the ordinates the stretching force, the area which gives the units of work will be a triangle. Hence:

Ex. 149.—There is a bar the length of which is L and section K; it is gradually elongated by a length l; if its modulus of elasticity be E, show that the work expended on its elongation will be given by the formula

$$U = \frac{l^2}{2L} \text{KE}.$$

Ex. 150.—The pumping apparatus of a mine is connected with the engine by means of a series of wrought-iron rods 200 ft. long; the section of each rod is ¾ of a square inch; the tension when greatest is estimated at 6 tons; how many foot-pounds of work are expended at every stroke upon the elongation of the bars? *Ans.* 830 ft.-pds.

---

* The curve given by the indicator is useful in other ways besides that mentioned in the text.—*Bourne on the Steam Engine*, p. 246.

Ex. 151.—A bar of wrought iron 100 ft. long with a section of 2 square inches has its temperature raised from 32° F. to 212° F.; how many foot-pounds of work has the heat done?   *Ans.* 3875 ft.-pds.

22. *The work expended in raising weights through various heights.*—The questions arising out of this important part of the present subject are solved by means of the following proposition.

Proposition 2.

When any weights are raised through different heights, the aggregate of the work expended is equal to the work that would be expended in lifting a weight equal to the sum of the weights through the same height as that through which the centre of gravity of the weights has been raised.

Let $w_1, w_2, w_3 \ldots \ldots$ be the weights of each separate body; conceive a horizontal plane to pass below them all; let $h_1, h_2, h_3 \ldots \ldots$ be the heights of those bodies above the plane before they are lifted, and let H be the height of their common centre of gravity; then (Prop. 16)

$$\text{H}(w_1 + w_2 + w_3 \ldots \ldots) = w_1 h_1 + w_2 h_2 + w_3 h_3 + \ldots \quad (1)$$

Also, let $k_1, k_2, k_3 \ldots \ldots$ be the heights of the weights respectively, after they have been lifted, and K the height of their common centre of gravity; then

$$\text{K}(w_1 + w_2 + w_3 \ldots \ldots) = w_1 k_1 + w_2 k_2 + w_3 k_3 + \ldots \quad (2)$$

hence, subtracting (1) from (2), we obtain

$$(\text{K} - \text{H})(w_1 + w_2 + w_3 \ldots) = w_1(k_1 - h_1) + w_2(k_2 - h_2) + w_3(k_3 - h_3) \ldots \quad (3)$$

Now, $w_1, w_2, w_3, \ldots \ldots$ are severally raised through the heights $k_1 - h_1, k_2 - h_2, k_3 - h_3 \ldots \ldots$; therefore the right-hand side of equation (3) gives the aggregate work expended in lifting them; hence that work is equal to

$$(\text{K} - \text{H})(w_1 + w_2 + w_3 \ldots \ldots),$$

i.e. to the work that must be expended in lifting a weight $W_1+W_2+W_3+\ldots$ through a height $K-H$. (Q. E. D.)

Cor.—In the case of the transport of bodies along any parallel lines, the principle enunciated in the theorem will hold good, since the resistances bear a constant ratio to the weights.

Ex. 152.—How many foot-pounds of work must be expended in raising the materials for building a column of brickwork 100 ft. high and 14 ft. square? and in how many hours would an engine of 2 horse-power raise them? *Ans.* (1) 109,760,000 ft-pds. (2) 27·71 hours.

[Since the material has to be raised from the *ground*, the common centre of gravity will have to be raised from the ground to the centre of gravity of the column, i.e. to its middle point 50 ft. above the ground.]

Ex. 153.—A shaft has to be sunk to the depth of 130 fathoms through chalk; the diameter of the shaft is 10 ft.; how many foot-pounds of work must be expended in raising the materials? In how long a time could this be done by a horse walking in a whim gin? How many men working in a capstan would do it in the same time? Determine the expense of the work supposing the horse to cost 3*s.* 6*d.* a day, and the wages of a labourer to be 2*s.* 6*d.* a day.

Ans. (1) 3457 million ft.-pds. (2) 409·6 days. (3) 5·62 men. (4) Cost of horse 71*l.* 14*s.* Cost of men 288*l.*

Ex. 154.—If the work in the last example is to be done in 24 weeks by a steam engine working 8 hours a day, 6 days a week, what must be the horse-power of the engine? *Ans.* 1·521 H.-P.

Ex. 155.—In *Ex.* 153 suppose the box in which the material is raised to weigh ½ cwt., the rope to be 3 in. in diameter, and each load to be 4 cwt. of chalk, also suppose the box to take as long in ascending as in descending and that ¼ of a minute is lost in unhooking and hooking at the bottom of the shaft and the same at the top; when the shaft is 100 ft. deep determine the time that elapses between the starting of one load and the starting of the next; the engine working at 1½ horse-power. *Ans.* 2·62 min.

Ex. 156 —Determine the same as in the last example when the shaft is x ft. deep. *Ans.* $\dfrac{112x+0\cdot045x^2}{5500}+0\cdot5$ min.

Ex. 157.—Determine the whole time of raising the materials of the shaft in *Ex.* 153 under the conditions of *Ex.* 155. *Ans.* 3331 hours.

Ex. 158.—Referring to *Ex.* 153, 155, suppose the drum of the winding machine to have two ropes wound round it in contrary directions, so that it unwinds one rope while winding up the other, and that consequently an empty box descends while a full one is being raised (as in *Ex.* 141); determine the time that must elapse between two consecutive lifts of 4 cwt. when the shaft is 100 ft. deep. *Ans.* 1·155 min.

Ex. 159.—Obtain a determination similar to that in the last example, when the shaft is x ft. deep. $Ans. \dfrac{448x}{49,500} + 0·25$ min.

Ex. 160.—Obtain the whole time of lifting the materials from the shaft under the circumstances of *Ex.* 158. *Ans.* 12·46 hours.

Ex. 161.—In how long a time would a 15 horse-power engine empty a shaft full of water, the diameter of the shaft being 8 ft. and the depth 200 fathoms? If the engine has a duty of 30 millions determine the amount of coal consumed in emptying the shaft.

Ans. (1) 76 hours.   (2) 75·4 bushels.

Ex. 162.—There is a certain railway 200 miles long; it may be assumed that in the course of 10 years there will be 50,000 tons of iron railing laid down, and that it will be equally distributed along the line. How many foot-pounds of work must be expended in conveying the rails (neglecting the weight of the trucks), if the depôt is at one end of the line? And how many if the depôt is in the middle of the line? The resistances being reckoned at 8 lbs. per ton.

Ans. (1) 211,200 million ft.-pds.   (2) 105,600 million ft.-pds.

Ex. 163.—How many journeys of 200 miles performed by a train weighing 50 tons does the difference of the results in the last example represent? Resistances 8 lbs. per ton. *Ans.* 250 journeys.

CHAPTER III.

ELEMENTARY STATICS.

23. *Mechanics.*—The science of Mechanics is that which treats of *the motion and rest of bodies as produced by force.* The words, 'as produced by force,' are added in order to exclude the science of *pure motion* or *mechanism*, which treats of the *forms* of machines, and in which machines are regarded merely as modifiers of *motion.* Into all questions which are properly mechanical the idea of *force* must enter.

Force may be defined to be any cause which puts a body in motion, or which tends to put a body in motion when its effect is hindered by some other cause. On this definition the following remark is to be made: Suppose a given weight is supported by a string passing over a pulley and fastened to a fixed point at the other end; next, suppose an equal weight to be supported by a man's hand; lastly, suppose an equal weight to be supported by the elastic force of a spring. Now, here we have three physical agents, viz. the reaction of the fixed point transmitted through the string, the muscular power of a man, and the elastic force of a spring, very different in many respects, but agreeing in their common capacity to support a given weight. They may clearly be regarded as equal, when viewed with reference to that capacity. In short, as, in geometry, we regard all bodies as equal which can successively fill the same space, without any regard to their physical qualities, such as weight, colour, &c., so in mechanics we regard all forces as equal which will severally balance by direct opposition a given weight irrespectively of their *physical origin*. By the weight of a body is meant the mutual attraction between the earth and that body; as this attraction has different amounts when the body is at different places, the weight of a body, when used as a standard of force, must be determined with reference to some assigned place. Thus:— there is kept in the Exchequer Office a piece of platinum called the standard pound (avoirdupois); the attraction of the earth on that body at London is a force of 1 lb., and any force which by direct opposition can support that body in London is also a force of 1 lb. If we suppose two forces each of 1 lb. to act in the same direction at a point and to be balanced by a single force, that force is one of 2 lbs.; and similarly a force of three, four, or more pounds can be defined.

24. *Statics and dynamics.*—It follows, from the definition, that, in Mechanics, we can consider a force either as producing motion, or as concurring with others in producing rest. Accordingly, the science of mechanics is divided into two distinct though closely connected branches, viz. statics and dynamics. Of these, statics is that science which determines the conditions of the equilibrium of any body or system of bodies under the action of forces. Dynamics is that science which determines the motion, or the change of motion, that ensues in a body or system of bodies subjected to the action of a force or forces that are not in equilibrium.

25. *Determination of a Force.*—From what has already been said, it appears that the *magnitude* of any force is assigned by considering the weight it would just support if applied directly upward; in other words, we arrive at the magnitude of any force by comparing it with the most familiar and measurable of forces, viz. weight. A little consideration will show that the effect of a force in any case depends not only on its amount, but

also on its point of *application*, and the *line* along which it acts. We may say, therefore, in general terms, that a force is completely determined when we know (1) its magnitude, (2) its point of application, (3) its line of action, and (4) its direction along that line.* A *line* is frequently said to *represent* a force; when this is the case it must be drawn *from* the point of application of the force along the line of its action, and must contain as many units of length (say inches) as the force contains units of force (say lbs.) It is of great importance that the student should attend to all the conditions which must meet when a line correctly represents a force. Suppose a force of P lbs. (fig. 11) to act on a body at the point A; if the force is a *pull*, as in the first figure, the line A B containing as many inches as P contains lbs. will represent the force; but if the force is a *push*, A B must be measured, as in the second figure.

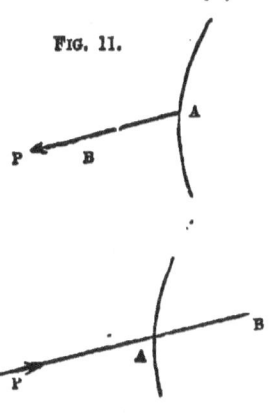

Fig. 11.

26. *Resultant and components.*—If we consider any forces that keep a body in equilibrium, it is plain that any one of them balances all the others: thus, if three strings be knotted together at A, and be pulled by forces of P lbs., Q lbs., and R lbs. respectively so adjusted as to balance one another, it is plainly a matter of indifference whether we consider that P balances Q and R, or that Q balances R and

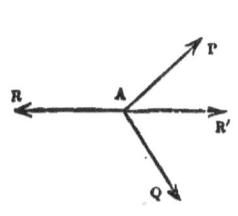

Fig. 12.

* The student must notice the distinction between the *line of action* and the *direction* of a force: e.g. in fig. 14 (p. 52) P and Q act in the *same* direction along different lines.

P, or that R balances P and Q. Let us consider that R balances P and Q; now R would of course balance a force R' exactly equal and opposite to itself: so that if we substitute R' for P and Q, or *vice versâ*, P and Q for R', in either case R is balanced, and the force R' is equivalent to P and Q; under these circumstances, R' is called the resultant of P and Q; and P and Q are called the components of R'. Hence we may state generally,

Def.—That force which is equivalent to any system of forces is called their resultant.

Def.—Those forces which form a system equivalent to a single force, are called its components.

27. *Resultant of forces acting along the same straight line.*—If the forces act in the same direction the resultant must be their sum. If some act towards the right and some towards the left, the first set can be formed into a single force (P) acting towards the right, the second set can be formed into a single force (Q) acting towards the left: the resultant of these two, and therefore of the original set of forces, will be equal to the difference between P and Q and will act in the direction of the greater. If the forces are in equilibrium, the sum of those acting towards the right must equal the sum of those acting towards the left.

Ex. 164.—If three men pull on a rope to the right with forces of 31, 20, and 27 lbs. respectively, and are balanced by two men who pull with forces of 40 and P lbs. respectively, find P. *Ans.* 38 lbs.

Ex. 165.—In the last example find the resultant of the 5 forces (1) if P = 30 lbs.; (2) if P = 40 lbs. *Ans.* (1) 8 lbs. acting towards the right. (2) 2 lbs. acting towards the left.

Ex. 166.—There is a rope AB and men pull along it in the following manner: the first with a force of 50 lbs. towards A; the second with a force of 37 lbs. towards B; the third with a force of 35 lbs. towards A; the fourth with a force of 20 lbs. towards A; the fifth with a force of 54 lbs. towards B; the sixth with a force of 27 lbs. towards A; the seventh with a force of 52 lbs. towards B; the eighth with a force of 30 lbs. towards B.

PARALLEL FORCES. 51

Determine the single force that must act along A B to balance them, and find whether it acts towards A or B. *Ans.* 41 lbs. acting towards B.

Ex. 167.—In the last example suppose the second force to act towards A, find the resultant. *Ans.* 33 lbs. acting towards A.

28. *The terms reaction, thrust, and tension* are of frequent occurrence in Mechanics, and it is important that their meaning should be distinctly understood. With regard to the first of them, it must be borne in mind that the only forces with which we are acquainted are exerted between different portions of matter; and if, for the sake of distinctness, only two bodies, A and B, are considered, the following statement is found to be universally true :—If A exerts a force on B, then B exerts an equal opposite force on A—a fact commonly expressed by saying that to every action there is an equal opposite reaction. Now let A B (fig. 13) be a body urged by a force T against a fixed plane A C, and let the motion which T tends to communicate to the body be prevented by the fixed plane; that fixed plane must supply a force (R) which exactly balances T; and the body A B is really compressed between two forces R and T, of which the former is the *Reaction* of the fixed plane, and the latter the *Thrust* along A B. A Thrust and a Reaction *compress* or *tend to compress* the body on which they act. If, on the contrary, the body (A B) had been acted on by two equal opposite forces T and R tending to produce *elongation*, it is said to sustain a *tension* T. One of the forces producing a tension may, of course, be a reaction; thus, if one end of a string is tied to a nail fast in a post, and the other end to a suspended weight of 10 lbs., the string is stretched by two forces each of 10 lbs., viz. the weight and the reaction of the nail, and the string is said to sustain a tension of 10 lbs.

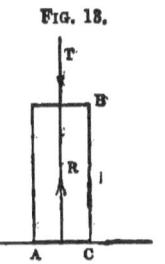

FIG. 13.

29. Resultant of two parallel forces.—First, let P and Q (fig. 14) be the two parallel forces acting in the same direction at the points A and B; join AB and divide it in C in such a manner that

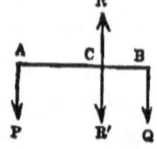
FIG. 14.

$$AC : CB :: Q : P$$

then the resultant (R') equals P+Q and acts through C along a line parallel to AP or BQ and in the same direction as P and Q.

If C rests on a fixed point P and Q will balance about C and the fixed point will sustain a pressure R'.

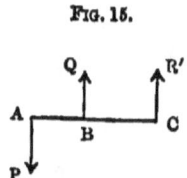
FIG. 15.

Secondly, let P and Q (fig. 15) be the two parallel forces acting at A and B in opposite directions. Suppose Q to be the greater. In AB produced take a point C such that

$$AC : CB :: Q : P$$

then the resultant (R') equals Q—P and acts through C along a line parallel to AP and BQ and in the same direction as Q.

If C rests on a fixed point P and Q will balance about C and the fixed point will sustain a pressure R'.

Ex. 168.—If weights of 12 lbs. and 8 lbs. are hung from A and B respectively, the ends of a rod 5 ft. long, and if the weight of the rod is neglected, determine the distance from A of the point round which these forces balance, and the pressure on that point. *Ans.* (1) 2 ft. (2) 20 lbs.

Ex. 169.—Let A B be a rod 12 ft. long (whose weight is neglected), from A a weight of 20 lbs. is hung, and an unknown weight (P) from B, it is found that the two balance about a point 3 ft. from A; determine P.
Ans. $6\frac{2}{3}$ lbs.

Ex. 170.—If a weight of 16 lbs. is hung from the end A, and 12 lbs. from the end B of a rod (whose weight is neglected), and if they balance about a point C, whose distance from A is $4\frac{1}{2}$ ft., what is the length of the rod?
Ans. $10\frac{1}{2}$ ft.

Ex. 171.—Draw a straight line A B, 8 ft. long; forces of 5 lbs. and 7 lbs. act at A and B respectively at right angles to AB and in opposite directions. Determine their resultant.

30. Conditions of equilibrium of three parallel forces.—In the last article we saw that the forces P and Q acting severally at A and B are equivalent to the force R' acting at C; now R' will clearly be balanced by an equal opposite force R; and therefore P and Q acting at A and B will be balanced by the force R acting at C. Hence the following conditions must be fulfilled by three parallel forces that are in equilibrium on a given body:—

(*a*) Two of the forces (P and Q) must act in the same direction, and the remaining force (R) in the opposite direction, the line along which the latter acts lying between those along which the former severally act.

(*b*) The sum of the former forces (P and Q) must equal the latter force (R).

(*c*) If any line be drawn cutting the lines of action of the forces (P, Q, R, in A, B, C, respectively) the portion of the line between any two forces is proportional to the remaining force, i.e.

$$BC : CA :: P : Q$$
$$CA : AB :: Q : R$$
$$AB : BC :: R : P$$

31. Centre of gravity.—Since each part of a body is heavy, it follows that the weight of a body is distributed throughout it; there exists, however, in every body a certain point called its *centre of gravity*, through which we may suppose the whole weight of the body to act, whenever that weight is one of the forces to be considered in a mechanical question. It admits of proof that the centre of gravity of any uniform prism or cylinder is the middle point of its geometrical axis: and as a uniform rod is merely a thin cylinder, its centre of gravity will be at its middle point.

Ex. 172.—Two men, A and B, carry a weight of 3 cwt. slung on a pole, the ends of which rest on their shoulders; the distance of the weight from A is 6 ft., and from B is 4 ft. Find the pressure sustained by each man.

If P is the pressure sustained by A and Q that sustained by B

$$P+Q=3 \text{ cwt.}$$

and $\quad 6:4::Q:P$

therefore $\quad P=1\tfrac{1}{5}$ cwt. and $Q=1\tfrac{4}{5}$ cwt.

Ex. 173.—There is a beam of oak 30 ft. long and 2 ft. square; at a distance of 1 ft. from one end is hung a weight of 1 ton; how far from that end must the point of support be on which the beam when horizontal will rest, and what will be the pressure on that point?

Ans. (1) 11·61 ft.   (2) 9245 lbs.

Ex. 174.—If a mass of granite 30 ft. long, 1 ft. high, and 3 ft. wide is supported in a horizontal position on two points each 3 inches within the ends (and therefore 29½ feet apart), find the pressure on each point of support. *Ans.* 7383 lbs.

Ex. 175.—If in the last example another mass of granite with the same section and half as long is laid lengthwise on the former, their ends being square with each other; determine the single force to which their two weights are equivalent, and the line along which it acts, and hence the pressure on the two points of support.

Ans. (1) Resultant acts 17·5 feet from one end.

(2) Pressures on points of support respectively 9197 and 12,950 lbs.

Ex. 176.—If in the last case the upper block is shifted round through a right angle in such a manner that the middle point of the upper block is exactly over a point in the axis of the lower, and the end of the lower in the same plane with one face of the latter, determine the pressures on the points of support. *Ans.* 7695 lbs. and 14,452 lbs.

Ex. 177.—A ladder A B, 50 ft. long, weighs 120 lbs.; its centre of gravity is 10 ft. from A; if two men carry it so that its ends rest on their shoulders, determine how much of the weight each must support. If the one of them nearer to the end B is to support the weight of 40 lbs., where must he stand?

Ans. (1) 96 lbs. and 24 lbs.   (2) 20 ft. from B.

32. *The parallelogram of forces.*—When two forces act at a point along different lines, their resultant is determined by the following rule, which is called the principle of the parallelogram of forces:—*If two forces act at a point, and if lines be drawn representing those forces, and on them as sides a parallelogram be constructed, that diagonal which passes through the point will represent the resultant of the forces.* The student, when applying this principle to any particular case, must bear in mind the meaning of the words a *line represents a force* (Art. 25).

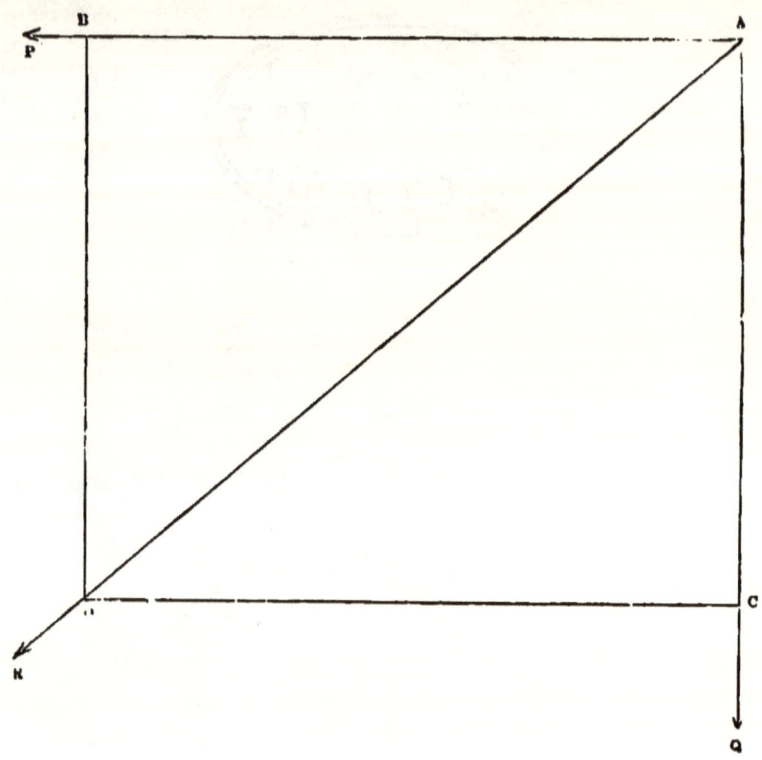

Fig. *a*, page 55.

PARALLELOGRAM OF FORCES.

Ex. 178.—If at a point A of a body two ropes A P and A Q are fastened and are pulled in directions A P, A Q at right angles to each other by forces of 120 and 100 lbs. respectively; determine the magnitude and direction of the resultant pull on the point A. (See fig. *a.*)

Along* A P measure on scale A B containing 120 units of length, and along A Q measure A C containing 100 units of length; complete the rectangle B C and draw the diagonal A D; this line represents the magnitude and direction of the resultant. In fig. *a* the scale employed is 1 in. for 40 lbs.; the results obtained by construction were the following R = 155·8 lbs. and P A R = 40° 5′; the measurement of the angle was made with a common ivory protractor, so that the number of minutes was determined by judgment: on calculating the parts of the triangle A B D, the results obtained were R = 156·2 lbs. and P A R = 39° 48′. It will be observed that when the construction is made on a small scale and with common instruments we can obtain by the exercise of moderate care a result that can be trusted to within the one-hundredth part of the quantity to be determined. The same remark applies to all the questions that were solved by the constructions from which the figures in the present volume were copied. If in this example the point A were to be *pushed* along the line A P by a force of 120 lbs., the resultant would of course be determined by the construction shown in the annexed figure.

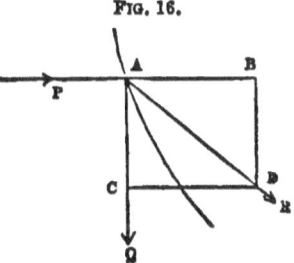

FIG. 16.

Ex. 179.—Draw A B and A C two lines at right angles to each other, a force of 50 lbs. acts from A to B, and one of 70 lbs. from A to C. Find their resultant by construction to scale.

Ex. 180.—Modify the construction of the last example, when the second force is made to act from C to A.

Ex. 181.—Draw an isosceles triangle, A B C, right-angled at C; forces of 60 lbs. act from A to B and from B to C respectively. Show by construction that the resultant is a force of about 46 lbs. and that the line representing it bisects the angle between C B and A B produced.

Ex. 182.—Draw two lines A B and A C at right angles to each other; a force of 50 lbs. acts along a line A D bisecting the angle B A C. Determine by construction the components of the force along A B and A C.

Ex. 183.—Draw A B and A C lines containing an angle of 135°; within this angle draw A D at right angles to A B; suppose a force of 100 lbs. to act from A to D. Show by construction that it is equivalent to forces of 100 and 141·4 lbs. acting respectively from A to B and from A to C.

---

* The examples in the present chapter may be worked by construction; if solved by calculation, some will be found to lead to very long arithmetical work, e.g. Ex. 184.

33. Resultant of more than two forces.—Since the rules of Arts. 27, 29, and 32 enable us to determine the resultant of any two forces (with one exception, explained in the next chapter) acting in the same plane, it is obvious that the resultant of three forces can be found, by finding the resultant of any two of the forces and then finding the resultant of that resultant and the third force, as shown in the following example. The same method can be extended to four or more forces.

Ex. 184.—Let A B C be an isosceles triangle, right-angled at C; a force of 100 lbs. acts from A to B, a force of 80 lbs. acts from A to C, a force of 80 lbs. acts from B to C. Find the resultant of the three forces.

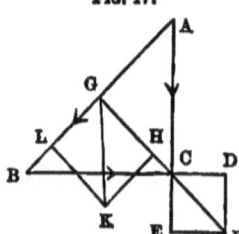

FIG. 17.

Draw the triangle A B C and mark by arrowheads the direction of the forces acting along the lines. On any scale take C D and C E to represent the forces acting from B to C and A to C respectively. Complete the parallelogram C D F E; the diagonal C F represents the resultant of the two forces of 80 lbs. Produce F C to meet A B in G, take G H equal to C F, and G L on the same scale to represent the force acting from A to B; complete the parallelogram G H K L. Then G K represents the required resultant; which is a force of 151 lbs. and acts along the line G K in the direction G to K.

Ex. 185.—Let A B C D be the corners of a square taken in order, produce A B to E, make B E equal to A B, and draw E F parallel to B C. If forces of 10 lbs. apiece act from A to B, B to C, and C to D respectively; show that their resultant will be a force of 10 lbs. acting along E F in the direction B to C.

Ex. 186.—In the last example if the force along C B has its direction reversed so as to act from C to B; show the resultant is still a force of 10 lbs. but acts along D A from D to A.

Ex. 187.—In *Ex.* 185 suppose an additional force of 15 lbs. to act from D to A; show that the resultant of the four forces is determined as follows :—produce B A to O, take A O equal four times A B, the resultant is a force of 5 lbs. acting through O parallel to D A and in the direction D to A.

Ex. 188.—Draw an equilateral triangle A B C, let a force of 20 lbs. act from A to B, one of 20 lbs. from B to C, and one of 30 lbs. from A to C; show that the resultant will be a force of 50 lbs. acting in the direction A to C along a line parallel to A C, drawn through a point P in B C such that B P is three-fifths of B C.

Ex. 189.—Determine the resultant in the last case when the direction of the force along A C is reversed ; the other forces remaining unchanged.

PARALLELOGRAM OF FORCES. 57

34. *Condition of equilibrium of three forces.*—If three forces, P, Q, and R, whose directions are not parallel, act on a body, it is necessary and sufficient for equilibrium that any one of them (P) be equal and opposite to the resultant of the other two (Q and R); the resultant of Q and R being found by the parallelogram of forces. It is worthy of remark that this condition involves the condition that the three forces act along lines which pass through a common point.

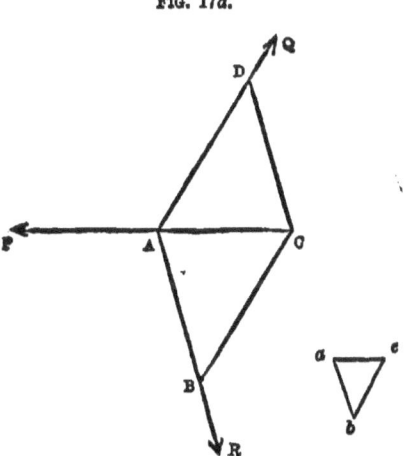

Fig. 17a.

Ex. 190.—Three ropes, P A, Q A, R A, are knotted together at the point A; on each a man pulls; the angle P A Q = 120°, Q A R = 132°, and therefore R A P = 108°; if the man who pulls on A P exerts a force of 24·5 lbs., find with what force the other men must pull that the three may balance each other.

[Produce P A to C and measure off on scale A C = 24½, this line must represent the resultant of Q and R, therefore drawing B C parallel to A Q and C D parallel to A R, the forces Q and R will be represented by the lines A D and A B respectively, and can be found by measuring them on scale or by calculating their lengths by trigonometry.]

Ans. Q = 31·35 lbs. R = 28·55 lbs.

Ex. 191.—If in the last example the rope A P were pulled with a force of 28 lbs.; A Q with a force of 35 lbs.; and A R with a force of 12 lbs., determine the angles P A Q, Q A R, and R A P.

Ans. Q A R = 134° 9'. R A P = 63° 46'. P A Q = 162° 5'.

Ex. 192.—If in *Ex.* 190 P A is pulled by a force of 28 lbs., Q A by a force of 40 lbs., and the angle P A Q is 135°, determine the magnitude of the force along R A, when they are in equilibrium, and the angles R A Q, and R A P.

Ans. Q A R = 135° 34' 30".
R A P = 89° 25' 30".
R = 28·28 lbs.

Ex. 193.—Let A B C D be a rectangle; A B is 7 ft. long, B C is 3 ft. long; join E F the middle points of A D and B C; at E act two forces, P and Q, in such directions that P E F = 45° and Q E F = 30°; the force P = 520 lbs.; find Q (1) when the resultant of P and Q acts through B, (2) when it acts through F, (3) when it acts through C.

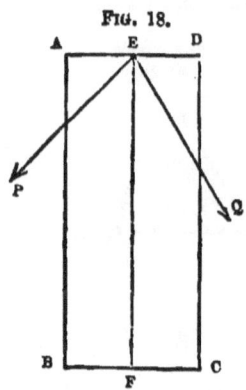

Fig. 18.

Ans. (1) 421 lbs. (2) 735 lbs. (3) 1420 lbs.

Ex. 194.—A boat is dragged along a stream 50 feet wide by men on each bank; the length of each rope from its point of attachment to the bank is 72 feet; each rope is pulled by a force of 7 cwt.; the boat moves straight down the middle of the stream; determine the resultant force in that direction. If, in the next place, one of the ropes is shortened by 10 ft., by how much must the force along it be diminished that the direction of the resultant force on the boat may be unchanged? What will now be the magnitude of the resultant force?

Ans. (1) 13·13 cwt. (2) $\frac{35}{36}$ cwt. (3) 12·08 cwt.

35. *Note.*—In a large number of questions the *solidity* of the bodies concerned does not enter the question, except so far as it affects the determination of their weight; it being manifest from the conditions of the question that all the forces act in a single plane; in many such cases a complete enunciation would be long and troublesome to the reader, while an imperfect enunciation is without any real ambiguity; wherever this happens the imperfect enunciation will be preferred; thus, in the next example all the forces are supposed to act in a vertical plane passing through the centre of gravity; and the diagram ought, strictly speaking, to be that given above, fig. 19, in which the dark lines are all that are shown in the figure which accompanies the example.

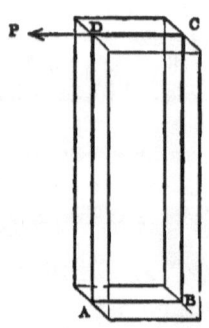

Fig. 19.

PARALLELOGRAM OF FORCES. 59

Ex. 195.—Let A B C D represent a rectangular mass of oak $2\frac{1}{4}$ ft. thick, A B and A D are respectively 4 ft. and 12 ft. long; it is pulled at D by a horizontal force P, and is prevented from sliding by a small obstacle at A; find P when the mass of oak is on the point of turning round A.   *Ans.* $1050\frac{3}{4}$ lbs.

FIG. 20.

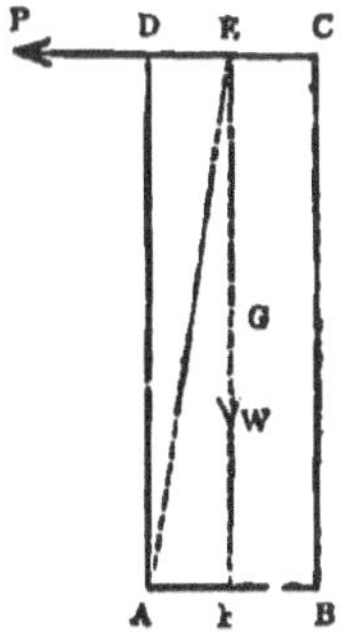

[Find G the centre of gravity of A B C D, and through it draw the vertical line B F meeting D C in E, the weight will act along the line E F, and the resultant of P and W must pass through A since the body is on the point of turning round A;—the remainder of the investigation is conducted as before.]

Ex. 196.—A B C D represents a block of oak 35 ft. long and 3 ft. square; the point A is kept from sliding; the body is held by a rope C E 60 ft. long in such a position that the angle D A E is 57°; determine the direction and amount of the pressure on the point A, and the tension of the rope.

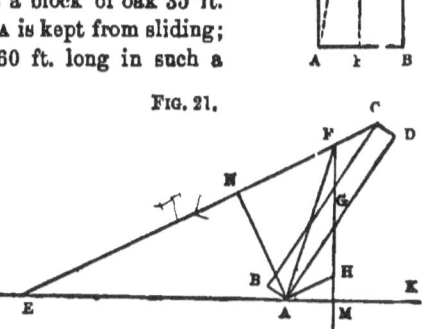

FIG. 21.

[Through G, the centre of gravity of the block, draw the vertical line G W, meeting B C in F; the forces that balance upon the block are the weight W, the tension T of the rope and the resistance of the ground at the point A; this force must pass through F, and then we have three forces acting in known directions through F; &c.]

Ans. (1) Tension 8453 lbs. (2) Pressure on ground, 23,900 lbs. making with vertical an angle of 17° 39'.

Ex. 197.—On every foot of the length of a wall of brickwork whose section is A B C D a force acts on the upper angle C, in a direction making an angle of 45° with the inner side B C; determine this force when the resultant of it and of the weight of the wall passes through the angle A at the bottom of the wall; the height of the wall being 20 ft. and its thickness 4 ft.   *Ans.* 1584 lbs.

Ex. 198.—If in the last example there were a bracket C E on the inside of the wall, C E being in the same line with D C, the top of the wall, and the force (inclined at the same angle as before) were applied at E, 2 ft. from the inside of the wall; what must be its magnitude if the resultant of it and of the weight of one foot of the length of the wall passes through the point A? determine also the point in which the resultant would cut A B, the base of the wall, if the force were the same as in the last example.

Ans. (1) 1810 lbs. (2) $2\frac{2}{5}$ in.

Ex. 199.—If A B are two points in the same horizontal line 10 ft. apart; A C and B C ropes 10 ft. and 5 ft. long respectively tied by the point c to a weight w of 3 cwt.; determine the tension of each rope.

Ans. Tension of A C = 86·8 lbs. Tension of B C = 303·6 lbs.

[The triangle A B C is, of course, fixed in position, the weight w will act vertically through c and be supported by the reactions of A and B transmitted along the ropes.]

36. *Triangle of forces.*—The reader will remark on reference to fig. 17a, that if lines be drawn parallel to the directions of P, Q, and R respectively, they will form a triangle abc similar to A B C, whose sides will therefore have to each other the same ratios as the forces, each side being homologous to that force to whose direction it is parallel. This fact is frequently of great importance. Thus in Ex. 195, if A E be joined, the sides of the triangle A E F are respectively parallel to the forces, so that

$$EF : FA :: W : P$$

and since E F, F A, and W are known, P is at once found. Again, in Ex. 196, if A H be drawn parallel to E C, the sides of the triangle A F H will be parallel to the forces, so that

$$FH : HA :: W : T$$

and

$$FH : AF :: W : R$$

from which T, the tension of the rope, and R, the pressure on the ground (or the reaction of the ground to which it is equal and opposite) are at once found. Hence also can be deduced a very simple construction for finding the resultant of any two forces P and Q. Referring to fig. 47, p. 83, draw any line bc parallel to A P in the direction A to P; from c draw ca parallel to A Q in the direction A to Q and of such a length that

$$P : Q :: bc :: ca$$

join ab; if R' is the required resultant we shall have

$$R' : P :: ab : bc$$

REACTION OF SMOOTH SURFACES. 61

and R' will act at A along a line parallel to ba and in the direction b to a. If the force at A had its direction reversed so as to act as R in fig. 47, viz. in the direction a to b, it will, as we have already seen, balance the forces P and Q.

37. *Reaction of smooth surfaces.*—We have already seen (Art. 28) that if a body is urged against a second body and thereby kept at rest, the second body reacts against the first. We have now to add that if we suppose the bodies to be perfectly smooth the reaction can only be exerted in the direction of the common perpendicular to the surfaces of contact. The supposition of perfect smoothness is commonly very far from the truth, but by making it we avoid a great deal of complexity in our reasoning and results. So long as both surfaces resist the tendency of the pressures to crush them any needful *amount* of reaction can be supplied, but, as already stated, only in the direction of the perpendicular, if the surfaces are perfectly smooth.

Ex. 200.—A body whose weight is w rests on a smooth plane A B inclined at a given angle B A C to the horizon; determine the force P which acting parallel to the plane will just support the body.

FIG. 22.

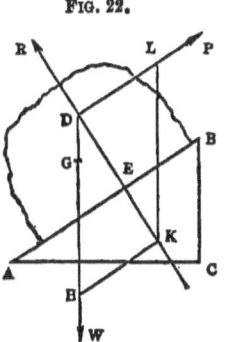

Find G, the centre of gravity of the body, and through it draw a vertical line G W, cutting in D the direction of P; through D draw D E at right angles to A B, then R, the reaction of the plane, must act along E D, and we have three forces P, W, and R in equilibrium acting in known directions; and since the magnitude of W is known, that of R and P can be found by the usual construction: viz. take D H to represent W, draw H K parallel to D P, and K L parallel to D H, then D K is proportional to R and D L represents P.

Ex. 201.—In the last example show that P : R : W : : BC : CA : AB.

Ex. 202.—In *Ex.* 200 if A were 45° and W were 1000 lbs., find P and R.
Ans. 707 lbs. (each).

Ex. 203.—In *Ex.* 200 if A were 30° and P were 200 lbs. what weight could P support? *Ans.* 400 lbs.

Ex. 204.—If a cylinder whose weight is w rests between two planes A B and A C inclined at different angles to the horizon (as shown in the figure); determine the pressures on the planes.

FIG. 23.

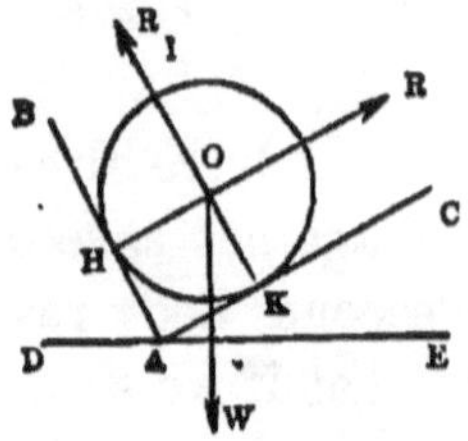

The weight w will act vertically through o, and will be supported by the reactions R and R_1 of the planes A B and A C : as these forces must act at right angles to the planes respectively, their directions will pass through o, and their magnitudes can be determined as usual. The pressures *on* the planes are, of course, equal and opposite to R and R_1 respectively.

Ex. 205.—In the last case if B A D and C A E are angles of 30° and w a weight of 112 lbs., determine the pressures.

Ans. 64·6 lbs. apiece.

Ex. 206.—Explain the modification that Ex. 204 undergoes if both A B and A C are on the same side of the vertical line drawn through A ; and determine the pressures when w equals 112 lbs. and C A B and B A C are each 30°.

Ans. R = 112 lbs., R_1 = 194 lbs.

38. *Transmission of force by means of a perfectly flexible cord.*—If a cord is stretched by two equal forces P and Q, one acting at each end, they will balance each other, and the tension of the cord is equal to either (Art.

FIG. 24.

28); suppose the cord to pass round a portion A B of a fixed surface, as shown in the figure, the portions A P and B Q of the cord will be straight, while A B will take the form of the surface (which is supposed to be convex), and if P and Q continue in equilibrium they must be exactly equal, provided the surface A B is perfectly smooth and the cord perfectly flexible ; conditions which are supposed to hold good unless the contrary is specified. Hence force is transmitted without diminution by means of a perfectly flexible cord which passes over perfectly smooth surfaces.

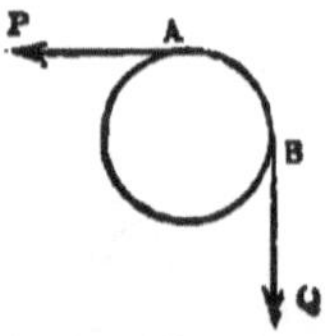

Ex. 207.—Let A and B be two perfectly smooth points in the same horizontal line, and let w be a weight of 100 lbs. tied at c to cords which pass over A and B, and let w be supported by weights P and Q tied to the ends of these cords respectively, and suppose the whole to come to rest in such a position that B A C equals 30° and A C B equals 90° ; find P and Q.

Since the forces P and Q are transmitted without diminution to C, W is supported by a force P acting along C A and by Q along C B. Hence draw C c vertically and such that on scale it represents the vertical force which balances W, and complete the parallelogram $a c b'c$, then $c a$ and $c b$ represent the transmitted forces that support W:—hence P equals 50 lbs., and Q equals 86·6 lbs.

FIG. 25.

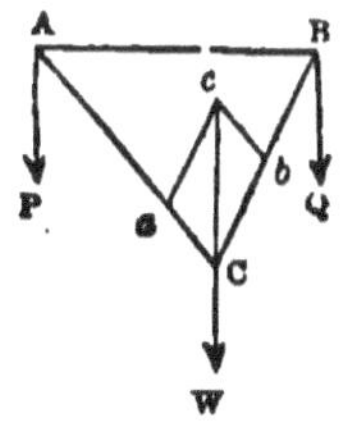

Ex. 208.—In the last example show that the pressures on A and B are equal to 86·6 lbs. and 167·3 lbs. and that their directions bisect the angles P A C and Q B C respectively.

39. *The principle of moments.*—A large class of questions has reference to the equilibrium of a body one point of which is fixed; in these cases it is frequently sufficient to determine the relation between the forces that tend to turn the body round the point, the actual amount and direction of the pressure on the point not being required; under these circumstances the relation sought is given at once by a principle called the PRINCIPLE OF MOMENTS. The definition of the moment of a force is as follows: If P represents any force, and A is any point, and A N is a perpendicular let fall on P's direction, then if the number of units of force in P is multiplied by the number of units of length in A N, the product is called the moment of the force P with reference to the point A. The principle of moments in its general form will be found in the next chapter; for present purposes the following statement will be sufficient. *If any number of forces acting in the same plane keep a body in equilibrium round a fixed point, and if their moments with reference to that point be taken, the sum of the moments of those forces which tend to turn the body from right to left round the fixed point, will equal the sum of the moments of those forces which tend to turn the body from left to right.*

The following case will exemplify the mode of applying

the principle of moments. In Ex. 196, let it be required only to determine the tension of the rope. Construct the figure to scale (see fig. *b*); determine G, the centre of gravity of the block, draw the vertical line G W, cutting A K in M; draw A N at right angles to C E; if T is the tension of the rope, and W the weight of the block which can be found to equal 18,388 lbs., then the moments of T and W are respectively A N × T and A M × 18,388; and the principle of moments assures us that these two are equal. In the construction from which fig. *b* was drawn, the scale employed was 1 inch to 10 feet; and it was found that A M equals 8·25 ft., and A N equals 18·1 ft.; hence was obtained for T a value of 8381 lbs.; the value of T as determined by calculation is 8453 lbs.

The student is recommended, as an exercise, to work by this method all the previous examples in the present chapter to which it can be readily applied, viz. Ex. 172, 173, 174, 175, 176, 177, 195, 197, 198.

40. *The lever.*—This is the name given to a rod capable of turning round a fixed point (called the fulcrum) and acted on by the reaction of the fixed point and by two other forces: as most machines are used for the purpose of moving bodies, one of these forces is to be overcome, or opposes motion, and this is called the *weight*, the other force which produces the motion is called the *power*. When the lever is in equilibrium the moments of the power and the weight with reference to the fulcrum must be equal; and, of course, those forces will tend to turn the lever in different directions round the fulcrum. Levers are sometimes classified as belonging to the first, second, and third orders respectively; those of the first order have the fulcrum between the power and the weight, as the beam of a pair of scales, or a poker when used to stir a fire; levers of the second order have the weight between the power and the fulcrum, as a crowbar

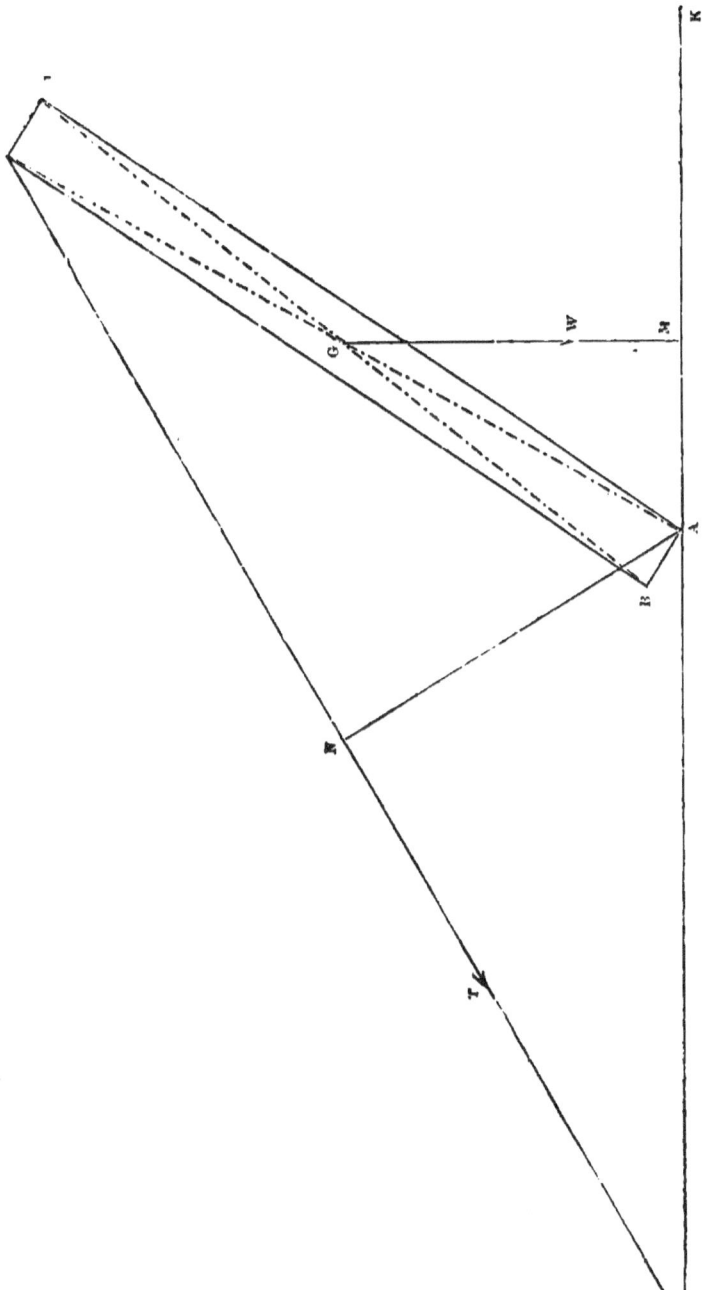

Fig. *b*, page 64.

THE LEVER.

when used to lift a weight one end resting on the ground, or an oar used in rowing, in which case the water is the fulcrum; levers of the third order have the power between the weight and the fulcrum, as the limbs of animals, e.g. when a man has a weight in his hand and extends his arm the forearm is a lever of which the elbow is the fulcrum and the power is the contractile force of the large muscle of the upper-arm acting by means of tendons fastened into one of the bones of the forearm—of course in such a case the *power* must be very much larger than the weight. Many simple instruments consist of two levers fastened together by, and capable of turning round, a common fulcrum; these are called double levers, and are classified as double levers of the first, second, and third orders respectively; a pair of scissors and of pincers are of the first order, a pair of nut-crackers of the second order, and a pair of tongs of the third order.

Ex. 209.—Let A B be a lever 16 ft. long; movable about a fulcrum D at a distance of 6 ft. from B; a weight of 28 lbs. is suspended from A and from B a weight of 336 lbs. Find the weight that must be hung at E (which is 7 ft. from D) to balance the lever. *Ans.* 248 lbs.

Ex. 210.—Let A B be a lever 8 ft. long, the end A resting on a fulcrum; a weight of 40 lbs. is hung at C, 3 ft. from A. The lever is held in a horizontal position by a force P, acting vertically upward at B. Find P and the pressure on fulcrum.
Ans. P = 15 lbs. Pressure on fulcrum 25 lbs.

Ex. 211.—Let A B and D E be levers turning on fulcrums B and F, connected by a bar D C, loosely jointed at D and C; A B and D E are respectively 5 and 6 ft. long, A C is 3 ft., and F E is 9 in. long; the weight P at E equals 1000 lbs. and is balanced by Q acting at A; find Q.

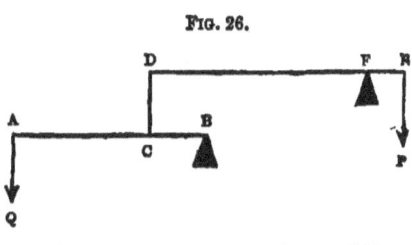

FIG. 26.

Ans. 57½ lbs.

Ex. 212.—A crane C B D is sustained in a vertical position by the tension
F

of a rope A E; its dimensions are as follows—B C, B D, B E, and A C respectively 19, 13, 1½, and 16 ft. long; the angle C B D equals 108°; a weight P of 7 cwt. is supported by a rope that passes over a pulley D and is fastened to C. Determine the tension of the rope A E, the weight of the crane and the dimensions of the pulley being neglected.   *Ans.* 7·329 cwt.

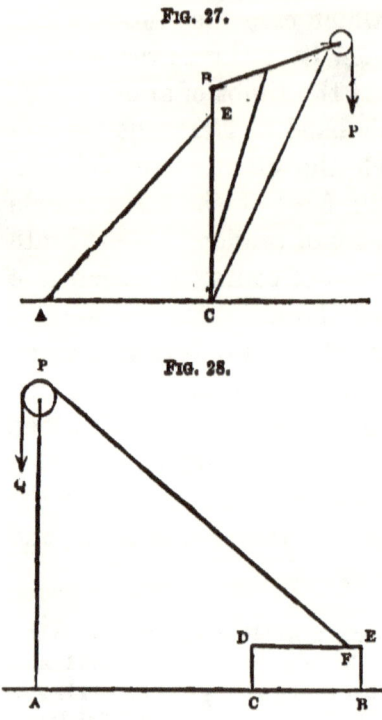

Fig. 27.

Fig. 28.

Ex. 213.—Let B C D E represent a block of Portland stone whose dimensions are 5 ft. long, 2 ft. high, and 2½ ft. wide; a rope F P Q is attached to it, which after passing over a pulley P, is pulled vertically downward by a force Q, which is just sufficient to raise the block: determine Q on the supposition that the dimensions of the pulley can be neglected, having given that E F equals 6 in. and B A and A P respectively 15 and 13 ft., the point A being vertically under P.
Ans. 1942 lbs.

Ex. 214.—In the last example determine the amount and direction of the pressure on the ground through the point C.

41. *The steel-yard.*—If a beam A B rests on a fine axis passing through its centre of gravity (G), and on the arm B G is placed a movable weight W, then if a substance equal in weight to W is suspended from A, the beam will balance when W is at a distance from G equal to A G; if the substance equals twice the weight of W, the beam will balance when W's distance from G equals twice A G; and so on in any proportion. Hence, if the beam is made heavy at the end A, so that G is very near A, the arm B G can be divided into *equal* divisions which shall indicate

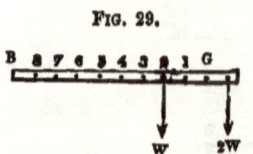

Fig. 29.

the weight of a substance suspended at A by means of the *position* occupied by W when it balances that substance. An instrument constructed on this principle is called a steel-yard, and is used when heavy substances have to be weighed, and extreme accuracy is not required; the advantage it possesses arises from the fact that the weights employed are much less heavy than the substance to be weighed. A very common application of the principle of the steel-yard can be seen in the weighing machines employed at most railway stations.

Ex. 215.—Show that the graduations of the steel-yard must be equal even if the centre of gravity of the beam do not coincide with the axis; but that the graduations must begin from that point at which the movable weight would hold the beam in a horizontal position.

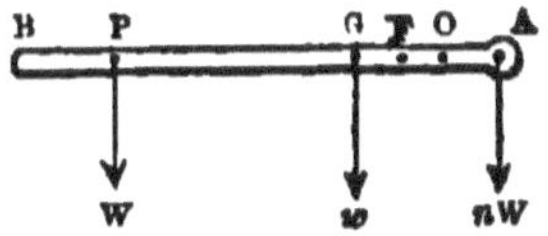

[Let F be the fulcrum, G the centre of gravity, and w the weight of the beam; suppose that O is so chosen that W at O balances w at G, then

$$F O \times W = F G \times w$$

FIG. 30.

Now, suppose that a body weighing n W is hung at A, and that the beam is kept horizontal by W at P; then, measuring moments round F, we have

$$F P \times W + F G \times w = F A \times n W$$

Therefore, by addition,

$$O P \times W = F A \times n W$$

Hence, if the weight of the body is W, OP must equal FA; if twice W, OP must equal twice FA, and so on in any proportion.]

42. *The equilibrium of walls.*—The question What is the force which, acting in a certain specified manner on a given wall, will be just sufficient to overthrow it? can be answered by an application of the Principle of Moments; the general method of considering this important question is as follows:—

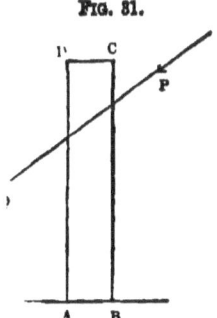

FIG. 31.

Let A B C D represent the section of a wall, the base A B being on the level of the ground; let it be acted on by a

F 2

force P along the line P Q: now, it is considered that a wall, to be stable, must be capable of standing irrespectively of the adhesion of the mortar;* hence, if we suppose B D to be a continuous mass, and simply to *rest* on the section A B, and determine the force P which will be on the point of turning the mass round the point A, we shall obtain the greatest force acting in the manner specified that the wall can support; the force is, of course, determined by the rule that its moment with reference to the point A equals the moment of the weight of the wall with reference to the same point.

Ex. 216.—A wall of brickwork 2 ft. thick and 25 ft. high sustains on the inner edge of its summit a certain pressure on every foot of its length; the direction of this pressure is inclined to the horizon at an angle of 60°; find its amount when it will just not overthrow the wall. (See fig. c.)

Draw the section of the wall A B C to scale; make the angle B A N equal to 30°, then the pressure P acts along the line P N; draw C N perpendicular to P N; through G, the centre of gravity, draw the vertical line G M, cutting C B in M; the principle of moments gives us

$$P \times CN = W \times CM$$

The weight W equals 5600 lbs.; C M equals 1 ft.; C N, as obtained by measurement, equals 10·8 ft.; whence P equals 518 lbs. When P is found by calculation it equals 520 lbs.

Ex. 217.—In the last example suppose the pressure to be applied by means of a bracket, at a horizontal distance of 3 ft. from the inner edge of the summit; determine its amount when it will just not overthrow the wall. *Ans.* 685 lbs.

43. *The effect of buttresses*.—Let fig. 32 represent the elevation of a wall, fig. 33 its plan, and fig. 34 its section made along the line A B; if now we neglect the

---

* 'Though ordinary mortar sometimes attains in the course of years a tenacity equal to that of limestone, yet, when fresh, its tenacity is too small to be relied on in practice as a means of resisting tension at the joints of the structure, so that a structure of masonry or brickwork, requiring, as it does, to possess stability while the mortar is fresh, ought to be designed on the supposition that the joints have no appreciable tenacity.'—Rankine, *Applied Mechanics*, p. 227.

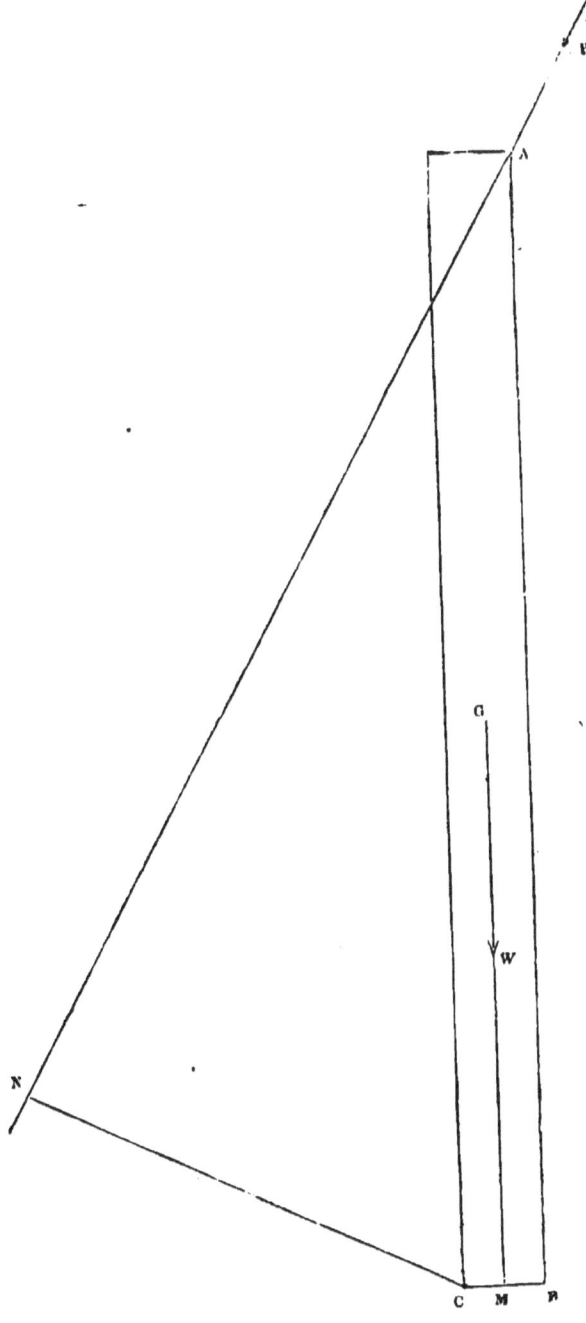

Fig. c, page 68.

weight of the buttresses their effect in supporting the wall will be understood by inspecting fig. 33; for it is manifest that the

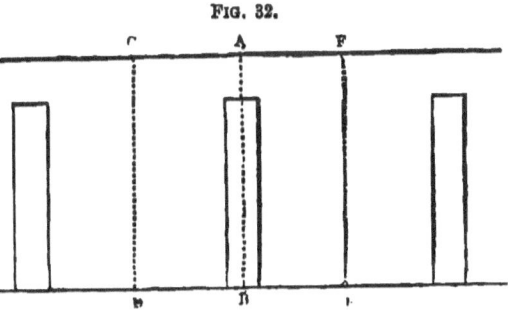

FIG. 32.

wall would fall by being caused to turn round the line X Y; but, if the buttresses were removed, by being caused to turn round the line xy; so that, in the former case, the moments must be measured round M (fig. 34), in the latter round K: in other words, the introduction of buttresses diminishes the moment of P, and increases that of the weight of the wall. Their useful effect is still farther increased by

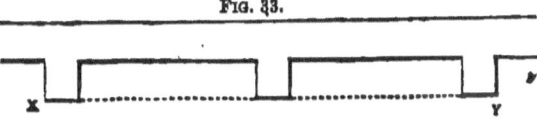

FIG. 33.

the fact that if the moment of the weight of the buttress is taken into account, it increases the moment of the weight of the wall.

It is to be observed that if C D (fig. 32) and E F be drawn at equal distances from A B, and at a distance from each other equal to the distance between the centres of two consecutive buttresses, then we may consider that the total pressure on C F is supported by the weight of the portion of the wall between C D and E F, and by the weight of the buttress.

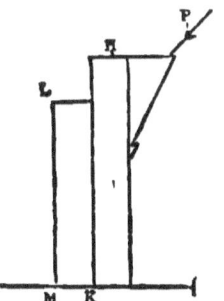

FIG. 34.

It must be remembered that the above explanation applies to the case in which the pressure is distributed uniformly along the top of the wall;

which in this case is supposed to be so strong as not to bulge between the buttresses. In many instances, however, particularly in large ecclesiastical buildings, the whole, or nearly the whole, weight of the roof and its lateral thrust act on the buttresses, and not on the portion of the wall between the buttresses; in such cases the wall serves as a curtain between the buttresses, and not as a support to the roof, and, of course, the moment of the lateral thrust must equal that of the weight of the buttress.

Fig. 35.

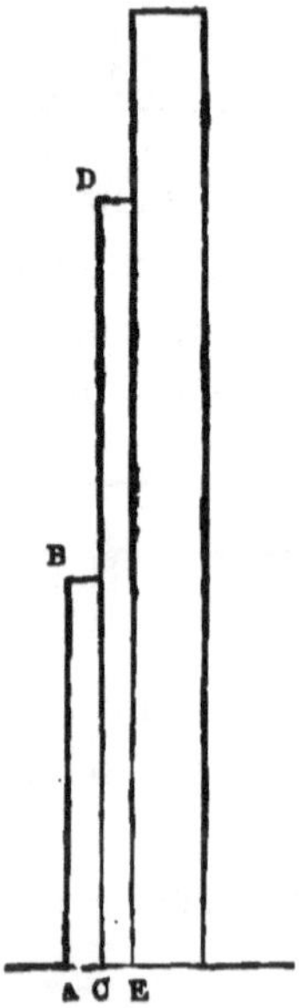

Ex. 218.—In the last example, if the wall were supported by buttresses 2 ft. thick,* to what can the pressure on each foot of the length of the wall be increased without overthrowing it—the weight of the buttresses being neglected? *Ans.* 2609 lbs.

Ex. 219.—In *Ex.* 216, suppose the wall to be supported by counterforts reaching to the top of the wall, 1 ft. thick, 1 ft. wide, and 10 ft. apart from centre to centre, determine the pressure on each foot of the length of the wall that can be supported—(1) when the direction of the pressure is inclined at an angle of 60° to the horizon; (2) when the direction is inclined at an angle of 30° to the horizon. *Ans.* (1) 1145 lbs. (2) 562·8 lbs.

Ex. 220.—In each case of the last example determine to what the pressure can be increased if the buttress assumes the form of a Gothic buttress, as indicated in the annexed diagram, where A C and C E are each a foot square, and C D and A B are respectively 20 and 10 ft. high. *Ans.* (1) 1903 lbs. (2) 875 lbs.

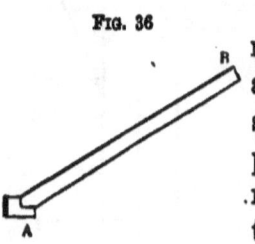

Fig. 36

44. *The thrust of props.*—Let A B represent a beam or prop resting on a fixed support at the end A; and suppose it to be acted on by certain pressures which are balanced by the reaction of the end A. That part of the reaction which acts along the

---

* The *thickness* of a pier or buttress is supposed to be measured in a direction perpendicular to the face of the wall.

axis of the beam A B is called the thrust of the prop, and is, of course, equal to the thrust produced by the pressures on the prop, the two being equal and opposite. If no pressure acts on the beam except at the end B, it is plain that the whole reaction from A must pass along the beam. In the following question, which concerns the thrust of props, it will be assumed that the *thickness* of the prop can be neglected, except so far as it affects its weight.

Ex. 221.—A wall of brickwork, 25 ft. high and 2 ft. thick, sustains on the inner edge of its summit a pressure of 1000 lbs. on every foot of its length, whose direction is inclined at an angle of 65° to the vertical; it is supported at every 5 ft. of its length by a prop 25 ft. long, resting against a point 3 ft. from the top; determine the thrust on the prop.

Ans. 7758 lbs.

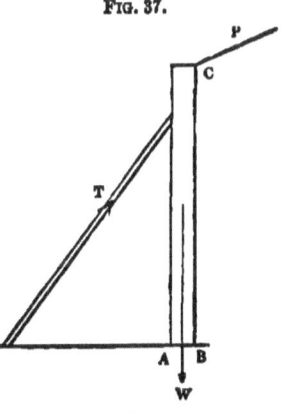

Fig. 37.

[If the annexed figure represent a section of the wall and prop, the forces acting are w, the weight of the wall, P, the pressure on the summit, and these are balanced by T, the thrust of the prop, and the reaction of the ground A B: now, unless the prop is wedged up against the wall, there will not be more reaction than is *just* sufficient to support the wall; consequently the resultant of P, W, and T must pass through A, at which point it will be balanced by the reaction of the ground; hence, by measuring moments round A we can find T.]

45. *The thrust along rods connected by a smooth hinge.*—Let A B be a rod capable of moving freely round a joint or hinge at A; if it were acted on by a force it would turn round A, unless the force acted through A. Suppose two such rods, A B and A C, to be connected by a perfectly smooth joint at A, while their ends B and C rest against immovable obstacles,

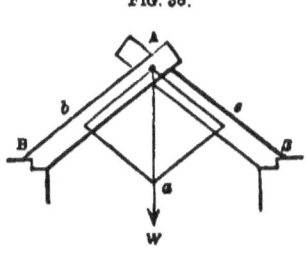

Fig. 38.

and let us suppose the rods to be geometrical lines and without weight; let a weight w be hung at A, and let it be required to determine the pressures against the fixed obstacles caused by w. Now (Art. 44), the reactions at B and C, which support w, must pass along B A and C A; hence, if we take A a to represent w and complete the parallelogram A b a c, the lines A b and A c will represent the *thrusts* caused by w along A B and A C, and these are respectively equal to the reactions by which they are balanced (Art. 28).

46. *The thrust along a rafter.*—The case which we have just explained enables us to determine the thrust produced on the summit of a wall by each rafter of an isosceles roof: let A B, A C, represent two of the principal rafters of such a roof, and let the whole weight sustained by each rafter (including its own weight) be represented by w; this weight will act at the middle point of the rafter, and therefore can be replaced by weights equal to ½w acting at each end of the rafter; so that the whole weight sustained by A B and A C may be distributed as shown in the figure, viz. it will be equivalent to w acting at A, ½w at B, and ½w at C; then the thrusts along the rafter (T) will be produced by w acting at A, and can be determined as explained above, viz. take A p to represent w, and complete the parallelogram A r p q, then A r and A q represent the thrusts in question: the total pressure on the wall at B will be found by compounding T with ½w. When the determination of the pressure is made for the purpose of ascertaining whether a certain wall will support the roof, it is much better not to compound the pressures T and ½w, but to regard the wall as acted on by the two uncompounded forces.

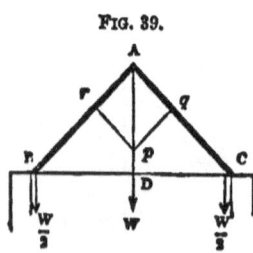

FIG. 39.

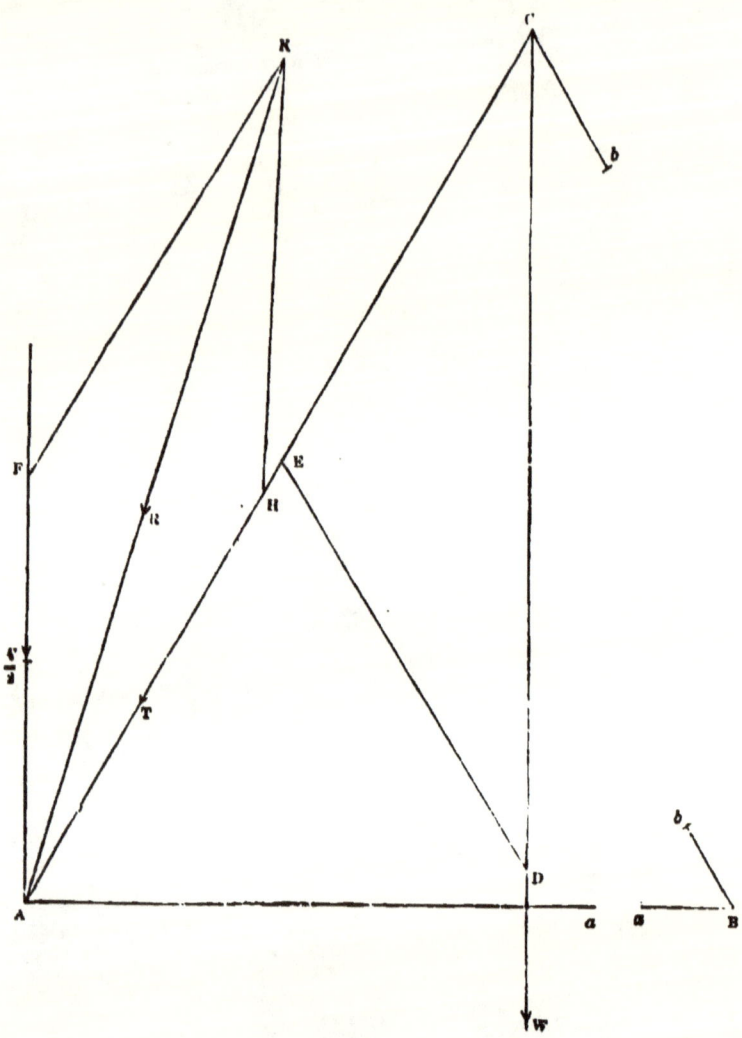

Fig. *d*, page 73.

THRUST ALONG RAFTERS. 73

Ex. 222.—There is a roof weighing 25 lbs. per square foot, the pitch of which is 60°; the distance between the side walls is 30 ft.; determine the magnitude and direction of the pressure on the foot of each rafter, the rafters being 5 ft. apart. (See fig. *d*.)

Let A B C represent the roof; then the weight (W) supported on each rafter equals 3750 lbs.; hence, when the weight is distributed, we have W at C, $\frac{W}{2}$ at A, and $\frac{W}{2}$ at B; draw C W vertical, and take C D to represent 3750 lbs.; draw D E parallel to B C [which is broken in the figure as indicated by the letters *a, a* and *b, b*]; then C E represents the thrust (T) along the rafter. The total pressure on the wall (R) is the resultant of $\frac{W}{2}$ and T acting at A; take A F to represent on scale 1875 lbs. and A H equal to C E; complete the parallelogram F H; then A K gives the magnitude and direction of the resultant R; it was found from fig. *d* that R equals 3885 lbs. and the angle K A F equals 16°; the results given by calculation are that R equals 3903 lbs., and that the angle K A F equals 16° 6′.

Ex. 223.—If in the last example the walls were 20 ft. high, 2½ ft. thick, and of Portland stone, would they support the roof?

Ans. The wall will stand—the excess of the moment of the weight of 5 ft. of its length over that of the thrust being 29,620.

Ex. 224.—If in the last example the walls be supported by buttresses 20 ft. apart from centre to centre, 15 ft. high, 2 ft. wide, and 2¾ ft. thick, would these support the wall if its thickness were reduced to 1½ ft.; and what would be the excess of the moment tending to support 20 ft. of the length of the wall over that which tends to overthrow it?

Ans. (1) Yes. (2) 221,000.

Ex. 225.—Show that the total pressure on each wall is equivalent to a vertical pressure W, and a horizontal pressure W × B C ÷ 4 A D. (Art. 46.)

Ex. 226. In the case of an equilateral roof show that the horizontal pressure equals 0·29 W.

FIG. 40.

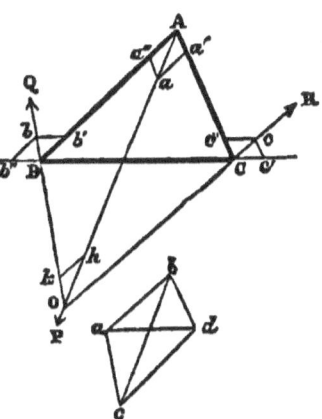

**47. *The equilibrium of a triangular frame.*—A triangular frame A B C consisting of rods loosely jointed at the angles is in equilibrium under the action of three forces acting one at each angle; it is required to find the thrust or tension to which each rod is subjected. Let the forces P, Q, R act at the angles A, B, C respectively, and suppose their directions to pass through a point O. In the first place they must be in equilibrium, or otherwise they would make the frame itself move; draw kh parallel to O R, then we know that the forces, P, Q, R are proportional to ho, ok, and kh. Take Aa, Bb, Cc equal to oh, ok, and kh respectively, and complete the parallelograms A $a'\,a\,a''$, B $b'\,b\,b''$, C $c'\,c\,c''$; it will be found that A $a''=$ B b', B $b''=$ C c', and C $c''=$ A a'. We see therefore that each rod is under the action of a pair of equal forces which tend to crush A B and A C, and to stretch B C; these forces are severally proportional to A a'', C c'', and B b''. These lines therefore measure the thrusts of A B and A C and the tension of B C. The magnitudes of these forces can be obtained by a more simple construction, thus:—Draw bc parallel to A O, and containing as many units of length as P contains units of force; draw ca, ab parallel to O B and O C respectively; draw ad parallel to B C, dc to A B, and join bd (the line bd is parallel to A C, as the student can prove), and we shall have ca, ab, ad, dc, bd containing as many units of length respectively as the forces Q, R, the tension of B C, and the thrusts of A B and A C contain units of force. This is plain from an inspection of the figure, since abc is the triangle of forces for the three forces in equilibrium at the point O, cda for the three forces at the point B, dab for the forces at C, and bcd for those at A. Referring to Art. 46; if the ends B and C of the rafters are connected by a beam B C (fig. 39), called a tie beam, they will constitute a triangular frame like that we have just considered; it can be easily shown that the tie beam is subject to a tension equal to the horizontal thrust of each rafter, i.e. equal to W × B C ÷ 4 A D (Ex. 225). Under these circumstances the roof

TRIANGULAR FRAME. 75

will act on the walls merely by its weight, and each wall will, of course, support half the whole weight of the roof.

Ex. 227.—If in fig. 40 the point o fall within the triangle, show that all the bars will be compressed or all stretched.

Ex. 228.—Two rafters A B and A C are each 20 ft. long, their feet are tied by a wrought-iron rod B C whose length is 35 ft., and a weight of 1 ton is suspended from A ; determine the tension it produces on the tie, the weight of the rafters, &c., being neglected. If the rod have a section of a quarter of a square inch, determine the weight that must be suspended at A to break it. *Ans.* (1) 2024 lbs. (2) 18,590 lbs.

Ex. 229.—There is a roof whose pitch is 22° 30', the rafters are 40 ft. long ; the weight of each square foot of roofing is 18 lbs. ; determine the diameter of the wrought-iron tie necessary to hold the feet of the principal rafters with safety, supposing them 10 ft. apart. *Ans.* 1·28 inches.

48. *Note.*—The foregoing remarks as to the thrusts of the rafters and the tension of the tie beam, apply to the cases in which the joints are perfectly smooth : as this is never the case, the thrusts, &c., may not equal the calculated amount ; but it is generally considered that reliance should never be placed on the resistance offered by a joint to the revolution of a rod round it. It will be instructive, however, to consider the case in which the rods and the joint at A (fig. 41) are perfectly rigid. Suppose two points, b and c, to be taken near to A, and joined by a rod bc; if this rod were inextensible, and if there were no tendency in the materials to give either by crushing or tearing at b and c, then would bc act the part of a tie beam, and there would be no horizontal thrust on the wall, which, as before, would merely have to support the *weight* of the roof.

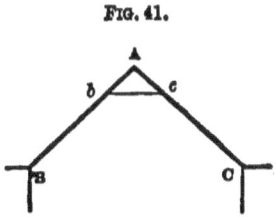

FIG. 41.

If we suppose the rod bc to be replaced by a metal plate firmly fastened to the beams, as shown by $abdc$ in fig. 42, this would tend to render the attachment of the beams rigid, the horizontal thrust being more or less neutralised by the resistance by the materials to crushing on the bolts, and to the tearing of the plate across ad. Hence, under all circumstances, the walls have to sustain the whole weight of the roof, and besides this, a horizontal thrust which will more nearly equal W × B C ÷ 4 A D, as the joint is less rigid

FIG. 42.

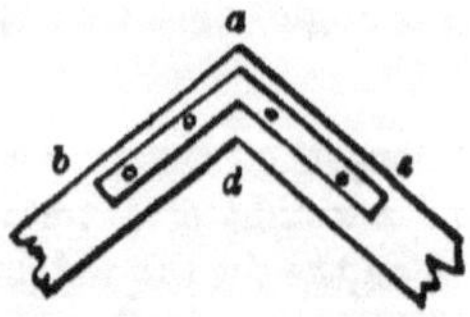

CHAPTER IV.

THE FUNDAMENTAL THEOREMS OF STATICS.

49. *Axioms.*—The following chapter contains demonstrations of the fundamental theorems of statics, so far as forces acting *in one and the same plane* on a rigid body are concerned. It may be well to invite the reader's attention to the order of proof adopted. In the first place the case of two forces and their resultant is fully discussed, together with the conditions of the equilibrium of three forces, and the case in which two forces do not have a resultant. In the next place, the results obtained for two forces are extended to any number of forces. Lastly, a peculiar property of parallel forces—the possession of a 'centre'— is proved. The demonstrations are of a very abstract character, and should be thoroughly mastered. Applications of several of the theorems have been already given in Chapter III., and many more will be found in the succeeding chapters. The demonstrations are based on certain assumed elementary principles or axioms. The assumption of these principles is, of course, not arbitrary, but justified by experience of the action of forces. The axioms are as follow:—

Ax. 1. The line which represents the resultant of two forces acting at a point, falls within the angle made by the lines that represent those forces. (See Art. 25.)

Ax. 2. If two *equal* forces act at a point, the line that represents their resultant bisects the angle between the lines that represent those forces.

Ax. 3. If a force acts upon a body, it may be supposed to act indifferently at any point in the line of its action, provided that point is rigidly connected with the body.

Ax. 4. It is *necessary* and *sufficient* for the equilibrium of any system of forces, that one of them be equal and opposite to the resultant of all the rest.

Ax. 5. If a system of forces in equilibrium be imposed on or removed from any system of forces, it will not affect the equilibrium of that system, if it be in equilibrium, nor its resultant, if it have a resultant.

Proposition 3.

The principle of the parallelogram of forces (Art. 32) *is true of the direction of the resultant of two equal forces.*

FIG. 43.

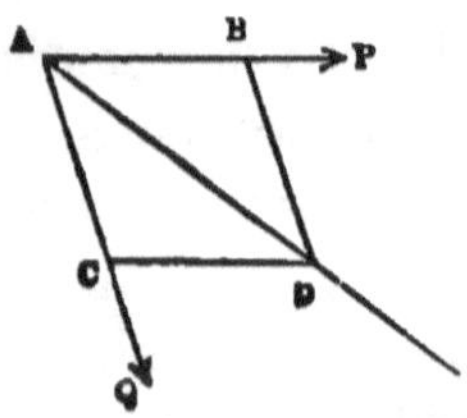

Let the equal forces P and Q act at the point A along the lines A P and A Q; let A B represent the force P, and A C the force Q, then will A B equal A C; complete the parallelogram A B C D, and draw the diagonal A D. We are to show that the resultant of P and Q acts along the line A D.

Since A C equals A B it equals C D, therefore the angle C A D is equal to the angle A D C, but since C D is parallel to A B, the angle A D C is equal to the angle B A D, therefore the angle B A D equals the angle C A D, and the line A D bisects the angle P A Q; but the line of action of the resultant of P and Q bisects the angle P A Q (Ax. 2), therefore the resultant acts along A D. Q. E. D.

50. *Remark.*—The following proposition may be regarded as the foundation of the science of statics; the demonstration generally seems obscure to readers who meet with it for the first time: this results from the somewhat unusual *form* of the proof; it may therefore be well to remark that the demonstration consists of two parts; in the first part it is shown that if the principle is true in two cases, viz. with regard to the pair of forces P and P_1 and the pair P and P_2, it must also hold good in a third case, viz. in regard to the pair of forces P and $P_1 + P_2$; this part of the proof is purely hypothetical, as much so as in the case of a demonstration by reduction to an absurdity; the second part of the proof takes up the argument, but as a matter of fact the proposition is true in two certain cases; therefore it must be true in a third case, therefore in a fourth case, and so on.

Proposition 4.

The principle of the parallelogram of forces is true of the direction of the resultant of any two commensurable forces.

Let the force P act at the point A along the line A B, and the forces P_1 and P_2 at the point A along the line A C: take A B, A C, C D, respectively proportional to P, P_1, and P_2, and complete the parallelograms B C, E D, then is the figure B D a parallelogram; draw the diagonals A E, C F, and A F, and suppose the points C, D, E, F to be rigidly connected with A.

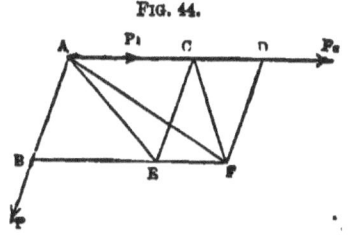

FIG. 44.

(*a*) The lines A B and A C represent the forces P and P_1; *assume* that their resultant acts along A E; then can P and P_1 be replaced by their resultant acting at A along A E, and, since A and E are rigidly connected, by that resultant acting at E along A E (Ax. 3); but this resultant acting at E can be replaced by its components acting at E, viz. by P_1 along B E, and by P along C E; and these again, since C and F are rigidly connected with E, by P_1 acting at F along B F, and P acting at C along C E.

(*b*) Since A and C are rigidly connected, P_2 may be supposed to act at C along CD; then CE represents the force P, and CD the force P_2; *assume* that their resultant acts along CF, then by reasoning in the same manner as in paragraph (*a*) it can be shown that the forces P and P_2 can be transferred to F.

(*c*) Thus it follows from our two *assumptions* that the forces P, P_1, P_2 may be supposed to act indifferently at A or F, therefore each of these must be a point in the direction of their resultant, i.e. their resultant must act along the line A F. Now A B represents the force P and A D the force $P_1 + P_2$; hence, if the proposition is true of the pair of forces P and P_1, and of the pair of forces P and P_2, it must also be true of the pair P and $P_1 + P_2$.

(*d*) But it appears from Prop. 3, that the proposition is true of equal forces, i.e. of any pair p and p, and of another equal pair p and p, therefore it will be true of the pair p and $p+p$, i.e. of p and $2p$; again, since the proposition is true of the pair p and p, and of the pair p and $2p$, it must be true of the pair p and $p+2p$, i.e. of p and $3p$; similarly it is true of p and $4p$, of p and $5p$, &c., and generally of p and mp.

(*e*) Again, since the proposition is true of the pair of forces mp and p, and of the pair mp and p, it must be true of the pair mp and $p+p$, i.e. of mp and $2p$; similarly it must be true of mp and $3p$, of mp and $4p$, and generally of mp and np.

(*f*) Now, any two *commensurable* forces P and Q must have a common unit (e.g. a pound, an ounce, &c.), and therefore can be represented by mp and np; hence the theorem is true of any two commensurable forces. Q. E. D.

Exercise.—The above demonstration may be put into a slightly different form, as follows : In the first place, suppose the forces P, P_1 and P_2 to be equal ; then the reasoning in § (*a*) and § (*b*) of Prop. 4 no longer proceeds from an assumption, but is based directly on Prop. 3 ; and the reasoning

PARALLELOGRAM OF FORCES. 81

in § (c) establishes the truth of the proposition in the case of the two forces P and 2P. The reasoning can be repeated for forces P, 2P and P, and the case P and 3P will be established; and by a repetition of the reasoning the cases P and 4P, P and 5P, and generally P and *m*P are established. A slight modification of the figure will then enable the reasoning to be extended to the case of *n*P, P, and P, so that the case *n*P and 2P will be established, then the cases *n*P and 3P, *n*P and 4P, and generally *n*P and *m*P. The student, having first mastered Prop. 4, will find it a useful exercise to write out the proof in this form.

Proposition 5.

The principle of the parallelogram of forces is true of the direction of the resultant of any two incommensurable forces.

Let P and Q be the two forces represented by the lines A B and A C; complete the parallelogram A B C D, then will the resultant (R) of P and Q act along the line joining A and D. For if not suppose R to act along any other line, this line must fall within the angle P A Q (Ax. 1), and therefore must cut either C D or D B; let it cut B D in the point E. Now, by continually

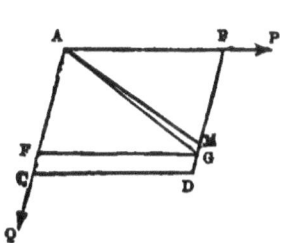

Fig. 45.

bisecting A B, a part can be found less than D E; set off distances equal to this part along A C, and let the last of them terminate at F (it cannot terminate at C, since A B and A C are incommensurable); therefore F C is less than this part, and therefore also less than D E; draw F G parallel to C D, this line will cut B D, in a point G between D and E, join A G. Suppose A F to represent a force Q′ and F C a force q, then will Q equal Q′ + q; now Q′ and P are commensurable, therefore their resultant (R′) will act along the line A G. But the resultant R of P and Q must equal the resultant of P, Q′, and q; i.e. of R′ and q; but R′ acts along A G, and q along A C, and therefore (Ax. 1) their resultant R must act *within* the angle G A Q; but by the

G

supposition it acts along A E *without* the angle G A Q; which is absurd. Therefore, &c. Q. E. D.

Proposition 6.

The principle of the parallelogram of forces is true of the magnitude of the resultant.

Let P and Q be the two forces acting at the point A, and let them be represented by the straight lines A B and A C, complete the parallelogram A B C D, and draw the diagonal A D; we have to prove that not only does the resultant (R) of P and Q act along the line A D, but also that it is represented in magnitude by that line. Suppose R' to be the force which balances P and Q, it must act along D A produced.

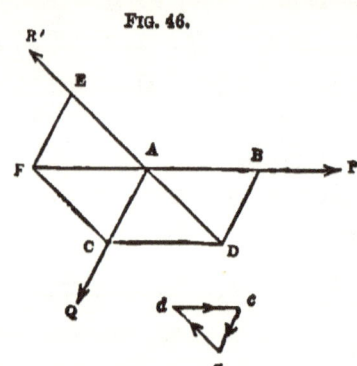

FIG. 46.

Let A E represent R'; complete the parallelogram C E, and join A F; the resultant of Q and R' must act along A F; but since P balances Q and R', it must act along F A produced; therefore F A B is one straight line, and is parallel to C D, so that F D is a parallelogram. Hence we have F C equal to A D, but F C equals A E, therefore E A equals A D. But R is equal and opposite to R', which is represented by A E, and therefore R is represented in magnitude by A D. Q. E. D.

51. *Application of trigonometry to statics.*—It is manifest that the sides of the triangle A C D (Prop. 6) are proportional to the three forces P, Q, R', which are in equilibrium. And hence if any triangle *a c d* be drawn similar to A C D, its sides will be proportional to the forces. Such a triangle will be formed by drawing lines respectively parallel to the directions of the forces, each force being an homologous term to the side parallel to its direction. The forces at A act in the directions *d c*,

PARALLELOGRAM OF FORCES. 83

ca, ad respectively, as shown by the arrow-heads. A similar remark applies to a triangle formed by drawing lines at right angles to the directions of three forces in equilibrium. The relations between three forces in equilibrium are thus reduced to the relations between the sides of a triangle; and of course all the trigonometrical relations between the sides and angles of that triangle will be analogous to relations between the forces and the angles between their directions. The two most important of these relations are proved in the following proposition :—

Proposition 7.

If three forces, P, Q, R, are in equilibrium, and act at a point A, to show that the following relations obtain :—

(1)   P : Q :: sin Q A R : sin R A P
      Q : R :: sin R A P : sin P A Q

(2) $R^2 = P^2 + Q^2 + 2$ P Q cos P A Q

(1) Draw the triangle abc whose sides bc, ca, ab are respectively parallel to the forces P, Q, R. Then it is evident that the angles a, b, c are respectively equal to $180° -$ Q A R, $180° -$ R A P, $180° -$ P A Q; now

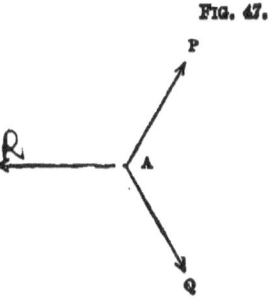

Fig. 47.

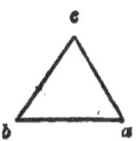

bc : ca :: sin bac : sin cba :: sin Q A R : sin R A P
ca : ab :: sin cba : sin acb :: sin R A P : sin P A Q

But by Art. 51—
bc : ca :: P : Q
ca : ab :: Q : R

G 2

therefore   P : Q :: sin Q A R : sin R A P
and   Q : R :: sin R A P : sin P A Q

These proportions are sometimes expressed by the rule, 'If three forces are in equilibrium, each force is proportional to the sine of the angle contained by the other two.'

(2) Employing the same figure, we have, by a well-known theorem in trigonometry,

$$ab^2 = bc^2 + ca^2 - 2\,bc\,.\,ca\,.\,\cos bca$$

Now bca is the supplement of P A Q, so that $\cos$ P A Q $= -\cos bca$;

therefore $ab^2 = bc^2 + ca^2 + 2\,bc\,.\,ca\,.\,\cos$ P A Q.

But bc, ca, ab are respectively proportional to the forces P, Q, R;

therefore $R^2 = P^2 + Q^2 + 2$ P Q $\cos$ P A Q   Q. E. D.

Ex. 230.—Show that when three forces are in equilibrium no one of them is greater than the sum of the other two.

Ex. 231.—Under what circumstances will three equal forces acting at a point balance each other?

Ans. Angle between directions of any two equals 120°.

Ex. 232.—Find the angle at which two forces of 8 lbs. must act so as to produce on a point a pressure of 12 lbs.   *Ans.* 82° 49'.

Ex. 233.—Let A B C be any triangle, D the middle point of B C; join A D; if A B and A C *represent* forces acting at A, show that their resultant will be represented by twice A D.

Ex. 234.—Explain the action of the forces by which a kite is supported in the air.

Ex. 235.—Explain the action of the forces by which a ship is made to sail in a direction nearly opposite to the wind.

Ex. 236.—The resultant of P and Q is 12 lbs. when their directions contain an angle of 60°, and 11 lbs. when they contain an angle of 90°: find P and Q.   *Ans.* 10·79 and 2·13 lbs.

Ex. 237.—There are two forces P and Q; when the lines representing them contain an angle θ, their resultant equals $\sqrt{\tfrac{5}{4}(P^2+Q^2)}$; but when those lines contain an angle $90° - \theta$, the resultant equals $\sqrt{\tfrac{3}{4}(P^2+Q^2)}$; find θ.

Ans. 63° 26'.

Ex. 238.—P and Q are two forces acting in directions at right angles to

RESULTANT OF TWO PARALLEL FORCES. 85

each other; their resultant equals $m(P+Q)$; if θ is the angle between the directions of their resultant and of P or Q, show that

$$m^2 \sin 2\theta = 1 - m^2$$

Ex. 239.—In the last example show that the ratio of the forces P and Q is

$$\frac{m^2 \pm \sqrt{2m^2-1}}{1-m^2}$$

Between what values must m lie? *Ans.* 1 and $\frac{1}{\sqrt{2}}$.

Ex. 240.—If P and $P+p$ are two forces very nearly equal, and if a is the angle between the lines representing them, then will the angle (in circular measure) between the direction of the resultant and of $P+p$ be very nearly

$$\frac{1}{2}\left(a - \frac{p}{P}\tan\frac{a}{2}\right)$$

Proposition 8.

To determine the resultant of two parallel forces acting in the same direction.

Let P and Q be the forces acting on a body at the points A and B; join A B; suppose any two equal and opposite forces T, T₁ to act at A and B respectively along the line A B; these forces being in equilibrium will not affect the resultant of P and Q (Ax. 5), therefore the required resultant will be that of T, P, Q, and T₁, i.e. of U and V, if U is the resultant of T and P, and V the resultant of Q and T₁. But since the line representing U falls within the angle T A P, and that representing V within the angle Q B T₁, these lines will meet when produced; let them be produced and meet in C; then if C be rigidly connected with the body, U and V may be supposed to act at C; through C draw C X parallel to A P or B Q; now U acting at C can be resolved into P, acting along C X, and T, acting parallel to B A, and similarly V can be resolved into Q acting along C X, and T₁ acting parallel to A B; hence the required resultant

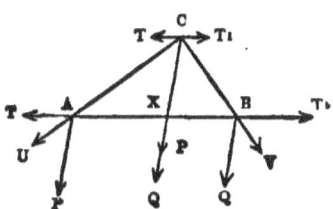

FIG. 48.

will be that of T, T₁, P, and Q, acting at C; or, since T and T₁ are in equilibrium, that of P and Q acting along C X, i.e. the resultant is a force equal to P + Q, acting along a line passing through X parallel to A P or B Q, and acting in the same direction as P and Q.

Next, to find the position of X. Since U is the resultant of P and T, those forces will be proportional to the sides of the triangle A X C;

therefore A X : X C :: T : P
similarly C X : X B :: Q : T₁
therefore A X : X B :: Q : P

i.e. the point X divides A B in the inverse ratio of the forces, which is the proof of the rule already given (Art. 29). Q. E. D.

Cor. 1.—Hence can be immediately deduced the conditions of the equilibrium of three parallel forces mentioned in Art. 30.

Cor. 2.—Hence, also, we can determine the resultant of two parallel forces acting in contrary directions. Thus

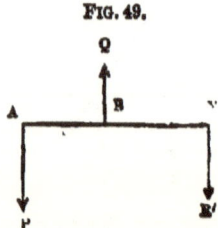

FIG. 49.

suppose P acting at A and Q acting at B to be the forces, and let Q be the greater; now if R′ is the force that balances P and Q, it must be equal and opposite to their resultant R; but R′ + P = Q, and A B : B X :: R′ : P, i.e. A B + B X : B X :: R′ + P : P, or A X : B X :: Q : P; i.e. the resultant equals Q − P, and acts in the same direction as Q through a point X, whose distances from A and B are inversely as the forces, and so taken that the greater force acts between the resultant and the lesser force.

Ex. 241.—Two parallel forces of 11 and 12 lbs. act in contrary directions at A and B respectively. The line A B is 6 ft. long, and is at right angles to the direction of the forces. Find the resultant.

Ans. A X = 72 ft. (fig. 49), R = 1 lb., acting in the same direction as Q.

USE OF THE POSITIVE SIGN.

Ex. 242.—A B is a straight rod 12 ft. long; C a point 4 ft. from B; the rod rests on a peg at C, and is kept horizontal by a peg placed over it at B; a weight of 20 lbs. is hung at A; find the pressure on each peg (neglecting the weight of the rod). *Ans.* Pressure on C 60 lbs., on B 40 lbs.

Ex. 243.—A and B are the pans of a pair of scales; a substance placed in A is balanced by P lbs. in B; when placed in B it is balanced by Q lbs. in A; find its true weight. *Ans.* $\sqrt{PQ}$ lbs.

Ex. 244.—A rod of uniform density A B is divided into any two parts at the point X; the middle points of A X, X B, and B A are P, Q, and R respectively; show that weight of A X : weight of X B :: Q R : R P.

Ex. 245.—A B C is an equilateral triangle, kept at rest by three parallel forces, P, 3P, and 2P, acting in the plane of the triangle at A, B, and C respectively. Determine the lines along which the forces must act.

Exercise.—Let A and B be two fixed points, let a force P act through A and a force Q through B, also let their directions intersect in a point X. Now, suppose the direction of Q to change in such a manner that the distance of X from A continually increases, and consequently the angle between the directions of P and Q continually diminishes. It is plain that the directions of P and Q will in the limit become parallel. It is required, by means of this consideration, to deduce the results of Prop. 8 from the previous Propositions.

52. *The use of the positive and negative signs to denote the directions of forces.*—Since a line can be taken to represent a force, and since if $+a$ be used to denote a line of a feet (or other units), measured to the right from a fixed point, then $-a$ must be used to denote a line of a feet measured to the left from that point, it should seem that the same principle ought to be applicable to forces, and that if $+P$ denote a force of P units acting to the right along a given line, then $-P$ must denote a force of P units acting towards the left along that line. That the principle so commonly used in geometry is correctly applied to forces, will be evident from a little consideration. Thus, if P and Q be two forces acting to the right along a line, and R their resultant, we have

$$R = P + Q \qquad (1)$$

If Q act to the left and be less than P, R will act to the right, and we have

$$R = P - Q \qquad (2)$$

If, however, Q be greater than P, R will act to the left, and we have

$$R = Q - P \qquad (3)$$

Here we have three equations to express a certain result; but if we suppose P+Q to be an *algebraical* sum, these three equations can be included in one, viz.

$$R = P + Q \qquad (4)$$

It is quite plain the (4) includes (1) and (2); it also includes (3), since that equation can be written

$$-R = P - Q$$

The same principle can be applied to the moments of forces. If we measure the moment of a force with reference to a certain point, we may agree to reckon it positive if the force tend to turn the body round that point in a direction contrary to that in which the hands of a watch move. If this assumption be made, then the moment of any other force must be reckoned negative which tends to turn the body in the contrary direction round the point. It will be remarked that in fig. 51 the moments of P, Q, R with respect to O are positive; in fig. 52 the moments of Q and R are positive, and that of P negative.

53. *Representation of a moment by an area.*—Let the line A B represent a force P, and from a point O let fall

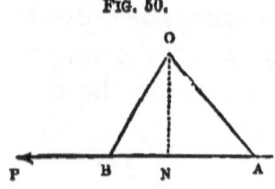

FIG. 50.

a perpendicular O N on A B or A B produced; join O A, O B; then twice the area of the triangle A O B equals the product of O N and A B, i.e. the product of the perpendicular on P's direction and the line that represents P; hence, twice the area of the triangle A O B represents the moment of the force P with respect to the point O.

Proposition 9.

The algebraical sum of the moments of two forces, whose directions are not parallel, taken with reference to any point in their plane, equals the moment of their resultant with reference to the same point.

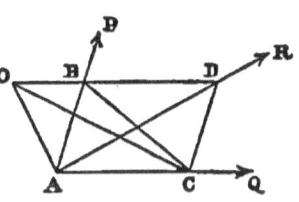

FIG. 51.

Let P and Q be two forces whose directions intersect in A; let O be the point with reference to which the moments are to be taken; draw O B (fig. 51) parallel to A Q, and take A C such that

$$AC : AB :: Q : P$$

then A B and A C will represent the forces P and Q; consequently if the parallelogram A B D C is completed, A D will represent the resultant (R) of P and Q. Join O A, O C, and B C. The moments of P, Q, and R are proportional to the areas of the triangles O A B, O A C, and O A D (Art. 53). Now O A C is equal to B A C, which being half of the parallelogram, is equal to A B D. But O A D is made up of O A B and B A D. Consequently,

moment of R = moment of P + moment of Q

In this case all the moments are positive; we will therefore take a case in which O is so situated that the moments of R and Q with regard to it are positive, and that of P negative, and in which consequently we have to show that

moment of R = − moment of P + moment of Q

In this case draw (fig. 52) O B parallel to A Q, and find A C from the proportion

$$AC : AB :: Q : P$$

so that A B and A C represent the forces P and Q; then on completing the parallelogram A B D C, A D will represent their resultant (R). Join O A, O C. We see that O A C is half the parallelogram, and consequently equals A D B, which is made up of O A D and O A B; hence

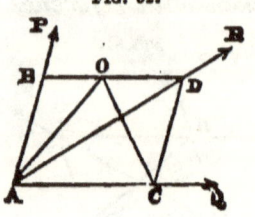

Fig. 52.

$$O\ A\ D = O\ A\ C - O\ A\ B$$

and as these triangles are proportional to the moments of the forces with respect to the point O, we have

moment of R = —moment of P + moment of Q.

Similar results are obtained for any other position of the point O; and the student will find it instructive to consider one or two other cases, e.g. that in which O falls within the angle R A Q, in which case the moments of P and R are both negative and that of Q positive. He will observe that, by the aid of the rule for the signs of the moments, all possible cases of two intersecting forces are included in the one statement given above.

Proposition 10.

The algebraical sum of the moments of two parallel forces with reference to any point in their plane is equal to the moment of their resultant with reference to the same point.

Let P and Q be the two forces, and let them act in the same direction, R their resultant, O the point about which the moments are measured; draw a line O B at right angles to the lines along which the forces act, and cutting them in A, B, and X respectively. Now in the case selected the moments of P, Q, and R are all positive, hence we have to show that

Fig. 53.

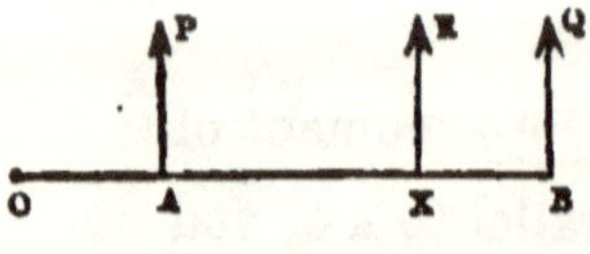

$$M^tR = M^tP + M^tQ$$

PRINCIPLE OF MOMENTS.

Since $R = P + Q$
we have
$$M^tR = O\ X.R$$
$$= O\ X.P + O\ X.Q$$
$$= O\ A.P + A\ X.P + O\ B.Q - B\ X.Q$$
but $A\ X.P = B\ X.Q$ (Prop. 8)
therefore
$$M^tR = O\ A.P + O\ B.Q$$
$$= M^tP + M^tQ$$

A similar proof will apply to every position of O, and to cases in which P and Q act in contrary directions. Hence, &c. Q. E. D.

Ex. 246.—If the point O (Prop. 9) be taken in the direction of the resultant, show that the moments of P and Q are equal and have opposite signs.

Exercise.—Prop. 3-10 can be proved by reasoning in the following manner:—*First.* Assume as an axiom that the resultant weight of a uniform rod acts through its middle point; and bearing in mind the remark in Article 23, that any force can be substituted for an equal force without reference to its physical origin, observe that *Ex.* 244 gives an independent proof of Prop. 8. *Secondly.* Observe that it follows from Axiom 2 (Art. 49), that when a body is acted on by two equal forces in the same plane, and has one point in the plane fixed, it will be at rest, provided the forces act at equal perpendicular distances from the point, and tend to turn the body round the point in opposite directions. This observation, combined with Prop. 8, will establish *Ex.* 246. *Thirdly.* The principle of the parallelogram of forces, so far as the direction of the resultant is concerned, can be easily deduced from *Ex.* 246. The student who has first mastered Prop. 3-10 will find it a most instructive exercise to write out proofs of the same propositions, adopting the method of proof above indicated.

54. *Statical couples.*—In Cor. 2 to Prop. 8 it was shown that if P and Q are two parallel forces acting at A and B in opposite directions, then if Q is greater than P their resultant R will be a parallel force acting in the same direction as Q through a point X given by the proportion

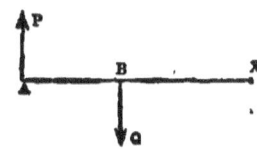

FIG. 54.

$$A X : B X :: Q : P$$

or
$$B X = \frac{A B.P}{Q - P}$$

Now, if we suppose Q to be gradually diminished, but A B and P to remain unaltered, the magnitude of R (or Q—P) will continually diminish and B X will continually increase, and in the limit when Q becomes equal to P, the magnitude of the resultant is zero, and X is removed to an infinite distance; in other words, two equal parallel forces acting in opposite directions have no resultant, and therefore cannot be balanced by any single force. Such a pair of forces constitute what is called a *statical couple*. If, in fig. 54, we suppose P and Q to be equal, and A B to be at right angles to their directions, A B is called the *arm* of the couple, and A B × P its *moment*. A little consideration will show that the sum of the moments of the forces with regard to *any* point in the plane of the couple will equal A B × P; and moreover, that if the sign of the sum of the moments with reference to one point is positive, it will be positive when taken with reference to *any* point in the plane of the couple; and if negative, negative; e.g. the couple represented in the diagram has a negative moment.

Proposition 11.

If two couples of equal moments and of opposite signs act in the same plane on a rigid body they will balance one another.

First. Let the forces which constitute the two couples not act along parallel lines, then must the four lines by their intersection form a parallelogram. Let A B C D be the parallelogram thus formed, and let the forces (P, P′) of the one couple act along A B and C D, then must the forces (Q, Q′) of the other couple act along A D and C B, since the moments of the couples

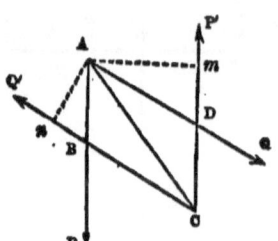

Fig. 55.

RESULTANT OF TWO COUPLES. 93

have contrary signs; draw A m and A n at right angles to
C D and C B, then since the moments of the couples are
equal
$$A m \times P = A n \times Q$$
also $\quad A n \times A D = A m \times A B$
since each product is the area of A B C D; therefore
$$A D \times P = A B \times Q$$
or $\quad P : Q :: A B : A D$

therefore A B and A D represent (Art. 25) the forces P
and Q, and therefore the diagonal A C represents
their resultant (R). In like manner P′ and Q′ are represented by C D, and C B respectively, and therefore C A
represents their resultant (R′). Hence, the four forces
P, Q, P′, Q′ are equivalent to a pair of equal opposite
forces R and R′, and therefore are in equilibrium.

Secondly. Let the four forces act along parallel lines;
draw a straight line cutting
those lines at right angles in
A, B, C, D, respectively; and let P
and Q act in the same direction,
and P′ and Q′ in the opposite direction, then the moments of the
couples will have contrary signs; now R the resultant
of P and Q equals P + Q, let it act through the point X,
then we have

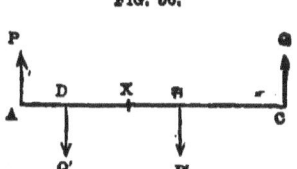

Fig. 56.

$$A X \times P = C X \times Q$$
also since the moments of the couples are equal
$$A B \times P = C D \times Q$$
therefore $\quad B X \times P = D X \times Q$
or $\quad B X \times P' = D X \times Q'$

hence the resultant (R′) of P′ and Q′ acts through the
point X, and as it equals P′ + Q′, the four forces P, Q, P′, Q′
are equivalent to two equal forces, R and R′ acting in

opposite directions along the same line, and therefore are in equilibrium.

Cor. 1. Hence two couples of equal moments and of the same sign and acting in the same plane on a rigid body are equivalent to one another, since either would be balanced by a couple of equal moment and of contrary sign. In other words, there will be no change produced in the effect of a couple by supposing it to act anywhere in its original plane, and by supposing its arm to be lengthened or shortened, provided the forces undergo a corresponding change, so that its moment remains unaltered in sign and magnitude.

Cor. 2. Hence, also, if M and N are the moments of two couples acting in the same plane, they will be equivalent to a single couple whose moment is their algebraical sum $M+N$. For let both couples be reduced to equivalent couples having arms of the same length a, then if P and P' are the forces of the one, and Q and Q' of the other, we shall have aP or aP' equal to M, and aQ or aQ' equal to N; now place the couples so that their arms coincide, then if both moments are positive, the couples will lie as shown in the figure, i.e. they are equivalent to a pair of parallel forces, $P+Q$ and $P'+Q'$ constituting a couple whose moment is $a(P+Q)$ or $M+N$. If the couples have contrary signs P and Q will act in contrary directions.

Fig. 57.

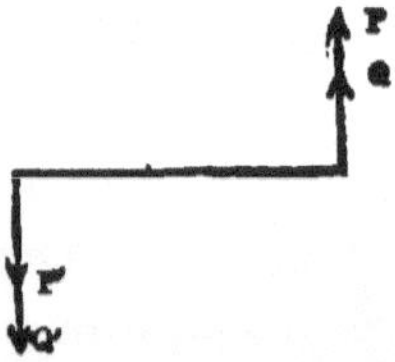

55. *Remark.*—In the previous propositions of the present chapter, we have completely discussed the relations which subsist between two forces acting in the same plane and their resultant; we have now to consider the case of any system of forces acting in one plane on a rigid body. It may be remarked that in general every such system will have a resultant; thus, if we have three forces, P_1, P_2, P_3, we can find the resultant R_1 of P_1 and

RESULTANT OF TWO COUPLES. 95

P_2, and then the resultant R of R_1 and P_3; the force R will be the resultant of P_1, P_2, and P_3; the same method can in general be applied to the determination of the resultant of any number of forces; two particular cases, however, may arise, *first*, when the system is in equilibrium, *secondly*, when the system reduces to a couple. A little consideration will show that no other exception can possibly arise in the case of a system of forces acting along lines in a common plane.

Ex. 247.—A, B, C, D are the angular points of a square taken in order. Forces of 5 lbs. each act respectively from A to B, from B to C, and from C to D. Find their resultant.

Ans. Produce A B to X, so that A X is twice A B, the resultant is a force of 5 lbs., acting through x parallel to and in the same direction as the force along B C.

Ex. 248.—If in the last example a force of 5 lbs. acted from D to A, to what would the four forces reduce?

Ans. A couple whose moment is 10 A B.

Ex. 249.—If, in *Ex*. 247, there are four forces of 10 lbs. apiece acting respectively from A to B, C to B, C to D, and A to D, to what can the four be reduced? *Ans*. They are in equilibrium.

Ex. 250.—Again, suppose that a force of 10 lbs. acts from A to B, 11 lbs. from C to B, 9 lbs. from C to D, and 10 lbs. from A to D, to what will the four forces reduce?

Ans. A force of $\sqrt{2}$ lbs. acting through C parallel to and in the same direction as a line drawn from D to B.

Ex. 251.—A B C is an equilateral triangle, three equal forces (P) act respectively from A to B, from A to C, and from B to C; what is their resultant?

Ans. A force 2P acting parallel to and in the same direction as A to C through the middle point of B C.

Ex. 252.—A, B, C, D are the angular points of a square taken in order; a particle at A is acted on by a force of 10 lbs. along A B from A to B, by a force of 20 lbs. along A C from A to C, and by a force of 25 lbs. along A D from A to D. Find the magnitude and direction of the resultant of the forces.

Ans. 46 lbs. acting in a direction within the right angle A, and making an angle of 58° 20′ with A B.

Ex. 253.—A B C is a triangle right-angled at C; B is an angle of 30°; a force of 4 lbs. acts along A B from A to B, of 3 lbs. along C B from C to B, of 2 lbs. along A C from A to C. Determine the magnitude and direction of the resultant.

Ans. Take D the middle point of B C, make B D E an angle of 31° 45′ (E and A on opposite sides of B C), the resultant is a force of 7·6 lbs. acting from B to E.

96   PRACTICAL MECHANICS.

Ex. 254.—When four forces acting in the same plane at a point are in equilibrium, show that a quadrilateral figure can be drawn, the sides of which are related to them in the same manner that the sides of the triangle in Art. 51 are related to three forces in equilibrium.

Ex. 255.—From the above example show that the resultant of three forces acting in the same plane at a point can be represented by a side of a quadrilateral, and state exactly how the quadrilateral must be drawn.

Ex. 256.—Extend the results in Ex. 254 and 255 to any number of forces (v. Art. 59).

56. *The resultant of any number of forces acting along the same straight line.*—Since the resultant of two such forces is their (algebraical) sum, the resultant of those two and a third force must be the (algebraical) sum of the three, and the same will be true of any number of forces; hence, *if any number of forces act along the same straight line their resultant will equal their algebraical sum.* If their algebraical sum is zero, the forces will be in equilibrium. In the following general theorems the term 'sum' means 'algebraical sum.'

57. *The resultant of any number of couples acting in the same plane.*—Since the moment of the resultant of two such couples is the sum of the moments of the two couples (Prop. 11, Cor. 2), that of the resultant of those two and a third will be the sum of the moments of the three, and the same will be true of any number of couples; hence, *if any number of couples act in the same plane, the moment of their resultant equals the sum of their several moments.* If the sum of the moments is *zero*, the couples will be in equilibrium; for if all the couples are reduced to equivalent couples with equal arms, and these arms are superimposed on each other, it is plain that the moment of the resultant couple can only become zero by each force of the couple becoming zero; i.e. the whole reduces to two systems of forces which are severally in equilibrium.

58. Extension of the principle of moments to any number of forces.—Let $P_1, P_2, P_3, \ldots P_n$ be any system of forces acting in one plane on a rigid body; let R_1 be the resultant of P_1 and P_2, R_2 of R_1 and P_3, and so on, and R the resultant of R_{n-2} and P_n. Now, if the moments are taken with respect to any one point in the plane, we shall have

$$m^tR_1 = m^tP_1 + m^tP_2$$
$$m^tR_2 = m^tR_1 + m^tP_3$$
$$\vdots \qquad \vdots \qquad \vdots$$
$$m^tR = m^tR_{n-2} + m^tP_n$$

therefore, by addition,

$$m^tR = m^tP_1 + m^tP_2 + m^tP_3 + \ldots + m^tP_n$$

Hence, if any forces act in a plane, the sum of their moments with respect to any point in that plane, will equal the moment of their resultant with respect to that point. A little consideration will show that if the forces reduce to a couple, the moment of the couple will equal the sum of the moments of the several forces.

Of course, if the point is taken in the direction of the resultant, its moment, and therefore the algebraical sum of the moments of the forces, will equal zero. Now, if a body acted on by any forces be kept at rest round a fixed point, the resultant must pass through that point; and therefore in this case the algebraical sum of the moments of the forces round that point will equal zero; a statement which coincides with that already given (Art. 39). It is plain that in this case the forces cannot be reduced to a couple; for if they could be so reduced they could not be balanced by the reaction of the fixed point.

Proposition 12.

To determine the resultant of any system of forces acting along parallel lines in one plane.

Let $P_1, P_2, P_3, \ldots$ be the forces; take any point O, and draw OA at right angles to the direction of the forces, and cutting the lines along which they act in $N_1, N_2, N_3, \ldots$ let $ON_1 = p_1$, $ON_2 = p_2$, $ON_3 = p_3 \ldots$;

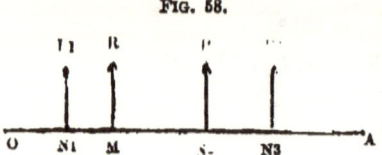

FIG. 58.

also let R, the resultant of the forces, act along a line cutting OA in M, and let $OM = r$; we have to find the magnitudes of R and r. Now the resultant of any two parallel forces equals their sum, therefore the resultant of those two and a third force will equal the sum of three, and so on for any number of forces, therefore their resultant must equal their sum, or

$$R = P_1 + P_2 + P_3 + \ldots$$

again, the moment of R round O must equal the sum of the moments of the separate forces, therefore

$$Rr = P_1 p_1 + P_2 p_2 + P_3 p_3 + \ldots$$

The former equation gives R and the latter r.

Cor. 1. Let the resultant of $P_2, P_3, \ldots$ be R′, and let its direction cut OA at a distance from O equal to r'; then it will be necessary and sufficient for the equilibrium of $P_1, P_2, P_3, \ldots$ that P_1 be equal and opposite to R′, i.e. that r' equal p_1, and that $P_1 + R'$ equal zero; but

$$R' = P_2 + P_3 + \ldots$$

and
$$R'r' = P_2 p_2 + P_3 p_3 + \ldots$$

Therefore it is necessary and sufficient for the equilibrium of the system of forces that

RESULTANT OF ANY NUMBER OF FORCES. 99

$$P_1+P_2+P_3+\ldots = 0$$
and
$$P_1p_1+P_2p_2+P_3p_3+\ldots = 0$$

By the words 'necessary and sufficient for equilibrium' is meant that on the one hand if the forces are in equilibrium the above equations will be satisfied, and on the other hand if the above equations are satisfied the forces will be in equilibrium.

Cor. 2. If the equations when formed lead to the following result,

$$P_1+P_2+P_3+\ldots = 0$$
and $P_1p_1+P_2p_2+P_3p_3+\ldots =$ a finite quantity,

the system of forces reduces to a couple.

Ex. 257.—A uniform rod is 3 ft. long and weighs 2 lbs.; weights of 1 lb., 3 lbs., 5 lbs., and 6 lbs. are suspended on it in order at distances of 1 ft. apart. Determine completely the resultant of the forces.

Ans. 17 lbs. acting along 5's line of action.

Ex. 258.—Let a horizontal line be drawn from a point A to the right, and let forces of 5 lbs., 12 lbs., and 19 lbs. act vertically upwards on it, and of 10 lbs. and 20 lbs. act vertically downwards on it, the former at distances of 2 ft., 5 ft., and 14 ft., and the latter at distances of 8 ft. and 20 ft. from A. Determine completely their resultant.

Ans. 6 lbs. acting upwards through a point 24 ft. to the left of A.

Ex. 259.—If in addition to the forces in the last example, one of 6 lbs. acts at a distance of 10 ft. to the right from A, determine the resultant (1) when the force acts vertically upwards; (2) when it acts vertically downwards.

Ans. (1) 12 lbs. acting vertically upwards 7 ft. to the left of A.
(2) A couple whose moment is -204.

59. *The resultant of any number of forces acting in one plane at a point* can be found by a very simple construction called the 'polygon of forces.'—Let the forces P, Q, R, S act at a point O in the directions O P, O Q, O R, O S, and let it be required to find their resultant. Draw any line A B proportional to P in the direction O P; from B draw B C proportional to Q and in the direction O Q, from

c draw C D proportional to R and in the direction O R, and finally D E proportional to S and in the direction O S. Then, if A E be joined, the resultant (T) of P, Q, R, and S will be a force acting at O in the direction A E, and proportional to A E. This is evident, since by the triangle of forces (Art. 36) the force at O represented by A E is equivalent to two forces at O represented by A D and D E; these to three forces at O represented by A C, C D, D E; and these in turn by four forces at O represented by A B, B C, C D, D E, i.e. P, Q, R, S. It is immaterial in what order we take the forces; for instance (fig. 59a), we may draw A B to represent P, then B C to represent R, then C D to represent Q, and finally D E to represent S; the resultant will be, as before, a force at O represented by A E.

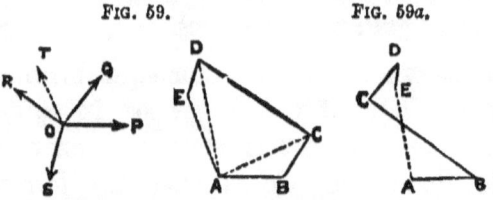

Fig. 59. Fig. 59a.

When the polygon is drawn, if it is found that E coincides with A, the magnitude of the resultant is zero, and the forces acting at O are in equilibrium.

To render the calculation of the magnitude and direction of the resultant intelligible it is necessary in the first place to explain what are the rectangular components of a force. Let O x, O y be two rectangular axes, and let P be a force acting at O along the line O P; let O A be the line which represents the force P, and let the angle it makes with the axis of x, viz. x O A, equal θ; now, if the parallelogram O B A C be completed, P will be equivalent to two forces respectively represented by O B and O C, and since these forces are at right angles to one another, they are called the rectangular components of P with respect to the axes O x and O y; again, since O C = O A sin θ

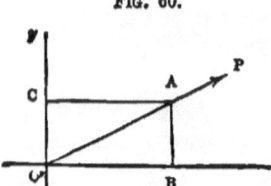

Fig. 60.

and $OB = OA \cos\theta$, it is plain that the rectangular components of P are $P\cos\theta$ along the axis Ox and $P\sin\theta$ along the axis Oy. If we always measure θ in the same direction, viz. upwards from Ox, so that it increases in a direction opposite to that in which the hands of a watch move, $P\cos\theta$ and $P\sin\theta$ will give not only the *magnitudes* of the components, but also the *directions* in which they act:—thus if we suppose P to act *towards* O, the line which represents the force is OA, so that θ is not xOP, but xOA, indicated by the dotted arc; and then, since θ lies between $180°$ and $270°$, both $P\sin\theta$ and $P\cos\theta$ will be negative, as they ought to be.

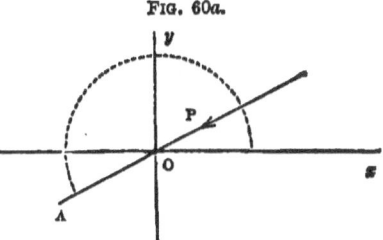

FIG. 60a.

Proposition 13.

To determine the resultant of any system of forces acting in one plane at a point: and to infer the conditions of equilibrium of such a system of forces.

(a) Let $P_1, P_2, P_3, \ldots$ be the forces acting at any given point O; through O draw two rectangular axes Ox and Oy, and let $\theta_1, \theta_2, \theta_3, \ldots$ be the angles that the lines representing the forces make with the axis of x. Then these forces can be replaced by their rectangular components along the axes of x and y, i.e. by

$P_1\cos\theta_1, P_2\cos\theta_2, P_3\cos\theta_3, \ldots$ along the axis of x, and
$P_1\sin\theta_1, P_2\sin\theta_2, P_3\sin\theta_3, \ldots$ along the axis of y.

Now, the former set is equivalent to a single force X acting along the axis of x, and the latter to a single force Y acting along the axis of y, provided

$$X = P_1\cos\theta_1 + P_2\cos\theta_2 + P_3\cos\theta_3 + \ldots$$
$$Y = P_1\sin\theta_1 + P_2\sin\theta_2 + P_3\sin\theta_3 + \ldots$$

Now, if R be the resultant of X and Y, and ϕ the angle which the line representing it makes with Ox, we must have

$$\text{R} \cos \phi = \text{X} \qquad (1)$$
and
$$\text{R} \sin \phi = \text{X} \qquad (2)$$

which equations determine R and ϕ. It will be remarked the determination is free from ambiguity, since the signs of X and Y will give the signs of $\cos \phi$ and $\sin \phi$, and therefore determine the *quadrant* in which the line representing R falls. Of course the magnitude of R is given by the equation

$$\text{R}^2 = \text{X}^2 + \text{Y}^2 \qquad (3)$$

(*b*) To obtain the conditions of equilibrium of $P_1, P_2, P_3 \ldots$

It must be remembered that it is necessary and sufficient for the equilibrium of these forces that P_1 be equal and opposite to the resultant of $P_2, P_3, \ldots$ (Ax. 4), so that the rectangular components of this resultant must be $-P_1 \sin \theta_1$ and $-P_1 \cos \theta_1$, therefore the required conditions are

$$-P_1 \sin \theta_1 = P_2 \sin \theta_2 + P_3 \sin \theta_3 + \ldots$$
and
$$-P_1 \cos \theta_1 = P_2 \cos \theta_2 + P_3 \cos \theta_3 + \ldots$$
or
$$P_1 \sin \theta_1 + P_2 \sin \theta_2 + P_3 \sin \theta_3 + \ldots = 0$$
and
$$P_1 \cos \theta_1 + P_2 \cos \theta_2 + P_3 \cos \theta_3 + \ldots = 0$$

That is to say—'It is necessary and sufficient for the equilibrium of any system of forces acting in one plane at a point, that the sums of their components along each of two rectangular axes be separately zero.'

FIG. 61.

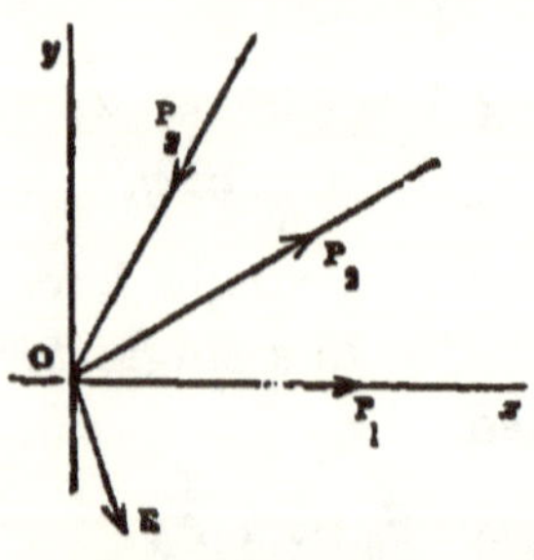

Ex. 260.—Let P_1, P_2, P_3 be three forces of 50, 30, and 100 lbs. respectively, acting at the point o, as shown in the figure; let the angle x o P_2 equal 30°, and x o P_3 equal 60°; it is required to determine their resultant by the method of Prop. 13.

In this case, $\theta_1 = 0$, $\theta_2 = 30°$, and $\theta_3 = 240°$, therefore,

RESULTANT OF n FORCES. 103

$R \cos \phi = 50 \cos 0° + 30 \cos 30° + 100 \cos 240°$
and $\quad R \sin \phi = 50 \sin 0° + 30 \sin 30° + 100 \sin 240°$
or $\quad R \cos \phi = 50 + 25\cdot 98 - 50 = \quad 25\cdot 98 \qquad$ FIG. 62.
and $\quad R \sin \phi = 15 - 86\cdot 60 \quad = -71\cdot 60$
hence $R = 76\cdot 17$ lbs. and $\phi = 289°\ 57'$, i.e. R acts as indicated in the diagram; this result may be verified by construction.

Ex. 261.—Let P_1, P_2, P_3 be three forces each of 100 lbs., let the angle $x \circ P_3$ be $135°$; find their resultant by the above method.

$\qquad Ans.\ R = 41\cdot 4$ lbs. $\phi = 315°$.

60. *Transfer of a force in a parallel direction.*—
Let A B and C D be two parallel lines, and p the length of the perpendicular O N drawn from O FIG. 63.
in A B to C D; then if a force P acts
from A to B along A B, it will be equivalent to an equal parallel force acting
along C D in the same direction, and a
couple whose moment is Pp, the sign of the couple being positive if O N is to the left of the direction of the force (as in the diagram), and negative if to the right. For if two opposite forces P′, P″, each equal to P, act along C D, they will be in equilibrium, and the three will be equal to P; but P and P″ constitute a couple with a positive moment Pp, hence P is equivalent to P′ and that couple.

Hence also we can determine the resultant of a force P, acting along a line A B, and a couple FIG. 64.
whose moment is M; for let M equal Pp,
from O in A B draw a perpendicular O N
equal to p, and to the right of P's direction, if the moment of the couple is
positive; make the arm of the couple coincide with O N then the couple will consist of the forces P′ and P″, each equal to P, acting as shown in the figure; hence the force and the couple are equivalent to the three forces P, P′, and P″, but P and P″ are in equilibrium, therefore the force P and the couple are equivalent to P′.

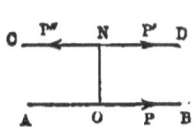

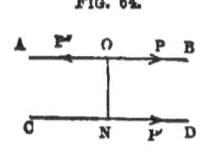

Ex. 262.—If A, B, C, D are the corners of a square taken in order, and if forces act along three of the sides, viz. P from A to B, P from A to D, and P from C to D, show that the three are equivalent to a single force P acting from B to C.

Proposition 14.

To determine the resultant of any system of forces acting in a plane.

Take ox, oy, any two rectangular axes, and let P_1, P_2, P_3, ... be the forces, acting along given lines; from o let fall perpendiculars p_1, p_2, p_3, ... on these lines; then P_1 is equivalent to an equal parallel force acting in the same direction through o, and a couple whose moment is $P_1 p_1$, the like is true of P_2, P_3, ...; let $\theta_1, \theta_2, \theta_3$, ... be the angles made with the axis of x by the lines representing the transferred forces.

Now, let R be the resultant of the transferred forces, and let ϕ be the angle which the line representing it makes with the axis of x. Therefore,

$$R \cos \phi = P_1 \cos \theta_1 + P_2 \cos \theta_2 + P_3 \cos \theta_3 + \ldots \quad (1)$$
$$R \sin \phi = P_1 \sin \theta_1 + P_2 \sin \theta_2 + P_3 \sin \theta_3 + \ldots \quad (2)$$

also let Rr be the moment of the resultant of the couples, therefore,

$$Rr = P_1 p_1 + P_2 p_2 + P_3 p_3 + \ldots \quad (3)$$

The equations (1) and (2) completely determine R. Hence the given system of forces is reduced to a known force and a couple of known moment; by compounding these we obtained the required resultant.

Cor. When equations (1)(2) and (3) are formed, if we obtain

$$P_1 \cos \theta_1 + P_2 \cos \theta_2 + P_3 \cos \theta_3 + \ldots = 0$$
$$P_1 \sin \theta_1 + P_2 \sin \theta_2 + P_3 \sin \theta_3 + \ldots = 0$$
$$P_1 p_1 + P_2 p_2 + P_3 p_3 + \ldots = \text{a finite quantity}$$

the system manifestly reduces to a couple.

RESULTANT OF n FORCES. 105

Ex. 263.—A B C is a triangle right-angled at A, its sides A B and A C are each 10 ft. long. The forces P_1, P_2, P_3, each of 100 lbs., act as shown in the figure: find their resultant by the method of Prop. 14.

FIG. 65.

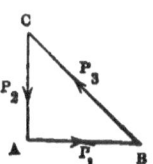

The force P_3 is equivalent to an equal parallel force whose direction passes through A, and a couple whose moment is $500\sqrt{2}$. Hence the three given forces are equivalent to the three forces of *Ex.* 261, and to the above couple. Now the latter three forces are equivalent to R, acting through A parallel to C B, where R equals $100(\sqrt{2}-1)$, and the couple is equivalent to the two forces R' and R" each equal to R acting as shown in the figure where the line A N is drawn at right angles to A R, and equals $500\sqrt{2} \div 100(\sqrt{2}-1)$ or $5(2+\sqrt{2})$ ft. in length. The required resultant is therefore the force R".

FIG. 66.

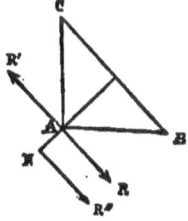

Ex. 264.—In the last case if P_3 equals 200 lbs., show, by the method of Prop. 14, that the resultant equals $100(2-\sqrt{2})$ lbs. and acts parallel to R' (fig. 66) along a line which cuts N A produced at a distance of $10(\sqrt{2}+1)$ ft. from A.

Ex. 265.—If A B C is a triangle, each of whose sides is 10 ft. long, and if a force P acts from A to B, an equal force from B to C, and another equal force from C to A, show that the three are equivalent to a couple whose moment is $5P\sqrt{3}$.

Ex. 266.—If A B C D is a square, and if a force equal to 2P acts from A to B, an equal force from B to C, 3P from C to D, and an equal force from D to A, show by the method of Prop. 14 that the resultant equals $P\sqrt{2}$, and acts in a direction parallel to the diagonal C A, along a line which cuts the diagonal B D produced in a point whose distance from D equals 2 B D.

Ex. 267.—Let A B C be an equilateral triangle, draw A D at right angles to B C, in B C produced take D E equal to D A, let equal forces (P) act from A to B, from B to C, from C to A, and from D to A respectively; show that their resultant equals P, and acts through B in direction parallel to D A.

Ex. 268.—In the last case determine the resultant if the fourth force had acted from A to D.

Ex. 269.—If three parallel forces are in equilibrium, they consist of two couples having equal moments of opposite signs.

Ex. 270.—If A B C is any triangle, and if a force P acts from A to B, Q from B to C, and R from C to A; and if $P : Q : R :: AB : BC : CA$, show that the resultant of the three forces is a couple whose moment is represented by twice the area of the triangle.

Proposition 15.

To determine the conditions of equilibrium of any system of forces acting in one plane on a rigid body.

Adopting the notation of Prop. 14, let R be the resultant of $P_2, P_3, \ldots$ Now, the necessary and sufficient condition of equilibrium is that P_1 shall be equal and opposite to R. But if we transfer P_1 to the point O, and then resolve it along Ox and Oy, we obtain a force $P_1 \cos \theta_1$ acting along Ox, a force $P_1 \sin \theta_1$ acting along Oy, and a couple whose moment is $P_1 p_1$: and in like manner by transferring R we shall obtain R $\cos \phi$ along Ox, R $\sin \phi$ along Oy, and a couple whose moment is Rr. But in order that P_1 and R may be equal and act in opposite directions along the same line, we must have $P_1 \cos \theta_1$ equal and opposite to R $\cos \phi$, $P_1 \sin \theta_1$ to R $\sin \phi$, and $P_1 p_1$ to Rr, i.e. it is necessary and sufficient for the equilibrium of the system that

$$P_1 \cos \theta_1 + R \cos \phi = 0$$
$$P_1 \sin \theta_1 + R \sin \phi = 0$$
$$P_1 p_1 \quad + R r \quad = 0$$

But by Prop. 14

$$R \cos \phi = P_2 \cos \theta_2 + P_3 \cos \theta_3 + \ldots$$
$$R \sin \phi = P_2 \sin \theta_2 + P_3 \sin \theta_3 + \ldots$$
$$R r \quad = P_2 p_2 \quad + P_3 p_3 \quad + \ldots$$

Hence the required conditions are

$$P_1 \cos \theta_1 + P_2 \cos \theta_2 + P_3 \cos \theta_3 + \ldots = 0 \quad (1)$$
$$P_1 \sin \theta_1 + P_2 \sin \theta_2 + P_3 \sin \theta_3 + \ldots = 0 \quad (2)$$
$$P_1 p_1 \quad + P_2 p_2 \quad + P_3 p_3 \quad + \ldots = 0 \quad (3)$$

These three conditions are sometimes stated thus: 'It is necessary and sufficient for the equilibrium of any system of forces acting in a plane that the sum of their horizontal components equal zero, the sum of their vertical components equal zero, and the sum of their moments with respect to any one point equal zero.'

61. *Remark.*—The determination of the resultant of any system of forces acting in a plane can also be effected by the following process: Resolve each force into components parallel to each of two rectangular axes, then the original system is replaced by two systems of parallel forces, viz. One parallel to ox, and the other parallel to oy. Find (by Prop. 12) the resultants R' and R'' of these systems respectively, and then the resultant of R' and R'' is the required resultant. The student will find it a useful exercise to work Ex. 252, 263, 264, 266, and 267 by this method; he may also prove that when the forces are in equilibrium the components parallel to ox generally constitute a couple, and likewise those parallel to oy, and that these couples have equal moments of opposite signs.

62. *The centre of parallel forces.*—If we conceive any system of Parallel Forces, and suppose that each force acts at a particular point, then if we suppose the lines along which the forces act to be turned round the points through any equal angles so that they still continue parallel, it will be found that there is a certain fixed point through which their resultant will always pass, whatever be the magnitude of the equal angles; the fixed point in the line of action of the resultant is called *the centre of that system of parallel forces*. If the parallel forces are the weights of the parts of a heavy body, or of the members of a system of heavy bodies, the centre of those parallel forces is the centre of gravity of the body or system of bodies.

If the parallel forces act at points which lie in a straight line, their centre can be found thus: Let P_1, P_2,

P_3 .... be the forces acting at N_1, N_2, N_3, ... in the line Ox, and let their lines of action make an angle θ with that line; also let $ON_1 = x_1$, $ON_2 = x_2$, $ON_3 = x_3$ ....; from O let fall a perpendicular Op cutting the lines of action of the forces in M_1, M_2, M_3, ... and let $OM_1 = p_1$, $OM_2 = p_2$, $OM_3 = p_3$, ... Let R be the resultant of $P_1, P_2, P_3, \ldots$ and let it act along a line cutting Ox in N and Op in M, also let $OM = p$, and $ON = \bar{x}$. Then (Prop. 12) pR or $p(P_1 + P_2 + P_3 + \ldots) = P_1 p_1 + P_2 p_2 + P_3 p_3 + \ldots$ But $p = \bar{x} \sin\theta$, $p_1 = x_1 \sin\theta$, $p_2 = x_2 \sin\theta$, ... Therefore by substitution we obtain, after dividing out $\sin\theta$,

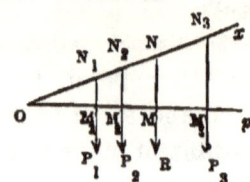

Fig. 67.

$$\bar{x}(P_1 + P_2 + P_3 + \ldots) = P_1 x_1 + P_2 x_2 + P_3 x_3 + \ldots \quad (1)$$

Now the value of $\bar{x}$ given by this equation is independent of θ, and therefore will be the same whatever value θ may have; hence the line of action of the resultant will always pass through N, when the lines along which the forces act are turned through any equal angles round $N_1, N_2, N_3 \ldots$ and continue parallel. The above equation therefore both proves the existence of a centre of parallel forces, and serves to determine it, in the case considered. If $P_1, P_2, P_3 \ldots$ are the weights of a number of particles arranged along a line, the above equation (1) serves to determine their centre of gravity.

Proposition 16.

To determine the centre of any system of parallel forces acting in one plane.

(1) Consider the case of two parallel forces, P_1, P_2; let them act at the points Q_1, Q_2, the co-ordinates of which are $ON_1 = x_1$, $N_1 Q_1 = y_1$, $ON_2 = x_2$, $N_2 Q_2 = y_2$. Divide $Q_1 Q_2$ in K, so that

$$Q_1 K : K Q_2 :: P_2 : P_1$$

CENTRE OF PARALLEL FORCES. 109

then the resultant R_1 of P_1 and P_2 will equal $P_1 + P_2$ and its direction will pass through K; let the co-ordinates of K be $OM = \bar{x}_1$ and $KM = \bar{y}_1$; through Q_1 and K draw lines $Q_1 n$ and Kk parallel to Ox, then by Eucl. (2—VI.) we have

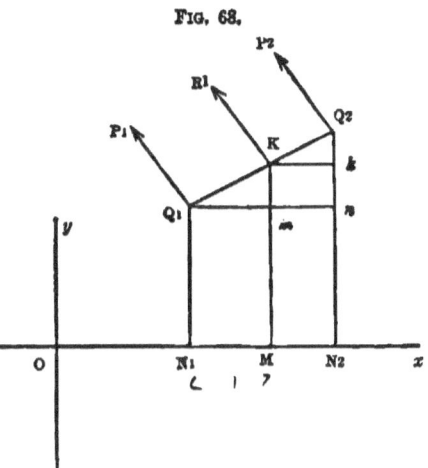

Fig. 68.

$$Q_1 K : K Q_2 :: Q_1 m : mn :: \bar{x}_1 - x_1 : x_2 - \bar{x}_1$$

therefore $\bar{x}_1 - x_1 : x_2 - \bar{x}_1 :: P_2 : P_1$

therefore $P_1 \bar{x}_1 - P_1 x_1 = P_2 x_2 - P_2 \bar{x}_1$

or $\bar{x}_1 (P_1 + P_2) = P_1 x_1 + P_2 x_2$

Again, since $Q_1 K : K Q_2 :: Km : Q_2 k$, we shall obtain, by reasoning in a precisely similar manner, that

$$\bar{y}_1 (P_1 + P_2) = P_1 y_1 + P_2 p_2$$

The position of K will not be affected if the lines of action of P_1 and P_2 be turned round Q_1 and Q_2 through equal angles so as to remain parallel; consequently K is the centre of P_1 and P_2 and its position is determined by $\bar{x}_1$ and $\bar{y}_1$.

(2) Suppose there are three forces, P_1, P_2, P_3. *First*, find R_1 the resultant of P_1 and P_2, acting at the point $\bar{x}_1 \bar{y}_1$; this, from the preceding paragraph, we do by the equations

$$R_1 = P_1 + P_2 \qquad (1)$$
$$\bar{x}_1 (P_1 + P_2) = P_1 x_1 + P_2 x_2 \qquad (2)$$
and $$\bar{y}_1 (P_1 + P_2) = P_1 y_1 + P_2 y_2 \qquad (3)$$

Secondly, find R the resultant of R_1 and P_3, acting at the point $\bar{x}\,\bar{y}$, for which we have the equations

$$R = R_1 + P_3 = P_1 + P_2 + P_3$$
$$\bar{x}(R_1 + P_3) = R_1 \bar{x}_1 + P_3 x_3$$

or
$$\bar{x}(P_1 + P_2 + P_3) = (P_1 + P_2)\bar{x}_1 + P_3 x_3 \quad (4)$$

and
$$\bar{y}(R_1 + P_3) = R_1 y_1 + P_3 y_3$$

or
$$\bar{y}(P_1 + P_2 + P_3) = (P_1 + P_2)\bar{y}_1 + P_3 y_3 \quad (5)$$

Hence, adding together (2) and (4), and also (3) and (5), we obtain

$$\bar{x}(P_1 + P_2 + P_3) = P_1 x_1 + P_2 x_2 + P_3 x_3 \quad (6)$$
$$\bar{y}(P_1 + P_2 + P_3) = P_1 y_1 + P_2 y_2 + P_3 y_3 \quad (7)$$

As before, $\bar{x}$ and $\bar{y}$ undergo no change if the lines of action of the forces are turned through equal angles and continue parallel. They are therefore the co-ordinates of the centre of the parallel forces.

The same proof can evidently be extended to four, five, or any number of forces. Q. E. D.

Cor. 1. If the points of application of the forces had been situated in space of three dimensions, and referred to three co-ordinate planes, a precisely similar proof would have given us

$$\bar{x}(P_1 + P_2 + P_3 + \ldots) = P_1 x_1 + P_2 x_2 + P_3 x_3 + \ldots$$
$$\bar{y}(P_1 + P_2 + P_3 + \ldots) = P_1 y_1 + P_2 y_2 + P_3 y_3 + \ldots$$
$$\bar{z}(P_1 + P_2 + P_3 + \ldots) = P_1 z_1 + P_2 z_2 + P_3 z_3 + \ldots$$

It will be remembered that the same values of $\bar{x}$, $\bar{y}$, $\bar{z}$ would be obtained in whatever order the forces had been taken, consequently a system of parallel forces cannot have more than one centre. It of course follows from this that a body or system of bodies cannot have more than one centre of gravity.

CENTRE OF PARALLEL FORCES. 111

Cor. 2. If the case should arise in which

$$P_1 + P_2 + P_3 + \ldots = 0$$
but $\quad P_1 x_1 + P_2 x_2 + P_3 x_3 + \ldots = A$
and $\quad P_1 y_1 + P_2 y_2 + P_3 y_3 + \ldots = B$

where one at least of A and B has some determinate finite value, the system reduces to a couple; and in this case there is *no* centre of parallel forces in finite space. If the forces are the weights of parts of a body they act in the same direction, and therefore their sum can never be zero, so that every body and system of bodies must have one, and only one centre of gravity, which can be determined by the above equations.

N.B.—For examples on this Proposition see Art. 69.

CHAPTER V.

OF THE CENTRE OF GRAVITY

63. *Definition of the centre of gravity.*—It has been already remarked that the weight of a body is an instance of a distributed force, and that it can be treated as a single force by supposing it to be collected at a certain point, called its centre of gravity. The centre of gravity of any system of particles is the centre of the system of parallel forces composed of the weights of those particles. If the particles form a solid body, it is plain that, if the centre of gravity be supported, the body will rest in any position under the action of gravity only, since the resultant of the applied forces will in all cases pass through the fixed point. It is also plain that no point but the centre of gravity has this property. That, as a matter of fact, every body has a centre of gravity, is shown in the corollary to Proposition 16. In determining the centre of gravity of any figure, it is assumed that a heavy line is made up of particles, a heavy plane of heavy parallel lines, and a solid of heavy parallel planes. It is also assumed that every figure is of uniform density, unless the contrary is specified.

Ex. 271.—Determine the centre of gravity of a uniform straight line A B.

The line A B may be conceived to be made up of a number of equal particles distributed uniformly along it (like beads on a wire); now if we take the two extreme particles, the resultant of their weights will pass through the middle point of A B, and in like manner that of each successive pair; consequently the weight of the whole will act through the middle point of A B, which is therefore the centre of gravity of the whole, or of the heavy line A B.

64. *Method of determining the centre of gravity of a plane area.*—Let A B C D be the plane area; we may

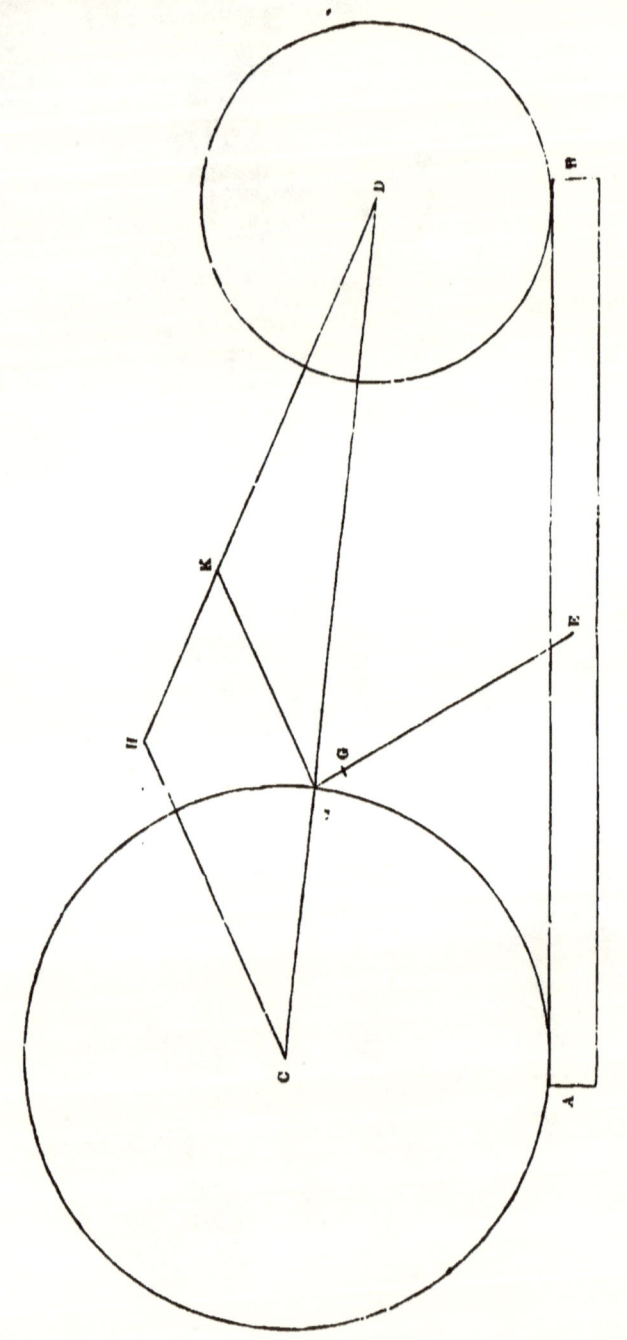

Fig. e, page 113.

conceive it to be made up of a set of parallel heavy lines, such as B D, E F ... drawn in any direction. If we can find a set of parallel lines all bisected by a single line A C, the centre of gravity of each line must be in A C, and therefore that of the whole figure must be in A C. If, moreover, we can determine a second line bisecting another set of parallel lines, we know that the centre of gravity must also be in this second line, and must therefore be at its point of intersection with A C. This method enables us to determine the centre of gravity of many simple figures: it also suggests a practical means of determining the centre of gravity of any plane area whatever. Suppose the figure to be cut out carefully to the required shape in cardboard or tin; suppose it to be suspended by a fine thread from any point B; now the forces in equilibrium are the tension of the string and the weight of the body; they must therefore act along the same line, so that the required centre of gravity must be in the prolongation B C of A B; this prolongation can easily be marked by suspending a plumb-line from A. Again, suspend the body by a fine thread D E fastened to any other point E, and draw the prolongation of this line, viz. E F; the centre of gravity must be in E F, and therefore at G, the point of intersection of E F and B C.

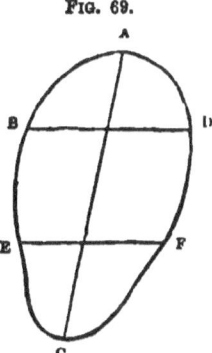
Fig. 69.

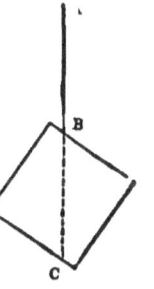
Fig. 70.

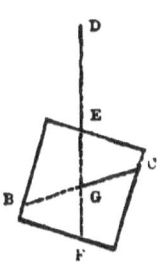

Ex. 272.—Show that the centre of gravity of the area of a circle is at its centre.

Since any diameter bisects all lines in the circle drawn perpendicularly to it, the centre of gravity must be in *any* diameter, and therefore at the centre of the circle.

Ex. 273.—Show that the centre of gravity of an ellipse must be at its centre.

Ex. 274.—Determine the centre of gravity of a triangle.

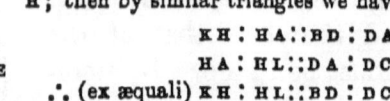

Fig. 71.

Let A B C be any triangle, bisect B C in D and join A D; draw any line K L parallel to B C cutting A D in H; then by similar triangles we have

$$KH : HA :: BD : DA$$
$$HA : HL :: DA : DC$$
∴ (ex æquali) $KH : HL :: BD : DC$

But B D is equal to D C, therefore K H is equal to H L, or K L is bisected by A D; and the same being true of any line drawn parallel to B C, the centre of gravity of the triangle must be in A D. Again, if A C is bisected in E and B E is drawn, the centre of gravity will be in B E, and therefore must be at G, the point of intersection of A D and B E.

It can be easily proved that $GD = \frac{1}{3}AD$. For join E D, then because A E = E C, and B D = D C we have

$$AE : EC :: BD : DC$$

and therefore E D is parallel to A B; hence the triangle D E G is similar to A B G and E D C to A B C:

therefore $\quad DG : DE :: GA : AB$
and $\quad DE : DC :: AB : BC$
therefore (ex æquali) $\quad DG : DC :: GA : BC$
But $\quad DC = \frac{1}{2}BC \therefore DG = \frac{1}{2}GA = \frac{1}{3}DA.$

Ex. 275.—Show that the centre of gravity of a parallelogram is at the intersection of the diagonals.

65. Centre of gravity of solids.

The above method can easily be extended to the case of solids; we may suppose them to be made up of heavy parallel planes: if we can show that the centres of gravity of these all lie along a line, we know that the centre of gravity of the solid must be in that line, and if two such lines can be found, the centre of gravity of the solid must be at their point of intersection.

Ex. 276.—Show that the centre of gravity of a sphere is at its centre.

Ex. 277.—Show that the centre of gravity of a cylinder is at the middle point of its axis.

[It may be regarded as evident that the same rule will hold good of any prism.]

Ex. 278.—Show that the centre of gravity of a parallelopiped is at the point of intersection of its diagonals.

CENTRE OF GRAVITY. 115

66. *Centre of gravity of a figure consisting of two or more simple figures.*—Let W_1 and W_2 be the weights of the simple figures and G_1, G_2 their centres of gravity, join $G_1 G_2$, divide it in G in such a manner that

Fig. 72.

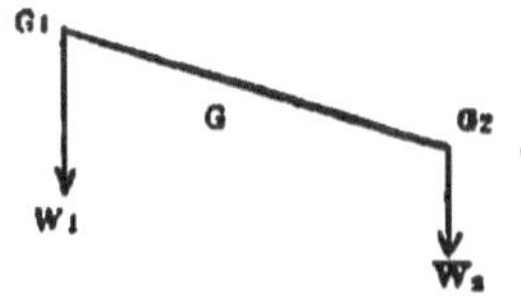

$$G_1 G : G G_2 :: W_2 : W_1$$

Then is G the required centre of gravity.

If there were a third body weighing W_3 whose centre of gravity is G_3, we can find the centre of gravity of the three bodies by joining $G G_3$ and dividing it into parts inversely proportional to $W_1 + W_2$ and W_3; and of course we could continue the same construction to a fourth or a fifth weight, &c.

Ex. 279.—Two spheres whose radii are respectively 4 and 5 in. touch one another; determine the distance of the centre of gravity from the centre of the smaller sphere when the former is of copper and the latter of cast iron.

Ans. 5·54 in.

Ex. 280.—A solid sphere 4 in. in radius touches a hollow sphere 5 in. in radius and 1 in. thick; they are of the same material; show that their centre of gravity is 4·392 in. from the centre of the solid sphere.

Ex. 281.—Determine by construction the centre of gravity of the bodies shown in fig. *e*, where A B is a beam 20 ft. long, and its section 1 ft. square: C and D the centres of two cylinders 1 ft. thick, the radii of whose bases are respectively 6 ft. and 4 ft.; they are of the same material as the beam, and rest with their centres of gravity vertically over the axis of the beam, at distances of 6 in. from A and B respectively.

Construct the figure to scale; this is done in fig. *e*, to the scale of 1 in. for 5 ft.—join C D, then the weights of the cylinders being in the proportion of 9 to 4, divide C D into parts D G, and G, C respectively proportional to 9 and 4; this will give the centre of gravity of the two cylinders. The construction may be made as follows, by Eucl., Bk. VI.—Take D H any line containing 13 equal parts (in the figure each part is ⅕th of an inch) and measure off D K containing 9 of them, join H C and draw K G, parallel to H C; then C G, : G, D :: H K : K D i.e. :: 4 : 9. Find E the centre of gravity of the beam, join E G, ; now the united weight of the cylinders is to the weight of the beam very nearly in the ratio 163 : 20, hence, divide E G, in G so that E G : G G, ::163 : 20, and the point G is the centre of gravity required.

I 2

Ex. 282.—At points 120° apart on the edge of a round table weights of 84 lbs. and 112 lbs. are respectively hung. Find where a weight of 224 lbs. should be placed so as to bring the centre of gravity of the three weights to the middle of the table.

Ex. 283.—A disc of cast iron 12 in. in radius and 2 in. thick rests on a disc of lead 24 in. in radius and 3 in. thick; the circumference of the upper disc passes through the centre of the lower; determine by construction the centre of gravity of the whole.

Ex. 284.—Show that the centre of gravity of any quadrilateral A B C D is given by the following construction:—take o the middle point of the diagonal B D; in O A take O P a third of O A, and in O C take O Q a third of O C; join P Q cutting D B in H; in P Q take P G equal to Q H; the centre of gravity is at G.

67. The centre of gravity of points lying in a straight line.

—The method above explained of finding the centre of gravity of a collection of two or more bodies can be applied to all cases; however, if there are only two bodies, or if the centres of gravity of three or more bodies lie in a line, it is commonly more convenient to determine its distance from some fixed point in that line. Let G_1, G_2 be the centres of gravity of the two bodies whose weights are W_1 and W_2 respectively; then the distance G O of the centre of gravity of W_1 and W_2 from O is determined by the equation

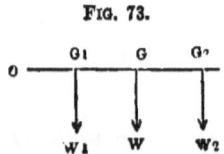
FIG. 73.

$$O G (W_1 + W_2) = O G_1 \times W_1 + O G_2 \times W_2$$

The method of treating three or more weights is exactly the same. It is also plain that if we know O G and O G_2, the same equation will give us O G_1.

Ex. 285.—How far from the one end of the handle is the centre of gravity of the hammer described in *Ex*. 9 situated, if we suppose the other end to fit square with the face of the hammer?

FIG. 74. [If the annexed figure represent the hammer, we have O A = 42 in. A B = 2 in., so that if G_1 is the centre of gravity of the handle and G_2 that of the head, we have O G_1 = 21 in. O G_2 = 41 in. Also the weight of the handle is 4·46 lbs. and of the head 8·37 lbs. Hence

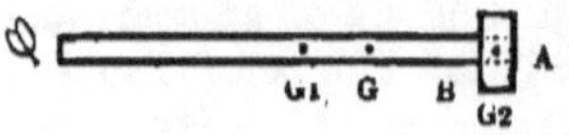

O G × 12·83 = 21 × 4·46 + 41 × 8·37

∴ O G = 34 inches]

CENTRE OF GRAVITY.

Ex. 286.—How far from the end of the handle is the centre of gravity of the hammer described in *Ex.* 12?   *Ans.* $72\frac{9}{11}$ in.

Ex. 287.—Let A B be the diameter of a circular disc of cast iron 12 in. in radius; out of the disc is cut a circular hole (whose centre is in A B) 4 in. in radius; the shortest distance between the circumferences is one inch; find the distance of G, the centre of gravity of the remainder, from A.   *Ans.* $11\frac{1}{5}$ in.

FIG. 75.

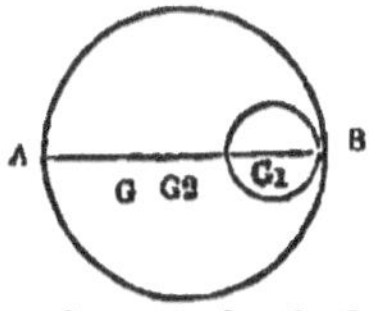

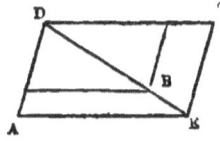

Ex. 288.—If in the last example the hole were filled up with lead, determine the distance of the centre of gravity of the body from A.   *Ans.* 12·42 in.

Ex. 289.—The gnomon A B C is cut out of a parallelogram A C; determine the distance of its centre of gravity from E; having given that D E and D B are respectively 20 and 15 ft. in length.   *Ans.* 6·786 ft.

FIG. 76.

Ex. 290.—If A B is the axis of a cross made up of six squares, the side of each being 3 in. long; find the distance of the centre of gravity from A.
   Ans. $6\frac{1}{2}$ in.

Ex. 291.—A rod capable of turning round a fixed point is kept in equilibrium by two weights suspended by strings of given length from the respective ends. Show that the centre of gravity of the weights is fixed whatever angle the rod makes with the horizon.

Ex. 292.—Weights of 7, 7, and 6 lbs. respectively are placed at the angular points of a triangle; find their centre of gravity relatively to that of the triangle.

Ex. 293.—Out of an isosceles triangle cut a square having two angles on the base and one on each of the equal sides. Find the centre of gravity of the remainder.

Ex. 294.—A piece of wire of uniform thickness is bent so as to form three sides of a triangle; show that the centre of gravity is the centre of the circle inscribed in the triangle formed by joining the middle points of the original triangle.

68. *Remark.*—The following examples of the determination of centres of gravity are similar to those contained in the former article, but involve somewhat greater geometrical difficulties; in many cases it will be well if the reader bear in mind, that when bodies are of the same substance their weights are proportional to their volumes, so that it frequently happens we may reason upon their *volumes* instead of their *weights*.

Ex. 295.—To find the centre of gravity of a triangular pyramid.

Let A B C D be the pyramid; bisect B D in H, join A H and H C; take $FH = \frac{1}{3}AH$ and $HE = \frac{1}{3}HC$; draw F C and A E, then these lines being in the same plane, viz. A C H, will intersect, let them do so in G; this point will be the required centre of gravity, and E G will equal $\frac{1}{4}$th part of A E. For draw any plane $b\,c\,d$ parallel to B C D cutting the plane A C H in $h\,c$, the line A E in e, and A H in h; then h is the middle point of $b\,d$; and it is evident by similar triangles that

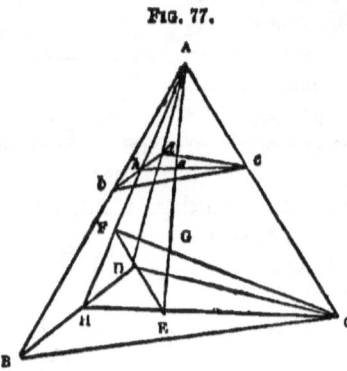

FIG. 77.

$$he : Ah :: HE : AH$$
and $$Ah : hc :: AH : HC$$
$$\therefore (ex\ æq.)\, he :: hc :: HE : HC$$

but $HE = \frac{1}{3}HC \therefore he = \frac{1}{3}hc$, and e is the centre of gravity of the triangle $b\,c\,d$; and the same being true of every other parallel section, the centre of gravity of the pyramid must be in A E; in the same manner it can be proved that the centre of gravity of the pyramid must be in C F; therefore it must be at G the point of intersection of A E and C F. Next, to show that $EG = \frac{1}{4}AE$. Join F E; then since $HE = \frac{1}{3}EC$ and $HF = \frac{1}{3}FA$, we have $HE : EC :: HF : FA$, and therefore F E is parallel to A C; hence the triangles G E F and G A C are similar, and we have

$$GE : GA :: EF : AC :: EH : CH$$

but $EH = \frac{1}{3}CH, \therefore GE = \frac{1}{3}GA = \frac{1}{4}AE$. Hence the centre of gravity of a triangular pyramid is found by the rule: Draw the line joining the centre of gravity of the base and the vertex of the pyramid, divide it into four equal parts; the first point of section above the base is the centre of gravity.

Ex. 296.—If the middle points of any two edges of a triangular pyramid which do not intersect are joined by a straight line, the middle point of that line is the centre of gravity of the pyramid.

Ex. 297.—Show that the centre of gravity of any pyramid or cone is found by the same rule as the centre of gravity of a triangular pyramid.

Ex. 298.—If out of any cone a similar cone is cut, so that their axes are in the same line and their bases in the same plane; show that the height of the centre of gravity of the remainder above the base equals $\frac{1}{4} \cdot \frac{h^4 - h'^4}{h^3 - h'^3}$ where h is the height of the original cone, and h' the height of that which is cut away.

Ex. 299.—If out of any right cylinder is cut a cone of the same base and height; show that the centre of gravity of the remainder is $\frac{5}{8}$ths of the height above the base.

CENTRE OF GRAVITY. 119

Ex. 300.—Find the centre of gravity of a trapezoid in terms of the lengths of the two parallel sides, and of the line joining their middle points.

Let A B C D be the trapezoid, of which A B and D C are the parallel sides; produce A D and B C to meet in E; bisect A B in F, join E F cutting D C in H, which is its middle point. Take $FG_1 = \frac{1}{3}FE$, $HG_2 = \frac{1}{3}HE$; then G_1 is the centre of gravity of the whole triangle A B E and G_2 of the part C D E; therefore G, the centre of gravity of the remainder, will lie in F H. Now, we have given $AB = a$, $DC = b$, and $FH = h$, and are to find $FG = x$.

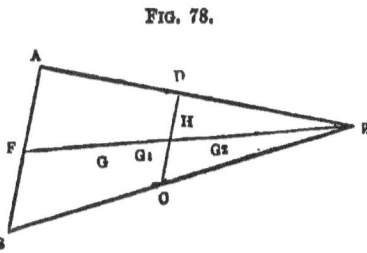

FIG. 78.

Since the weights are in the same proportion as the areas of the triangles A B E and C D E, we have

$$FG_1 \times ABE = FG \times ABCD + FG_2 \times CDE$$

Now, $FG_1 = \frac{1}{3}FE$ and $FG_2 = h + \frac{1}{3}HE = h + \frac{1}{3}(FE - h) = \frac{2}{3}h + \frac{1}{3}FE$

therefore $x \times ABCD = \frac{1}{3}FE \times ABE - (\frac{2}{3}h + \frac{1}{3}FE) \times CDE$

But by similar triangles (Euc. 19—VI.)

$$ABE : CDE :: a^2 : b^2$$

therefore $ABCD : CDE :: a^2 - b^2 : b^2$

therefore $x(a^2 - b^2) = \frac{1}{3}FE \times a^2 - (\frac{2}{3}h + \frac{1}{3}FE)b^2$

$$= \frac{1}{3}FE \times (a^2 - b^2) - \frac{2}{3}hb^2$$

Again, by similar triangles,

$$FE : HE :: AE : DE :: a : b$$

therefore $FE : FE - HE :: a : a - b$

and $FE = \dfrac{ha}{a-b}$

therefore $x(a^2 - b^2) = \frac{1}{3}ha(a+b) - \frac{2}{3}hb^2$

$$= \frac{h}{3}(a^2 + ab - 2b^2)$$

$$= \frac{h}{3}(a + 2b)(a - b)$$

therefore $x = \dfrac{h}{3} \cdot \dfrac{a + 2b}{a + b}$

Ex. 301.—Show that the centre of gravity of the frustum of a pyramid is situated in the line joining the centres of gravity of the ends and at a distance from the lower end, given by the formula $x = \dfrac{h}{4} \cdot \dfrac{a^2 + 2ab + 3b^2}{a^2 + ab + b^2}$, where a and b are any pair of homologous sides of the ends, and h is the length of the line joining the centres of gravity of the ends.

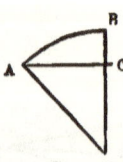

Fig. 79.

Ex. 302.—If a segment of a sphere is described by the revolution of A B C round B O, show that the centre of gravity of the surface of the segment is in the middle point of B C.

[It can be easily proved that if B C is divided into any number of equal parts, and planes are drawn perpendicularly through the points of section, they will divide the surface of the segment into equal zones—the weight of each can be collected in B C; and as these equal weights will be uniformly distributed along B C, the required centre of gravity will be in its middle point.]

Ex. 303.—Show that the centre of gravity of the spherical sector formed by the revolution of the sector A B O (fig. 79) round B O is at a distance from $O = \frac{3}{4} OB - \frac{3}{8} BC$ or $\frac{3}{8}(OB + OC)$

[It must be remembered that the spherical sector may be conceived to be made up of an indefinitely great number of equal pyramids having a common vertex O, whose bases form the spherical surface; the weights of each of these can be collected at its centre of gravity, distanced $\frac{3}{4}$ O B from O, and the question is reduced to a case of the last example.]

Ex. 304.—Determine the position of the centre of gravity of the volume of the spherical segment formed by the revolution of A B C round B O. And when A B C is a quadrant, show that the centre of gravity of the hemisphere generated by its revolution is at a distance of $\frac{3}{8}$ths of the radius from the centre of the sphere.

69. *Applications of the formulæ of Prop.* 16.—When a body consists of parts, and we know the weights of the several parts, and the co-ordinates of their centres of gravity; the co-ordinates of the centre of gravity of the body will be found by means of the formulæ of Prop. 16.

Ex. 305.—A, B, C, D are the angular points taken in order of a square (one of whose sides is a) and E the intersection of its diagonals; weights of 3, 8, 7, 6, and 10 lbs. are placed at these points respectively. Find their centre of gravity.

Ans. If A B and A D are the axes of x and y, $34\bar{x} = 20a$, $34\bar{y} = 18a$.

Ex. 306.—Weights of 1, 2, 3, 4, 5, and 6 lbs. are placed respectively at the angular points of a regular hexagon (one of whose sides is a) taken in order. Find their centre of gravity.

Ans. If the lines joining the points at which 1 and 2 and 1 and 5 are placed be the axes of x and y, $14\bar{x} = 5a$, $14\bar{y} = 9a\sqrt{3}$.

Ex. 307.—A B C is an isosceles triangle right-angled at C; parallel forces

APPLICATIONS OF PROP. XVI. 121

of 4, 6, and 8 lbs. act at A, B, and C respectively. Find their centre when the two former act in the same direction and the latter in the opposite direction. Likewise when the third force is 10 lbs.

Ans. (1) If C A and C B are the axes of x and y, $\bar{x} = 2a$, $\bar{y} = 3a$.
(2) Centre at an infinite distance, forces reducing to a couple.

Ex. 308.—A, B, C, D are the angular points taken in order of a square, one of whose sides is a; parallel forces of 5, 9, 7, and 3 lbs. act at the angular points respectively. Find their centre—(1) supposing 5 and 9 to act in the same direction, and 7 and 3 in the opposite direction; (2) supposing 5 and 7 to act in the same direction, and 9 and 3 to act in the opposite direction.

Ans. (1) If A B and A D are the axes of x and y, then $2\bar{x} = a$, $2\bar{y} = -5a$. (2) Centre at an infinite distance, forces reducing to a couple.

Ex. 309.—Parallel forces of 5 lbs. apiece act in the same direction through the angular points of a square, and a parallel force of 20 lbs. acts through the intersection of the diagonals in the opposite direction. Find the centre.

Ans. Centre indeterminate, forces being in equilibrium.

Ex. 310.—Find the co-ordinates of the centre of gravity of the trapezoid A B C D, having given O B = 7 ft., O C = 19 ft., A B = 12 ft., D C = 18 ft.; the angles at B and C being right angles.

FIG. 80.

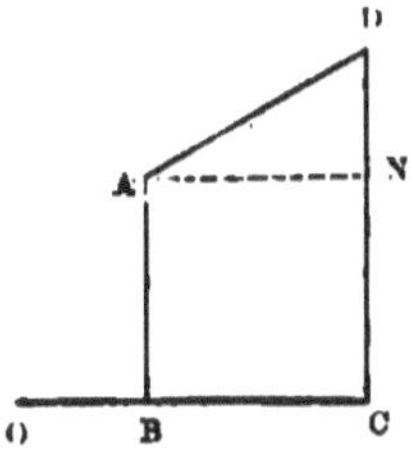

[If A N is drawn parallel to B C dividing the figure into a triangle and a square, the co-ordinates of the centre of gravity of each can be easily found, and if $\bar{x}$ and $\bar{y}$ are the required co-ordinates, it will appear that they are determined by the equations

$$180\,\bar{x} = 13 \times 144 + 15 \times 36$$
$$180\,\bar{y} = 6 \times 144 + 14 \times 36]$$

Ans. $\bar{x} = 13\tfrac{2}{5}$, $y = 7\tfrac{3}{5}$.

Ex. 311.—Let A B C D represent the section of a ditch: the breadth A D is 20 ft. and the depth 8 ft.; the slope of A B is 1 in 1 and of D C is 2 in 1; determine the horizontal distance from A of the centre of gravity of the section.

FIG. 81.

Ans. $10\tfrac{4}{9}$ ft.

Ex. 312.—If in the last example the breadth A D is a feet, the depth of the ditch h feet, and if A B has a slope of m in 1 and D C of n in 1, show that if $\bar{x}$ be the horizontal distance of the centre of gravity of the section from A; then $\bar{x}$ will be found by the formula

$$\bar{x}\left\{2a - \left(\frac{1}{m} + \frac{1}{n}\right)h\right\} = a^2 - \frac{1}{n}\,ah - \tfrac{1}{3}\left(\frac{1}{m^2} - \frac{1}{n^2}\right)h^2$$

Ex. 313.—If A B C D represents the section of a wall of which B C is vertical and equal to h, A B $= a$ and D C $= b$; then if w is the weight of a cubic foot of th material, the moment of 1 foot of the length of the wall round A and B respe tively are given by the formulæ

$$M = \frac{wh(2a^2 + 2ab - b^2)}{6}$$

and

$$M = \frac{wh(a^2 + ab + b^2)}{6}$$

Ex. 314.—The engine-room of a steam-vessel is 30 ft. long, 20 ft. wide, and 15 ft. high; at 10 ft. from one side, 6 ft. from one end, and 5 ft. from the floor, is situated the centre of gravity of the boiler, the weight of which is 2 tons; at 4 ft. from the same side, 11 ft. from the same end, and 7 ft. from the floor, is the centre of gravity of the beam of the engine, which weighs $\frac{1}{2}$ a ton; at 9 ft. from the side, 7 ft. from the end, and 3 ft. from the floor, is the centre of gravity of the furnace, which weighs $1\frac{1}{2}$ ton; at 5 ft. from the side, 11 ft. from the end, and 10 ft. from the floor, is the centre of gravity of the cylinder, which weighs 1 ton; where is the centre of gravity of the whole?

Ans. 8·1 ft. from the side, 7·8 ft. from the end, 5·6 ft. from the floor.

70. *On stable and unstable equilibrium.*—Bearing in mind that when forces are in equilibrium any one of them is equal and opposite to the resultant of all the rest, it is plain that when a heavy body is supported by any forces their resultant must act vertically upward through the centre of gravity. Suppose, then, that a body is supported at one point, the reaction of the fixed point and the weight of the body are in equilibrium, therefore the direction of the reaction must pass vertically through the centre of gravity, consequently the conditions of equilibrium are fulfilled when the line joining the centre of gravity and the fixed point is vertical, or, which comes to the same thing, when the centre of gravity is vertically under or vertically over the fixed point.

Practically speaking, there is the greatest possible difference between these two cases, for a body could scarcely be made to rest in the latter position, and could be displaced from it by the smallest possible force and caused to take up the former position. In fact, the former case—

STABLE AND UNSTABLE EQUILIBRIUM. 123

centre of gravity *under* the point of support—is said to be a position of *stable* equilibrium, while the latter—centre of gravity *above* the point of support—is said to be one of *unstable* equilibrium. The distinction between *stable* and *unstable* equilibrium is thus stated: Suppose a body to be in equilibrium under the action of given forces, and suppose it to be slightly displaced, if the forces tend to bring the body *back* again to the original position, that position was one of *stable* equilibrium, but if they tend to make it move *farther* from its original position, that position was one of *unstable* equilibrium. If the student will draw a figure of a body suspended from a point he will see at once that the two positions of equilibrium are *stable* and *unstable* according to the terms of the definition.

It is obvious that there may be an intermediate case in which, after the body has been displaced, the forces have no tendency to move it either backward or forward. In this case the body is said to have been in a position of *neutral* equilibrium. If, in the example already given, the point supported had been the centre of gravity the equilibrium would have been neutral. A sphere of uniform density on a horizontal plane is in a position of neutral equilibrium; if it be *loaded* at the top of a vertical diameter its position becomes one of *unstable* equilibrium, if loaded at the lower end of a vertical diameter it is in a position of *stable* equilibrium.

Ex. 315.—A hemisphere (whose radius is r) and a cone (the radius of whose bar is r and height h) of equal and uniform density are fastened together so that their bases coincide. They are placed on a horizontal plane, and are in equilibrium resting on the lowest point of the hemisphere; show that the equilibrium is *stable, neutral,* or *unstable,* according as

$$r\sqrt{3} > \text{ = or } < h.$$

Our limits will not allow us to develop this subject fully, but one other point must not be passed over. A body may be in stable equilibrium in two or more positions, but the *degree* of stability in the two cases may be very different:—

Thus, referring to fig. 19 and supposing the force P not to act, if the body were placed on one of its edges upon a horizontal plane and with either diagonal (joining A and C or B and D) vertical, it would be in a position of *unstable* equilibrium; but if it is placed with the face containing A B or A D on a horizontal plane the equilibrium is *stable*; but manifestly far more *stable* in the latter case than in the former; indeed, if A D were many times (e.g. a hundred times) greater than A B the degree of stability in the former case would be so small that practically the body would not retain its position without support.

71. *Geometrical applications of the properties of the centre of gravity.*—The most important of these are proved in Prop. 17, 18, 19; but before considering them one class of applications may be noticed. Suppose it can be proved by any means that the centre of gravity of a figure or collection of points lies in two or more lines, then, as there can be only one centre of gravity, it will follow that those lines must pass through a common point, e.g. in any triangle the lines joining each angle with the middle point of the opposite side must pass through one point. This admits of independent geometrical proof; it also follows at once from the fact that the centre of gravity of the triangle is in each of the lines.

Ex. 316.—Draw any quadrilateral, show that the lines joining the points of bisection of opposite sides mutually bisect each other.

[Suppose equal weights to be placed at each angle of the quadrilateral, and find their centre of gravity.]

Ex. 317.—In any triangular pyramid the three lines, joining the middle points of each pair of edges which do not meet, pass through a common point.

Ex. 318.—If A B C is any triangle, and points X, Y, Z are taken on the sides B C, C A, A B respectively, in such a manner that

$$BX \cdot CY \cdot AZ = XC \cdot YA \cdot ZB,$$

the lines A X, B Y, and C Z will pass through a common point.

GEOMETRICAL APPLICATIONS. 125

Proposition 17.

If a surface be described by the revolution of a plane curve round an axis fixed in its plane, its area is found by multiplying the length of the curve into the length of the path described by its centre of gravity.

Let A B be the curve, C D the axis of revolution; G the centre of gravity of the curve; draw G M at right angles to C D; we have to show that the area of the surface described by the revolution of A B round C D is found by multiplying the length of A B into the length of the path described by G, i.e. into 2π G M.

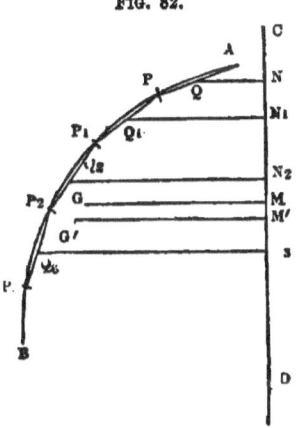

FIG. 82.

In A B place any number of equal chords, viz. A P, P P_1, $P_1 P_2$, &c. Take Q, Q_1, Q_2, ... their middle points, and draw Q N, $Q_1 N_1$, $Q_2 N_2$ ... at right angles to C D; also find G' the centre of gravity of the chords, and draw G' M' at right angles to C D; now when the curve revolves round C D, the chords will describe frustums of cones, the surfaces of which, by a well-known rule of mensuration, will be respectively $2\pi \times$ A P $\times$ Q N, $2\pi \times$ P P_1 $\times Q_1 N_1$, $2\pi \times P_1 P_2 \times Q_2 N_2$, &c., and therefore the sum of the surface of these frustums will equal

$$2\pi \left(\text{A P} \times \text{Q N} + \text{P } P_1 \times Q_1 N_1 + P_1 P_2 \times Q_2 N_2 + \ldots \right)$$

But by a property of the centre of gravity (Prop. 16) we have

$$\text{G}'\text{M}' \left(\text{A P} + \text{P } P_1 + P_1 P_2 + \ldots \right) = \text{A P} \times \text{Q N} + \text{P } P_1 \times Q_1 N_1 + P_1 P_2 \times Q_2 N_2 + \ldots$$

Therefore the sum of the surfaces of the conic frustums will equal

2π G'M' $\times$ the sum of the chords A P, P P_1, $P_1 P_2$ . . . .

Now this being true, however great the number of chords,

will be true in the limit; but the surface of the solid cf revolution is the limit of the sum of the surfaces of the conic frustums; the length of the curve is the limit of the sum of the chords; and since G′ must ultimately coincide with G, the limit of G′M′ is G M. Therefore, area of surface described $= 2\pi$ G M × length of curve A B.   Q. E. D.

Cor.—It is manifest that the above proof includes the case of the figure described by the revolution of an area bounded by straight lines. It is also obvious that the same rule applies to any portion of the area contained between two given positions of the revolving curve.

Proposition 18.

If a plane curve revolve about an axis fixed in its plane, the volume of the solid described is found by multiplying the area of the curve by the length of the path of its centre of gravity.

Let A B C D be the plane curve; the lines A C and B D are perpendicular to C D, the axis about which the curve revolves;

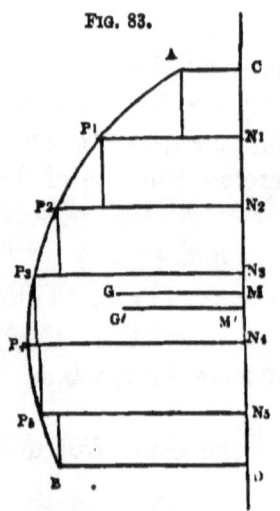

Fig. 83.

find G its centre of gravity, and draw G M at right angles to C D: we have to show that the volume of the solid described by the revolution of A B C D equals the length of G's path multiplied by the area of A B C D.

Divide C D into any number of equal parts in $N_1, N_2, N_3, \ldots$ and from these points draw ordinates to meet the curve in $P_1, P_2, P_3, \ldots$ and complete the rectangles A N_1, $P_1 N_2$, $P_2 N_3$, ...; when the figure revolves round C D, these rectangles will describe cylinders, and the united volumes will equal

$$\pi \left(A C^2 \times C N_1 + P_1 N_1{}^2 \times N_1 N_2 + P_2 N_2{}^2 \times N_2 N_3 + \ldots \right)$$

Let G' be the centre of gravity of these rectangles, draw G'M' at right angles to CD; now, the centre of gravity of AN$_1$ is at a distance from CD equal to $\frac{1}{2}$ AC, that of P$_1$N$_2$ is at a distance from CD equal to $\frac{1}{2}$ P$_1$N$_1$, and similarly of the others. Hence by Prop. 16 G'M' × sum of rectangular areas equals

$$\tfrac{1}{2}AC \times AC \times CN_1 + \tfrac{1}{2}P_1N_1 \times P_1N_1 \times N_1N_2 + \tfrac{1}{2}P_2N_2 \times P_2N_2 \times N_2N_3 + \ldots$$

and therefore 2π G'M' × sum of rectangular areas equals

$$\pi A\,C^2 \times C\,N_1 + \pi P_1N_1^{\,2} \times N_1N_2 + \pi P_2N_2^{\,2} \times N_2N_3 + \ldots$$

that is, equals the sum of the above-mentioned cylinders, and this, being true whatever be the number of parts into which CD is divided, will be true in the limit; now, the volume of the solid of revolution is the limit of the sum of the cylinders; the curvilinear area is the limit of the sum of the rectangles; and since G' must ultimately coincide with G, the limit of G'M' is G M. Hence the volume of the solid of revolution is found by multiplying the area of the curve by the length of the path described by its centre of gravity.

Cor.—The remarks contained in the corollary to the last are applicable, *mutatis mutandis*, to the present Proposition.

Proposition 19.

If a right prism or cylinder be cut by any plane, the volume of the frustum is found by multiplying the area of the base into the length of a line drawn perpendicularly to the base through its centre of gravity, and terminated by the cutting plane.

Let ABCD be the frustum of the right prism or cylinder, standing on the base ABE, whose centre of gravity is G; through G draw GQ at right angles to the plane of the base ABE and terminated by the cutting plane DCF; we have to show that the volume of the frustum is found by multiplying the area of A E B into the length of G Q. Suppose the plane of the paper to be perpendicular to the planes of the ends, and to cut them in A B B' C D; if the planes of the two ends are produced, they will intersect in a line K K' perpendicular to the plane of the paper; hence A B' D is the angle of inclination of the cutting plane to the base; we will denote this angle by θ. Draw G M at right angles to K K' and join Q M.

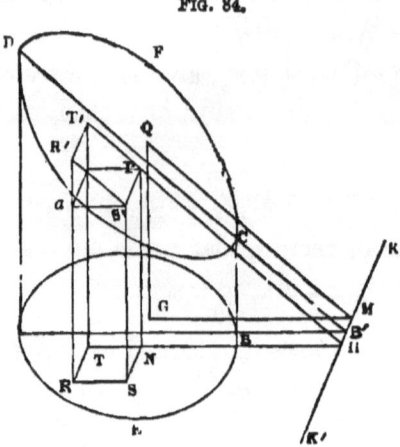

FIG. 84.

Suppose the base A E B to be divided into a large number of small rectangular areas (such as N S R T), then ultimately the sum of these rectangles will equal the area of the base. On the rectangles describe rectangular parallelopipeds such as P a N R, then ultimately the sum of their volumes will equal the volume of the frustum. Let N S R T be denoted by p_1 and N H by y_1, then the volume of P a N R is

$$p_1 y_1 \tan \theta$$

since P N plainly equals N H $\times \tan \theta$. Adopting a similar notation for the other parallelopipeds, the sum of their volumes will equal

$$(p_1 y_1 + p_2 y_2 + p_3 y_3 + \ldots) \tan \theta$$

and this by Prop. 16 equals

$$(p_1 + p_2 + p_3 + \ldots) \bar{y} \tan \theta$$

Now in the limit

$$p_1 + p_2 + p_3 + \ldots = \text{A E B}$$

and $\quad \bar{y} \tan \theta = \text{G M} \tan \theta = \text{G Q}$

Therefore the volume of the frustum equals A E B $\times$ Q G.

GEOMETRICAL APPLICATIONS. 129

Cor.—It is evident that if the prism or cylinder is cut by another plane inclined at any angle to the base, the volume contained between the cutting planes equals the area of the perpendicular section multiplied into the part contained between the planes of a line drawn through the centre of gravity of the perpendicular section at right angles to its plane.

Ex. 319.—Show that Prop. 17 and 18 are true in the case when the curve is a closed curve and revolves round an axis wholly without it.

Ex. 320.—In Prop. 19 show that Q is the centre of gravity of D C F.

Ex. 321.—An equilateral triangle revolves round its base, whose length is a; find the area of the surface and volume of the figure described.

$$Ans. \ (1) \ \pi a^2 \sqrt{3}. \quad (2) \ \frac{\pi a^3}{4}$$

Ex. 322.—An equilateral triangle revolves round an axis parallel to the base, the vertex of the triangle being between the axis and the base; the base is 6 in. long and the distance from the vertex to the axis is 9 in.; determine the volume of the ring described. *Ans.* 1220·7 cub. in.

Ex. 323.—Determine the volume of a ring formed like that in the last example, having given that each side of the triangle is 6 in. and the external diameter of the ring 3 ft. *Ans.* 1593·4 cub. in.

Ex. 324.—The section of a ring is a trapezoid, its height is 3 in. and its parallel sides are respectively 7 in. and 3 in. long, they are parallel to the axis, the shorter being the nearer to the axis and at a distance of 11 in.; find the volume of the ring. *Ans.* 1196·9 cub. in.

Ex. 325.—In the last example, if the longer side of the trapezoid had been the nearer to the axis, the external diameter of the ring being the same in both cases, what would have been the volume?

Ans. 1159·2 cub. in.

Ex. 326.—Determine the volume and surface of a ring with a circular section whose internal diameter is 12 in. and thickness 3 in.

Ans. (1) 333·1 cub. in. (2) 444·1 sq. in.

Ex. 327.—Determine the volume and surface of a ring whose section is a regular hexagon, whose circumscribing circle has a radius a, and whose centre is at a distance b from the axis of revolution.

Ans. (1) $3\pi b a^2 \sqrt{3}$. (2) $12\pi a b$.

Ex. 328.—Find the centre of gravity of the arc of a semicircle.

$$Ans. \ \text{Distance from centre} = \frac{\text{diam.}}{\pi}$$

K

Ex. 329.—Find the centre of gravity of the area of a semicircle.

Ans. Distance from centre $= \frac{2}{3} \cdot \frac{\text{diam.}}{\pi}$

Ex. 330.—A cylindrical shaft is cut off obliquely at an angle of 45° to the axis, its radius is 6 in. and its extreme height is 2 ft. 6 in.; find its solid contents. *Ans.* 1·5708 cub. ft.

Ex. 331.—A cylindrical shaft is cut obliquely at an angle of 60° to the axis, the radius of the base is 10 in., the extreme height of the shaft 3 ft.; find its volume. *Ans.* 9497 cub. in.

Ex. 332.—A right prism stands on a triangular base, the angles of which are A, B, C, the angles of the other end being D, E, F; the sides AB, AC are each 15 ft. long, BC is 18 ft. long; the edges AD, BE, CF are each 30 ft. long; through the edge BC passes a plane making an angle of 60° with the base; determine the volumes of the parts into which the prism is divided. Also if the prism were cut by a plane parallel to the former and cutting AD at a distance of 24 ft. above A, find the volumes of the two parts.

Ans. (1) 748·3 and 2491·7 cub. ft. (2) 1095·6 and 2144·4 cub ft.

Ex. 333.—Show that if any triangular prism be cut by a plane so that the edges perpendicular to the base are respectively a, b, c, and the area of the base A, then the volume of the frustum will be $\frac{1}{3}$ A $(a + b + c)$.

FIG. 85.

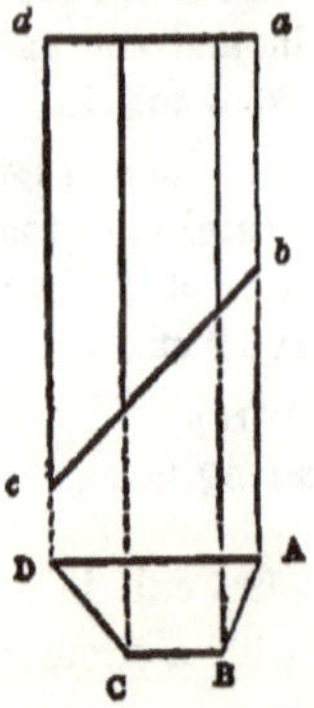

Ex. 334.—Let $a\,b\,c\,d$ represent the plan and ABCD the section of a portion of a ditch; AD = 20 ft.; depth of ditch 8 ft.; slope of AB is 2 in 1, and that of DC is 1 in 1; ab and cd are respectively 20 and 40 ft. long. Find the volume; and determine the error that would be committed if we had found the volume by multiplying the area of the section by half the sum of ab and dc.

Ans. (1) 3264 cub. ft. (2) Error 96 cub. ft.
[Compare *Ex.* 311.]

Ex. 335.—Let ABCD be the plan of a square redoubt, each side of which is 150 ft., the corners of the ditch are quadrants of circles whose centres are respectively A, B, C, D. So that the ditch has a uniform width which is 24 ft., its depth is 9 ft., the inside slope is 3 in 1 and the outside 1 in 1. Find the volume of the ditch. *Ans.* 108,057 cub. ft.

Ex. 336.—If the ditch in the last example were surrounded with a glacis 3 ft. high whose outside slope is 1 in 10 and inside slope 1 in 1; find its volume. *Ans.* 40,897 cub. ft.

CHAPTER VI.

FRICTION OF PLANE SURFACES—
INCLINED PLANE, WEDGE SCREW.

SECTION I.

72. *Reaction of surfaces.*—It nearly always happens that amongst the forces which keep a body at rest is the reaction of one or more surfaces; to explain the nature of this reaction let us consider a particular case; suppose a mass M to rest on a table, A B, and suppose it to weigh 1000 lbs.; that weight must be supported by the table, which must therefore exert upwards a force of 1000 lbs. in a direction opposite to the direction of the weight. If we consider the case particularly we shall see that this reaction is an instance of a *distributed* force, for the under surface of C D will be in contact with the table at many points, and at each point there will be a reaction; what are the magnitudes of the reactions respectively at the points we do not commonly know; they must, however, be such that their resultant shall act vertically upward through the centre of gravity of M and shall equal 1000 lbs. And, in general, if a body is at rest when pressed against a surface, the various points of that surface must supply reactions whose resultant is equal and opposite to the resultant of the forces by which the body is urged

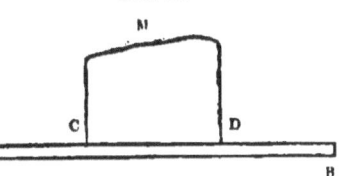

FIG. 86.

against the surface; this resultant reaction is called *the reaction* of the *surface*.

73. *The limiting angle of resistance.*—The question now arises—Under what circumstances is the plane capable of supplying the reaction necessary to produce equilibrium? There will be equilibrium if the plane do not break, if the body do not turn over, and if the reaction keep the body from sliding; it is with the last condition we are here

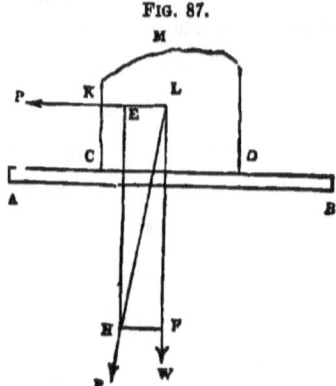

FIG. 87.

concerned. Let us revert to the example discussed in the last article, and let us suppose a rope to be fastened to the point K by means of which the body is pulled horizontally by a force P; we know that if P have a certain magnitude it will just make the body slide, but if it be less than that certain magnitude the body will continue at rest; suppose that a force of 190 lbs. will just not make the body slide; produce P K to meet the vertical line through the centre of gravity in L, let L E represent P (190) and L F represent W (1000), complete the parallelogram and draw the diagonal L H, this must be the direction of the resultant R, and its direction makes with a perpendicular to A B an angle of 10° 45'; now, if the force P is less than 190 lbs. the direction of the resultant will fall within the angle R L W; but if P is greater than 190 lbs. the direction of the resultant will fall without the angle R L W; in the former case the surface A B can supply a reaction which prevents motion, in the latter it cannot; and thus in the case we have supposed the surface A B can supply a reaction in any required direction which makes an angle less than 10° 45' with the normal, i.e. the perpendicular, to the surface; and when the body is in a state bordering on motion, the direction of the reaction will make an angle equal to 10° 45' with the normal.

Now it appears from experiment that if the surface A B were of cast iron, and the mass M of wrought iron, a force of 190 lbs. would be required just not to produce motion in the case above discussed; and it also appears from experiment that within very considerable limits, the same proportions are preserved, irrespectively of the *extent* of the surface pressed and the amount of the force; so that we may state as a fact of experience, that when wrought iron rests on cast iron the former will exert a reaction in any direction required to produce equilibrium that does not make with the normal an angle greater than 10° 45'; and when motion is about to ensue, the direction of the reaction will make an angle with the normal equal to 10° 45'; this angle is therefore called the *limiting angle of resistance*, or *the angle of friction* in the case of cast iron upon wrought. It further appears from experiment, that in the case of any two surfaces whatever, there is a limiting angle of resistance proper to those surfaces, and depending on their physical character; for instance, in the case of wrought iron on oak, the angle is 31° 50', and similarly in other cases. Values of this angle in several cases are given in Table XI.

Hence if a body is urged against a fixed surface by any force or forces, the direction of the reaction of that surface can never make with the normal an angle greater than a certain angle. That angle is called the limiting angle of resistance or the angle of friction; its magnitude is fixed by the physical nature of the surfaces in contact.

If the resultant of the forces which urge the body against the fixed plane be found, the body will continue at rest, provided the direction of the resultant makes with the normal an angle less than the limiting angle of resistance; for under these circumstances the reaction can act in a direction opposite to the resultant and balance it. If the resultant make with the normal an angle equal to the limiting angle of resistance the body will still be in equilibrium, but will now be *in the state bordering on motion*, for if the angle between the resultant and normal be increased by ever so small an amount, the reaction can no longer act in a direction opposite to the resultant, and therefore can no longer balance it. Under all circumstances the reaction will *oppose* the motion of the body. In the following pages ϕ will be used to denote the limiting angle of resistance.

Ex. 337.—If a mass whose weight is w rests on a horizontal plane A B, and is pulled by a force P whose direction (c P) makes an angle a with the horizon, determine P when it is on the point of making the body slide.

FIG. 88.

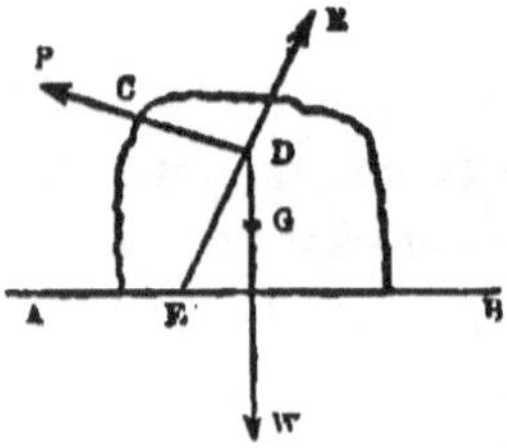

Find G the centre of gravity, and draw G W a vertical line; produce P C to cut G W in D: then since the body is held at rest by P, W, and the reaction of the plane (R), the direction of R must pass through D, also since the body is on the point of sliding from B to A, the direction of R must make with D W an angle B D W equal to ϕ. Then we have W D R = 180° $-\phi$, R D P = 90 $-a+\phi$, and P D W = 90 $+a$, therefore (Prop. 7) P : W : R :: sin ϕ : cos $(a-\phi)$: cos a.

Ex. 338.—In the last example determine P and R if the mass M, weighing 750 lbs., is of wrought iron, on oak, and the direction of P inclined to the horizon at an angle of 15°.   *Ans.* P = 413·3 lbs. R = 756·9 lbs.

Ex. 339.—What would be the required force P in the last case if its direction were horizontal?   *Ans.* P = 465 lbs.

Ex. 340.—Show that when a body rests on a horizontal plane the smallest force that will bring it into the state bordering on motion will act in a direction inclined upwards from the horizon at an angle equal to the limiting angle of resistance.

74. *Conditions under which a body acted on by certain forces will neither be overthrown nor slide.*—Let a mass A B rest on a horizontal plane C D, and let the forces concerned be its weight acting vertically along the line E W

EQUILIBRIUM OF A BODY ON A PLANE. 135

and a force P acting along the line E P : find R the resultant of these forces ; in order that the body may be at rest it is necessary that R be balanced by a reaction equal and opposite to it; this cannot happen if the direction of R cuts C D outside the base ; hence the condition that the body be not overthrown is that the direction of the resultant fall within the base; if this condition be fulfilled, the body will slide or not, according as the direction of R makes with the normal to the point where it cuts the surface, an angle greater or less than the limiting angle of resistance. The question may be asked, if A B be pulled along the line E P by a continually increasing force, will it slide before it topples, or *vice versâ?* This is readily answered by joining A E ; then if A E W be less than the limiting angle of resistance, the body will topple before it slides, since R's direction will fall without the base before its direction makes with the perpendicular an angle greater than the limiting angle of resistance ; if, however, A E W be greater than the limiting angle of resistance, the body will slide before it topples. In the intermediate case, when A E W equals the limiting angle of resistance, the body will be on the point of toppling and sliding for the same value of P. It obviously follows from the above reasoning that when a body stands on a horizontal plane a vertical line drawn through its centre of gravity must cut the plane within its base. If a body rest upon points its base is the polygon formed by joining the points in succession. It is to be observed, however, that if any points would fall *inside* the polygon formed by joining the rest, they are not to be reckoned.

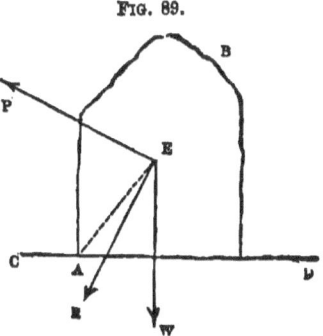

Fig. 89.

Ex. 341.—A rectangular mass of oak the base of which is 2 ft. square and height 7 ft. rests endwise on a floor of oak, a rope is fastened to it at a certain height above the floor and is pulled by a force in a direction inclined upward at an angle of 20° to the horizon; it is found to be on the point both of toppling and sliding; find the height of the point of attachment from the floor, and the magnitude of the force.

Ans. (1) 2·68 ft. (2) 648·7 lbs.

[It is manifest, referring to fig. 89, that B will be found by making the angle B A W equal to the complement of the limiting angle of resistance when the circumstances are those mentioned in the question.]

Ex. 342.—A cylinder of copper the radius of whose base is 2 in. and height 3½ in. rests on a horizontal oak table, it is pulled by a horizontal force whose direction coincides with a radius of the upper end; find the force that will just make the body move, and determine whether the motion will be one of sliding or toppling.

Ans. (1) 8 lbs. (2) The body will topple.

Ex. 343.—Work the last example supposing the cylinder to be of oak, the fibres being parallel to the axis of the cylinder.

Ans. (1) 10·2 oz. (2) The body will slide.

Ex. 344.—A round table stands on four legs, one at each angle of the inscribed square. It weighs 120 lbs.; find the least weight which hung from its edge would overthrow it. *Ans.* 290 lbs.

Ex. 345.—A rectangular box is overthrown by turning round a horizontal edge; given the lengths of the edges; determine the height through which its centre of gravity must be raised.

75. *Friction and the laws of friction.*—Let A B be a table; M a mass which, in consequence of the action of certain forces, is on the point of sliding in the direction B A; then the reaction R' will be equal to their resultant, and its direction will be inclined to the perpendicular to A B at an angle φ equal to the limiting angle of resistance; let C R' be the direction of this reaction; draw C D at right angles to A B, then the angle D C R is equal to φ; take C E to represent R', and complete the

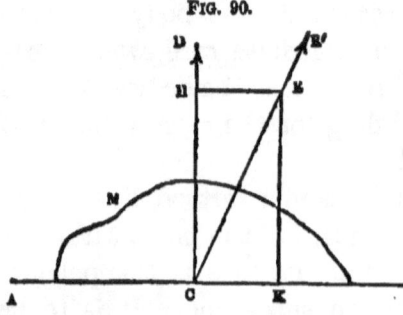

FIG. 90.

LAWS OF FRICTION.

rectangle H K; we may replace R' by two components R and F, of which R acts along C D and F along A B; these components are represented by C H and C K respectively; now it is evident that $\tan \phi = \dfrac{HE}{CH}$ i.e. $\tan \phi = \dfrac{F}{R}$

therefore $\qquad$ F = R $\tan \phi$

The tangential reaction F is commonly called the *Friction*, and $\tan \phi$ (which is generally denoted by the letter μ) is called the *coefficient of friction*; so that when a body resting on a plane is in the state bordering on motion, the friction equals the normal reaction multiplied by the coefficient of friction; it will be remarked that unless the body is in the state bordering on motion the whole of the friction is not called into play, but only so much of it as is sufficient to produce equilibrium.

If in any particular case we are required to determine the relation between the forces which keep a body in the state bordering on motion, and amongst these forces is the reaction of a rough surface, we may treat this reaction in either of two ways:—First, we may consider the reaction (R') to be a single force making an angle ϕ with the normal; or, secondly, we may replace that reaction by two forces, viz. a reaction R acting along the normal, and a friction μ R acting along the tangent; the former way of looking at the question is generally more convenient when the body is acted upon by only three forces, the latter when it is acted on by more than three forces, and when, consequently, it is necessary to have recourse to the general equations of equilibrium.

In order to complete our remarks on this subject, it is to be observed that when the body actually slides, its motion is opposed by a constant friction which is properly represented by μ times the normal reaction; it appears, however, that the numerical value of μ for the same substances is different in the cases of motion and of rest. The difference is most conspicuous in the case of soft substances (e.g. various kinds of wood) that have been some time in contact; wherever a difference exists the value of μ for substances at rest is larger than the value for the same substances in motion.

The chief general results that have been elicited by experiments on the friction of surfaces, are called the *laws of friction*, and may be thus stated:—

(1) Friction is proportional to the normal pressure.
(2) It is independent of the extent of the surfaces in contact.
(3) In the case of motion it is independent of the velocity.
(4) If unguents are interposed so as to form a continuous stratum between the surfaces of contact, the friction depends mainly on the nature and quality of the unguent.

It must be added that these laws depend entirely on experimental evidence, and that the first of them ceases to be true when the pressure per square inch becomes very great. The accurate determination of the values of μ, the coefficient of friction for different substances, is due to General Morin, on whose authority the results rest that are registered in the following table.*

---

* The establishment of the laws of friction appears to be due to Coulomb, whose Memoir on Friction was published in A.D. 1785; a very full abstract of the paper is given in Dr. Young's *Natural Philosophy*, vol. ii. p. 170 (1st ed.) General Morin's Tables are very extensive: they have been several times printed. A sufficient account of the experiments on which they are based, together with the Tables themselves, will be found in his work, *Notions Fondamentales de Mécanique*. To enable the reader to form some conception of the limits within which the laws of friction hold good, the following (somewhat favourable) instance may be adduced. The coefficient of friction is given in the tables as 0·54 in the case of oak resting in the state bordering on motion on oak with the fibres perpendicular to

LAWS OF FRICTION.

Table XI.
COEFFICIENTS OF FRICTION

AND LIMITING ANGLES OF RESISTANCE OF SUBSTANCES BETWEEN WHICH NO UNGUENTS ARE INTERPOSED.

Substance	Disposition of Fibres	State bordering on Motion			State of Motion		
		ϕ	μ or $\tan \phi$	$\sin \phi$	ϕ	μ or $\tan \phi$	$\sin \phi$
Oak on oak . .	Parallel	31°50′	0·62	0·53	25°40′	0·48	0·43
,, ,, . .	Perpendicular	28°20′	0·54	0·47	18°45′	0·34	0·32
,, ,, . .	Endwise	23°20′	0·43	0·40	10°45′	0·19	0·19
Oak on elm . .	Parallel	20°50′	0·38	0·35			
Elm on oak . .	Parallel	34°40′	0·69	0·57	23°20′	0·43	0·40
,, ,, . .	Perpendicular	29°40′	0·57	0·50	24°15′	0·45	0·41
Wrought iron on oak . . .	Parallel	31°50′	0·62	0·53	31°50′	0·62	0·53
Cast iron on oak .	Parallel	33°0′	0·65	0·55			
Copper on oak .	Parallel	31°50′	0·62	0·53	31°50′	0·62	0·53
Wrought iron on cast . .	—	10°45′	0·19	0·19	10 10′	0·18	0·18
Cast iron on cast	—	9°5′	0·16	0·16	8°30′	0 15	0·15
Oak on calcareous oolite * . .	Endwise	32°10′	0·63	0·53	20°50′	0·38	0·35
Wrought iron do.	—	26°10′	0·49	0·44	34°40′	0·69	0·57
Brick do. . .	—	33°50′	0·67	0·56			
Calcareous oolite on do. . .	—	36°30′	0·74	0·59	32°40′	0·64	0·54

each other. The experimental results from which this value was deduced are as follows :—

Surface of Contact	Normal Pressure	Pressure on point of causing Motion	Coef. Friction μ
0·947 ft.	121 lbs.	67 lbs.	0·55
	283 ,,	151 ,,	0·53
	495 ,,	252 ,,	0·51
	1995 ,,	1171 ,,	0·58
	2525 ,,	1287 ,,	0·51
0·043 ft.	389 ,,	204 ,,	0·52
	403 ,,	213 ,,	0·53
	1461 ,,	855 ,,	0·52

* The stone employed in M. Morin's experiments seems to have been a soft oolitic stone from the quarries at Jaumont near Metz.

It is to be observed that in the above Table the numerical values of μ were ascertained by experiment; the values of ϕ and sin ϕ have been obtained by calculation. General Morin's Tables give the values of μ corresponding to various unguents; of these, the following comprehensive results will be sufficient for our purposes: any two of the following substances, oak, elm, cast iron, wrought iron, bronze, pressed against each other, tallow being employed as an unguent, have for the coefficient of friction $\mu = 0.10$, and therefore $\phi = 5° 40'$ and sin $\phi = 0.10$. The same substances when in motion, and the unguent is either tallow, hog's lard, or any similar substance, have the coefficient of friction equal to 0·07, and therefore $\phi = 4°$ and sin $\phi = 0.07$.

76. *The inclined plane.*—The principles which regulate the equilibrium of a body resting on a plane inclined to the horizon are the same as those which regulate the equilibrium of a body resting on a horizontal plane—a case which has been already considered;—the applications of the former case are, however, very numerous and very important, it will therefore be discussed at some length. It is scarcely necessary to observe that the inclined plane is commonly reckoned amongst the 'Mechanical Powers.'

Ex. 346.—A mass whose weight is W rests on a plane A B (fig. *f*), inclined to an angle α to the horizon A C; it is acted on by a force P in a direction (N P) making an angle β with A B: determine the relation between the forces P and W when P is on the point of making the body slide up the plane.

Take G the centre of gravity of the body, and through it draw the vertical line G W, cutting P N in D, both lines being produced if necessary. Now, the only forces acting on the body are its weight W along D W, the force P along D P, and the reaction (R) of the plane A B; R's direction must pass through D, and must be inclined to a perpendicular to A B at an angle equal to ϕ, the limiting angle of resistance; draw D M at right angles to A B, and make M D E equal to ϕ; then R will act along the line E D. (The line E D is drawn as in the figure, since the reaction R tends to oppose the sliding of the body.) Hence we have

$$P : W :: \sin WDR : \sin RDP :: \sin WDE : \sin EDP$$
But $\quad WDE = \alpha + \phi$, and $EDP = 90 + \beta - \phi$
Therefore $\quad P : W :: \sin(\alpha + \phi) : \cos(\beta - \phi)$
In the same manner it can be shown that
$$W : R :: \cos(\beta - \phi) : \cos(\alpha + \beta)$$
If the question is solved by the general equations of equilibrium (Prop. 15)

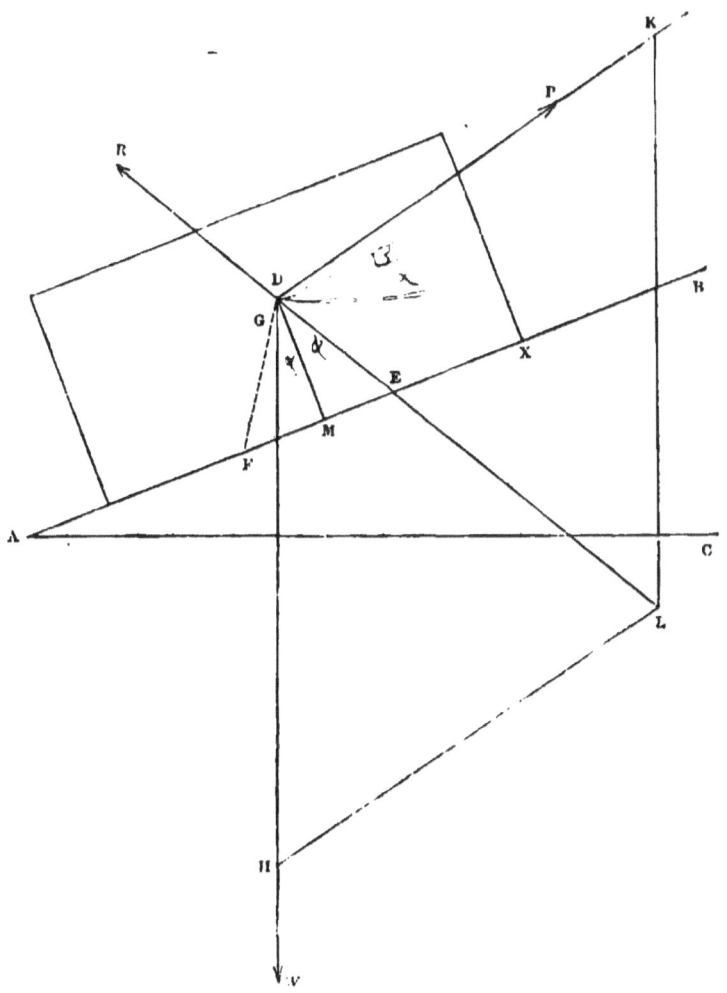

Fig. *f*, page 140.

THE INCLINED PLANE. 141

we may call R' the normal reaction, acting at a point whose distance from M is x; the friction will be μ R', acting from B to A; also we may represent D M by p. Then, if we resolve the forces along and at right angles to A B, and measure moments round D, we shall obtain

$$W \sin \alpha + \mu R' - P \cos \beta = 0 \qquad (1)$$
$$-W \cos \alpha + R' + P \sin \beta = 0 \qquad (2)$$
$$x R' - \mu p R' = 0 \qquad (3)$$

Equations (1) and (2), when solved, give relations between P and W, and between W and R', equivalent to those already obtained; equation (3) shows that R' will act through the point E.

With numerical data, a solution can be obtained by construction, as indicated in the diagram, by the parallelogram H K, in which, if D H represents the given weight, D K will represent the required force, and L D the reaction.

Ex. 347.—If α be greater than ϕ, show that when the body is on the point of sliding down the plane

$$P : W :: \sin(\alpha - \phi) : \cos(\beta + \phi)$$
$$W : R :: \cos(\beta + \phi) : \cos(\alpha + \beta)$$

Ex. 348.—Show that if $\alpha < \phi$ the body will remain at rest without support.

Ex. 349.—A mass of wrought iron weighing 500 lbs. rests on a plane of oak inclined at an angle of 20° to the horizon, a force P acts upon it so as just not to pull it up the plane in a direction inclined to the plane at an angle of 12°; find P. *Ans.* 417·9 lbs.

[In fig. f the construction is shown by which this example was solved, the scale being 1 in. to 200 lbs.; the result obtained by the construction was 415 lbs., the correct answer being 417·9 lbs.]

Ex. 350.—In the last example suppose P to act along P D as a pushing force; find its magnitude that it may just not push the body down the plane. *Ans.* 142·1 lbs.

Ex. 351.—Referring to *Ex.* 349 and 350: first, if P had been a force of 200 lbs. acting up the plane; next, if P had been a force of 100 lbs. acting down the plane; and, lastly, if there were no force P; find the magnitude and direction of the reaction of the plane.

Ans. (1) 428·7 lbs. P D R = 81° 18'. (2) 559·4 lbs. P D R = 130° 42'.
(3) 500 lbs. acting vertically upward.

Ex. 352.—Show that the direction of the smallest force which will make a body slide either up or down an inclined plane makes an angle ϕ with the plane.

Ex. 353.—What is the least force that will draw a cubic foot of cast iron down a plane of oak inclined to the horizon at an angle of 14°?

Ans. 146·7 lbs.

Ex. 354.—In the last example what would have been the least force necessary to support the mass had the plane been of cast iron?

Ans. 38·6 lbs.

Ex. 355.—What would be the horizontal force that would just push the body *up* the inclined plane in the last case?   *Ans.* 192 lbs.

Ex. 356.—If the body represented in fig. f is a cylinder the radius of whose base is r and height $2h$, and if P acts at a point N so chosen that for the same value of P the body is on the point of turning round X when it is also on the point of sliding up the plane, show that

$$x_N = \frac{(r \cos \alpha + h \sin \alpha) \cos (\beta - \phi)}{\cos \beta \sin (\alpha + \phi)}$$

and transform the expression into one adapted for logarithmic calculation.

Ex. 357.—A rectangular mass of cast iron rests on an inclined plane of oak; it is on the point both of sliding down and of overturning; its base is 2 ft. square, what is its height?   *Ans.* 3·08 ft.

Ex. 358.—In the last example what force acting parallel to the inclined plane would be just sufficient to draw the mass of iron up it? Could this force be applied at any point of the body so far above the plane as to overturn the body before making it slide up the plane?

Ans. (1) 6100 lbs.   (2) It will overturn the body if applied at a point more than 1·54 ft. above the plane.

Ex. 359.—If 2A is the vertical angle of a cone standing on a plane whose inclination to the horizon is ϕ (the limiting angle of resistance), show that $4 \tan \text{A} = \tan \phi$, if the cone is such as to be on the point both of toppling and sliding.

Ex. 360.—The earliest experiments on friction were made in the following manner: The substances were formed into rectangular blocks—shaped like bricks—and were placed on planes of various substances; the planes were then gradually raised, and the angles noted at which sliding commenced; it was found that for the same substances this angle was the same whatever the weight of the block, and whether it rested on a broad or narrow face; what conclusions could be inferred from these facts as to the nature of friction?

Ex. 361.—Given an incline of 1 in n (i.e. 1 ft. vertical to n ft. horizontal), and that a body weighing w lbs. rests upon it; given also that the friction is 1 lb. in m: show that the force which, acting parallel to the plane, will be on the point of making the body move up the plane very nearly equals $w\left(\frac{1}{n} + \frac{1}{m}\right)$.

Ex. 362.—Let CA and CB be two equally rough planes inclined downward from C on opposite sides of the vertical through C, and let AB be horizontal;

THE INCLINED PLANE.

let a weight w_1 be placed on C A, and a weight w_2 on C B, and let them be connected by a fine smooth cord passing over C: if w_1 is on the point of sliding down C A, and thereby dragging w_2 up C B, show that

$$w_1 \sin(A-\phi) = w_2 \sin(\phi+B)$$

77. Many questions arise out of cases in which a body rests on two planes inclined at a certain angle to each other; in most of them it is convenient to have recourse to the general equations of equilibrium (Prop. 15); a few such examples are here added.

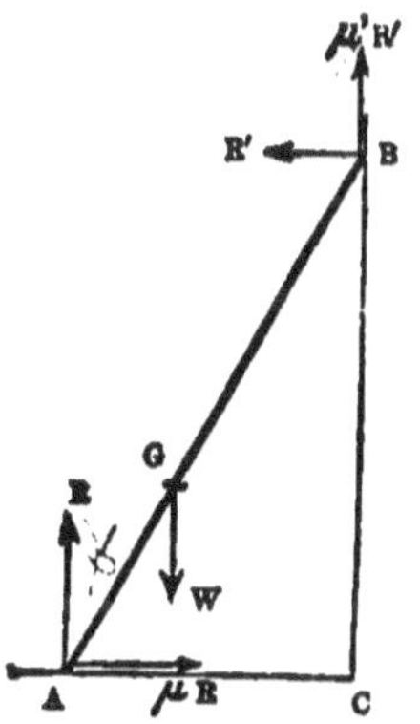

Ex. 363.—A B represents a ladder, one end of which rests on the ground at A, and the other against a vertical wall at B; its length is a, the distance from its foot to its centre of gravity (A G) is b, its weight is w: determine the angle B A C, or θ, at which it will just slide.

The point A must just be sliding outward, and B downward; hence the forces act on the ladder as shown in the figure, and, taking the horizontal and vertical components, and measuring moments round A, we have the following equations:—

$$R + \mu'R' - w = 0$$
$$\mu R - R' = 0$$
$$aR' \sin\theta + a\mu'R' \cos\theta - bw \cos\theta = 0$$

Hence $(1+\mu\mu') R = w$, $(1+\mu\mu') R' = \mu w$
and $\mu a \tan\theta = b - (a-b)\mu\mu'$

The ladder will stand in every possible position if

$$b(1+\mu\mu') < a\mu\mu'$$

FIG. 91.

It may be remarked that, though any point may be chosen from which to measure moments, it is generally advantageous to choose a point through which the directions of one or more of the unknown forces pass—e.g. in the above question A or B should be chosen.

Ex. 364.—In the last example, if the ladder is placed in a known position, determine at what distance (x) from A a weight w_1 must be placed that the ladder may be on the point of sliding $(\mu = \mu' = \tan\phi)$.

$$Ans. \quad x = a\left(1 + \frac{w}{w_1}\right)\frac{\sin\phi \sin(A+\phi)}{\cos A} - \frac{bw}{w_1}.$$

Ex. 365.—In *Ex.* 363, suppose C to be an obtuse angle $(=180°-\gamma)$, and suppose $\mu = \mu'$; find θ, and find the condition of the ladder resting in all positions.

$$Ans. \quad (1) \; \mu \tan\theta = 1 - \frac{(a-b)(1+\mu^2)\sin\gamma}{a(\sin\gamma - \mu\cos\gamma)}$$

$$(2) \; b\sin\gamma < a\Big(\sin\gamma - \sin(\gamma-\phi)\cos\phi\Big)$$

78. *The wedge*.—In the above examples the inclined plane, though reckoned as one of the mechanical powers, can hardly be regarded as a machine; in many cases, however, the inclined plane is itself movable, and is employed to separate bodies that are urged together by great forces; in this case it is correctly spoken of as a *machine*. The simplest instance of this use of the inclined plane is the wedge, which is, in fact, nothing but a movable inclined plane.

Ex. 366.—To determine the relation between the resistance and the force which is on the point of urging forward an isosceles wedge.

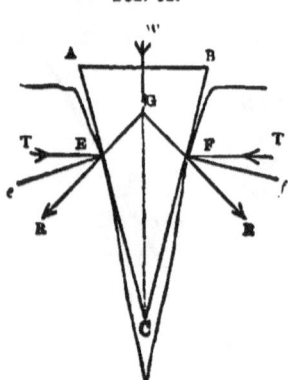

FIG. 92.

Let A B C be the wedge, W the force acting along the axis G C, E and F the points of contact of the sides of the wedge with the obstacle; draw eE and fF at right angles to A C and B C respectively; make the angles eER and fFR each equal to ϕ (the limiting angle of resistance between the sides of the wedge and the obstacle): then, since the wedge is on the point of moving forward, the mutual action between the surfaces of contact at E and F will act along these lines, and the wedge is kept at rest by W and reactions R' and R', equal and opposite to R and R; the directions of these three forces must pass through a common point G; therefore

$$R' : W :: \sin CGR : \sin RGR$$

Now, if A C G equals α, we have C G R equal to $90 - (\phi + \alpha)$, and R G R equal to $180 - 2(\phi + \alpha)$; therefore

$$W = 2R' \sin(\alpha + \phi) \tag{1}$$

Now, suppose that T, the tendency of the obstacles to collapse, acts along T E, and let T E e equal ι; then the resolved part of R along E T must equal T, the remaining part of R being transmitted to the ground. Hence

$$R \cos(\iota + \phi) = T \tag{2}$$

Therefore, remembering that R and R' are equal,

$$W \cos(\iota + \phi) = 2T \sin(\alpha + \phi)$$

The angle ι is commonly unknown and very small; it is therefore generally neglected.

Ex. 367.—If W is the force required to keep the wedge from starting, show that

$$W \cos(\iota - \phi) = 2T \sin(\alpha - \phi)$$

THE WEDGE.

Ex. 368.—Show that if w is the force that has driven a wedge into a given position and w_1 the force required to extract it, then ($\iota = 0$)
$$w_1 = w \frac{\sin(\phi - \alpha)}{\sin(\phi + \alpha)}.$$

Ex. 369.—An iron wedge whose vertical angle is 13° is driven into a mass of oak by a force of 1 cwt. :—what force will be necessary to extract it? *Ans.* 77·27 lbs.

Ex. 370.—Show that the wedge will start if the force be withdrawn, provided the angle of the wedge be greater than 2ϕ.

Ex. 371.—An iron wedge whose angle is 7° is driven into a mass of oak; find what fraction of the driving force is consumed by friction.

Ans. If w′ is the force on the smooth edge which exercises the same normal pressure on the block as that produced by w on the rough edge, then $w' = 0.09\,w$.

Ex. 372.—In *Ex.* 366, if A B C is not isosceles, and if the limiting angles of resistance at E and F are ϕ and ϕ_1, and if R is the pressure caused by w at E, show that
$$R \sin(c + \phi + \phi_1) = w \sin(B - \phi_1)$$

Ex. 373.—In the annexed figure, D C is a horizontal table, H K a fixed obstacle, A B C D, A B E F two movable inclined planes, having a surface of contact A B, inclined at an angle α to the horizon; the former is urged forward by a force P, the latter downward by a force w; ϕ, ϕ_1, ϕ_2, are the limiting angles of resistance at A B, H K, and D C respectively: show that when the horizontal force P is about to overcome the vertical force w

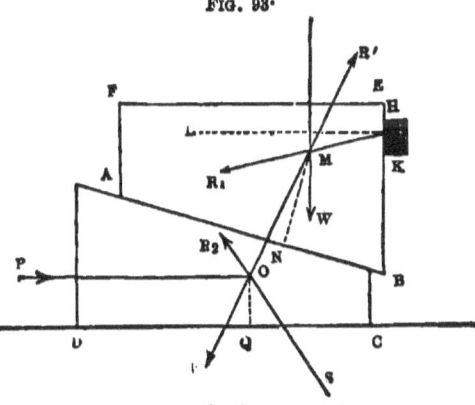

FIG. 93.

$$P \cos \phi_2 \cos(\alpha + \phi + \phi_1) = w \cos \phi_1 \sin(\alpha + \phi + \phi_2)$$

[The diagram shows how the various reactions act. The student must bear in mind that the upper plane is in equilibrium under the action of w, R_1, and R′; the lower plane under the action of P, R, and R_2, and of these R equals R′. He will find it a useful exercise to determine independently the relation between P and w, when H K and D C are smooth.]

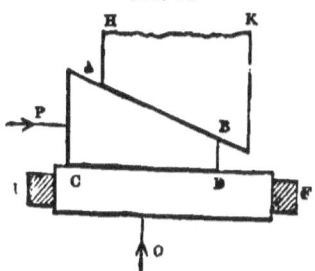

FIG. 94.

Ex. 374.—In the annexed figure let A B K H be fixed, C D a horizontal plate capable of moving up and down between the

L

guides E and F: if the inclination of A B to C D is a, and all the surfaces are smooth except A B, show that when the horizontal force P is about to overcome the vertical force Q

$$P = Q \tan(a + \phi)$$

Ex. 375.—In the last example, if all the surfaces are rough, show that

$$P \cos(a + \phi) \cos(\phi_1 + \phi_2) = Q \cos \phi_2 \sin(a + \phi + \phi_1)$$

Ex. 376.—In the last example, if Q_1 is the force that will just drive P out, show that

$$Q \sin(a + \phi + \phi) \ \cos(a - \phi) \cos(\phi_1 - \phi_2) = Q_1 \sin(a - \phi - \phi_1) \cos(a + \phi) \cos(\phi_1 + \phi_2)$$

What is the smallest slope of A B at which it will be possible for this to happen?

79. *The form of the helix or the thread of the screw.*

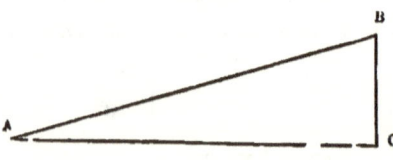
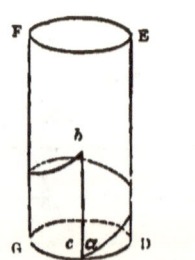

FIG. 95.

—Let A B C be a right-angled triangle, and DEFG a cylinder, the circumference of whose base is equal to the base (A C) of the triangle; if we suppose this triangle to be wrapped round the cylinder so that A and C come together, as indicated by the small letters $a c b$, the hypothenuse A B will take the form of a curve called the helix, i.e. the curve to which the thread of a screw would be reduced if it became merely a line.

Ex. 377.—If the distance measured parallel to the axis between two turns of a thread of a screw (or its pitch) is h and the radius of the cylinder is r, show that the length of n turns of the thread is $n\sqrt{4\pi^2 r^2 + h^2}$.

Ex. 378.—Show that if h is the pitch and r the radius of the cylinder, then if θ is the angle of inclination of the thread of the screw we shall have

$$\tan \theta = \frac{h}{2\pi r}.$$

Ex. 379.—The length of a screw is $1\frac{1}{2}$ ft., in which space the screw makes 36 turns, the radius of the cylinder is $1\frac{1}{4}$ in.; determine the angle of inclination of the thread and its length. *Ans.* (1) 3° 2′ 12″. (2) 339·7 in.

80. *The form of screw with a square thread.*—In

THE SCREW.

the last Article we considered the form of the geometrical curve called the helix. If we suppose that instead of the triangle A C B we have a solid, such that, when it surrounds the cylinder, its upper face projects at right angles to the cylinder at every point, as shown in the annexed figure; this upper surface will have the form of the upper surface of the square-threaded screw; if now the lower part of this projection be cut away, so as to leave a projecting piece of uniform thickness, we shall obtain a screw with a square thread, as shown in fig. 97,* a section of which made by a plane passing through the axis of the cylinder is shown in fig. 98. The student will remark that the thread of a screw, though a very common object, has a very remarkable form; for instance, the curve

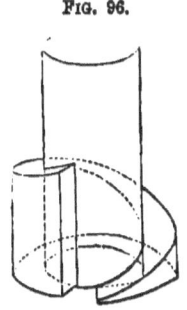

FIG. 96.

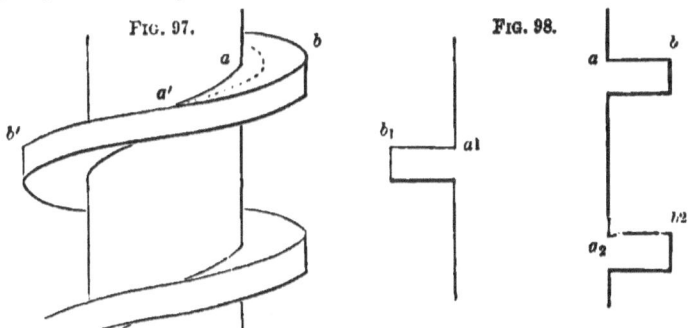

$a\,a'$ (fig. 97), which when prolonged passes through the points a, a_1, a_2 (fig. 98), is a helix, as also is the curve $b\,b'$ (fig. 97), which when prolonged will pass through the points b, b_1, b_2 (fig. 98). Now imagine a cylinder to be described whose axis coincides with that of the screw, and

---

* When there is a considerable distance between two consecutive turns of the thread, as is the case with the screw represented in the figure, it is usual to have a second intermediate thread running round the cylinder. This is done for the purpose of distributing the pressure exerted between the thread and its companion over a larger area, and thereby decreasing the risk of breaking the thread.

whose surface cuts the thread between a and b (fig. 97), the curve of section will be a helix, as indicated by the dotted line; the triangles whose hypothenuses form these helices will all have the same height, viz. $a\,a_2$ or $b\,b_2$ (fig. 98), but their bases will be the circumferences of the bases of their respective cylinders.

Ex. 380.—If h is the height between two turns of the thread of a screw (or its pitch), r and r_1 the radii of the external and internal cylinders, and θ and θ_1 the angles of inclination of the external and internal helices, show that

$$\tan(\theta_1 - \theta) = \frac{2\pi h\,(r - r_1)}{4\pi^2 r r_1 + h^2}$$

and show that the formula gives a correct result when $r_1 = 0$.

Ex. 381.—If the thread of the screw in *Ex.* 379 were cut half an inch deep, determine the difference between the lengths of the interior and exterior helices, and the inclination of the mean helix. *Ans.* (1) 112·8 in. (2) 3° 38′.

Ex. 382.—The external and internal radii of the thread of a square-threaded screw are r and r_1; its thickness (measured parallel to the axis) is a; show that the volume of one turn of the thread is $\pi\,(r^2 - r_1^2)\,a$.

Ex. 383.—A wrought-iron screw is 1 ft. long, and 1½ in. in radius, the thread makes 3 turns in 2 in., its thickness is ⅓ in., its depth ½ in.; find its weight, and the weight of the part cut away when the screw was made.

Ans. (1) 276·1 oz. (2) 106·2 oz.

FIG. 99.

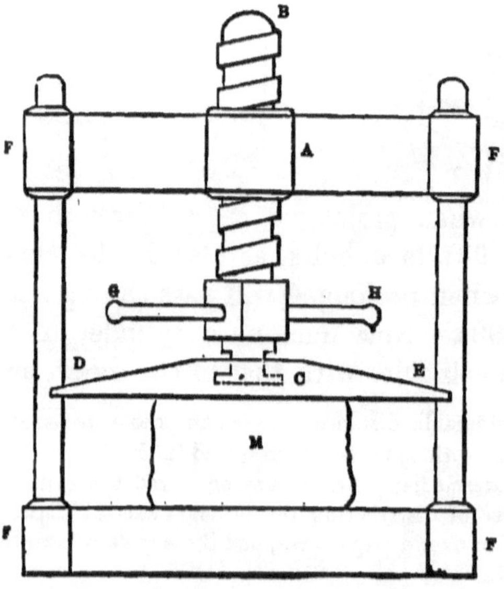

31. *The screw-press*.—The most familiar application of the screw occurs in the screw-press, and as it is very desirable that the student should get a clear conception of the mode of action of the forces in the case of the screw, he will do well to examine a screw-press; its most usual form is represented in the annexed figure, and can be sufficiently described as follows: F F F F is a strong frame; at A in the middle of the cross piece is a hollow nut, on whose interior surface is cut a groove, called the companion screw, which the thread of the screw B C exactly fits; the end C of the screw is fixed to the piece D E in such a manner that the screw is free to turn, while the piece D E can only move in a vertical direction in consequence of the guides F F and F F; it moves downward when the screw is turned by the handle G H in one direction, and upward when the screw is turned in the opposite direction; in the former case a pressure is exerted on the mass M which it is the purpose of the machine to compress. The action of the forces in this case will be understood by considering the annexed figure, in which A A A A represents a section of the nut, B C of the screw, F F the guides, D E the movable piece, Y Y the thread of the screw, X X the groove of the companion; the force P is equivalent to the pressure at the end of the arm which tends to turn the screw; Q is the reaction against D E which balances P; the frictions called into play in this case are the following: (1) between the thread and the groove, (2) between the end of the screw and the piece D E, (3) between the guides F F and the sides of the piece D E, (4) between the cylindrical surfaces of B and A. It is not easy to obtain the relation between P and Q in the

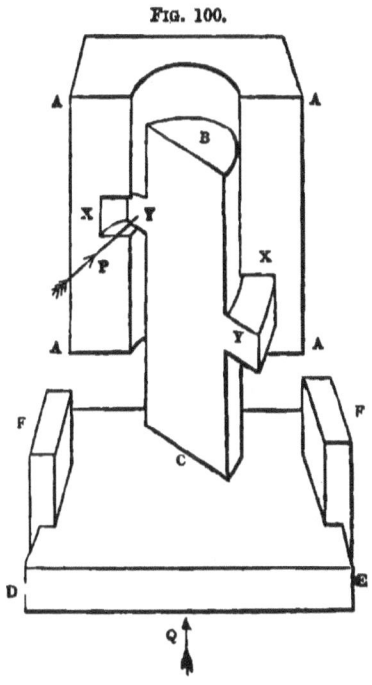

Fig. 100.

state bordering on motion when all the frictions are taken into account;* the frictions marked (3) and (4) are, however, small, and in the following pages will be neglected.

Ex. 384.—Show that in the case of the screw-press the relation between P and Q is given by the formula

$$P a = Q r \tan (a + \phi)$$

where a is the length of the arm on which P acts, r the radius of the screw, a the angle of inclination of the thread, and ϕ the limiting angle of resistance between the thread and groove; all other frictions being neglected.

If (referring to fig. 94) C D is a horizontal table, movable in a vertical direction between guides E F; and if A B H K is a fixed inclined plane, and A B C D a movable inclined plane, then if a is the inclination of A B, and if all the surfaces are smooth except A B,

$$P = Q \tan (a + \phi)$$

If A B C D is wrapped round a cylinder, and A B H K round a hollow cylinder, we obtain the same arrangement of pieces that exists in a screw working against a fixed nut; but Q acts along the axis of the cylinder, and P acts, not tangentially to the cylinder, but at the end of an arm a.

Suppose the force Q to cause pressures $q_1, q_2, q_3, \ldots$ at different points of the thread of the screw, and suppose $p_1, p_2, p_3 \ldots$ to be the forces which acting horizontally in directions touching the surface of the cylinder at those points would be on the point of overcoming $q_1, q_2, q_3, \ldots$ respectively, then the relation between p_1 and q_1 must be the same as that between P and Q given above. Hence

similarly
$$p_1 = q_1 \tan (a + \phi)$$
$$p_2 = q_2 \tan (a + \phi)$$
$$p_3 = q_3 \tan (a + \phi)$$

and therefore $p_1 + p_2 + p_3 + \ldots = (q_1 + q_2 + q_3 + \ldots) \tan (a + \phi)$.

Now $p_1, p_2, p_3, \ldots$ have the same tendency as P to turn the screw round its axis, and therefore the principle of moments gives us

$$P a = p_1 r + p_2 r + p_3 r + \ldots$$

---

* If $2b$ equals D E, ρ the radius of the end of the screw, μ, μ', μ'' the coefficients of friction between screw and nut, screw and D E, and guides and D E respectively, and the remaining notation the same as that employed in *Ex.* 384, the following, it is believed, will be found to be the correct formula for the relation between P and Q:—

$$P \left(a - \frac{\mu r \cos (2a + \phi)}{\sqrt{1 + \mu^2 \cos^2 a} \cos (a + \phi)} \right) = \frac{b Q}{b + \tfrac{2}{3} \mu' \mu'' \rho} \left(r \tan (a + \phi) + \tfrac{2}{3} \mu' \rho \right)$$

evidently differs but little from the formula of *Ex.* 393.

SCREW-PRESS. 151

also since the pressures $q_1, q_2, q_3, \ldots$ are all parallel to Q's direction, we have

$$Q = q_1 + q_2 + q_3 + \ldots$$

therefore
$$Pa = Qr \tan(\alpha + \phi)$$

Ex. 385.—Show, by a method similar to that employed in the last example, that when all the frictions are neglected

$$Pa = Qr \tan \alpha$$

and that P : Q :: the pitch of the screw : the circumference of the circle described by the point at which P acts.

Ex. 386.—There is a screw with a square thread the radius of which is 1 in., the pitch is $\frac{1}{3}$ in., the nut is of cast iron and the screw of wrought iron, their surfaces are well greased; determine the pressure that would be produced on the substance in the press if we neglect all the frictions but that between the thread and the groove, when the screw is turned by a force of 150 lbs. acting at a distance of 3 ft. from the axis of the screw.

Ans. 35,275 lbs.

Ex. 387.—In the last example determine Q if the screw is not greased.

Ans. 22,007 lbs.

Ex. 388.—Find the number of turns per foot which the thread of a perfectly smooth screw will make whose power is the same as that of the screw described in *Ex.* 386. *Ans.* $12\frac{1}{2}$ nearly.

Ex. 389.—If in any screw reckoned perfectly smooth a force P were required to compress a substance with a force Q, and if P' were the additional force required in consequence of the friction between the thread and the groove, show that

$$P' = \frac{2\mu P}{\sin 2\alpha}, \text{ very nearly,}$$

where α is the angle of inclination of the thread of the screw, and μ the coefficient of friction—neither being large.

Ex. 390.—If the screw described in *Ex.* 386 has to exert a pressure Q, find both from first principles and from the formula in the last example the value of $\frac{P'}{P}$. *Ans.* (1) 1·885. (2) 1·880.

Ex. 391.—The diameter of the screw of a vice is 1 in. and the thread makes 4 turns to the inch, the whole is of cast iron and the screw is well greased; the handle by which it is turned is 6 in. long and is urged by a force of 100 lbs.; the jaws of the vice hold an ungreased piece of wrought iron by friction only; find the force requisite to extract it. *Ans.* 2530 lbs.

82. *Friction on the end of the screw.*—Let A B C be a cylinder or pivot, the end of which is urged against a rough plane by a force Q acting along its axis O C; the cylinder

is supposed to be on the point of turning round the axis, and is opposed by the friction; it is required to determine the moment of the frictions with respect to the axis O C.

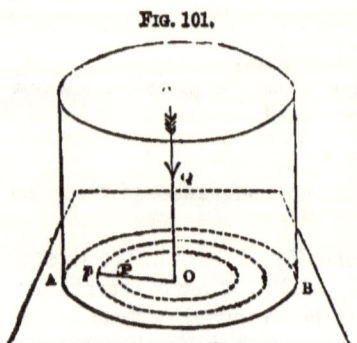

Fig. 101.

It may be assumed that the inequalities of the surfaces will wear away, and that the pressure will be equally distributed; consequently if ρ is the radius of the pivot (say in inches), $\dfrac{Q}{\pi \rho^2}$ will be the pressure per square inch, and consequently $\dfrac{\mu Q}{\pi \rho^2}$ will be the friction per square inch; hence if we consider a small ring enclosed between two circles, whose radii O P and O p are respectively r and $r + \delta r$, its area will ultimately equal $2\pi r \delta r$, and the friction on it will equal $\dfrac{2\mu Q}{\rho^2} r \delta r$. Now the friction at every point of this ring acts in a direction perpendicular to the radius at that point, and hence the sum of the moments of the frictions on this ring with respect to the axis will ultimately equal $\dfrac{2\mu Q}{\rho^2} r^2 \delta r$; the same will be true of any other ring, and therefore we shall obtain the required moment if we divide the area into a great number of rings, and ascertain the limit of the sum of the moments of the frictions on each ring; this can be done as follows:

Take D E $= \rho$ and at right angles to it draw E F $= \rho$, perpendicularly to both draw E H $= \dfrac{2\mu Q}{\rho}$, complete the rectangle E F G H, and complete the pyramid D E F G H; take D P $= r$ and P $p = \delta r$, and through P and p draw planes parallel to the base enclosing the lamina P R S; then it is

FRICTION ON END OF SCREW. 153

plain by similar triangles that $PS = r$ and $PR = \dfrac{2\mu Q}{\rho^2} r$, consequently the volume of the lamina is ultimately equal to

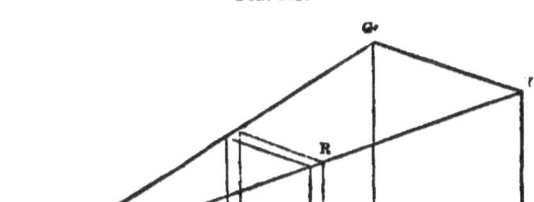

Fig. 102.

$\dfrac{2\mu Q}{\rho^2} r^2 \delta r$, i.e. the moment of the friction on the ring is correctly represented by the volume of the lamina, and the same being true of any other lamina, we shall have the moment of the whole correctly represented by the volume of the pyramid,* i.e. the moment equals $\tfrac{1}{3} \rho \times \rho \times \dfrac{2\mu Q}{\rho}$ or moment of friction $= \tfrac{2}{3} \rho Q \mu$.

Ex. 392.—If the screw rests on a hollow pivot whose internal and external radii are respectively ρ_1 and ρ, show that the moment of the friction round the axis of the screw is given by the formula

$$\tfrac{2}{3} . \dfrac{\rho^3 - \rho_1^3}{\rho^2 - \rho_1^2} . Q\mu$$

and show from this formula that when ρ_1 is very nearly equal to ρ the friction is very nearly equal to $\rho Q\mu$.

Ex. 393.—In the screw when the friction on the end as well as the friction on the thread is taken into account we have

$$P = \dfrac{rQ}{a} \tan(\alpha + \phi) + \tfrac{2}{3} . \dfrac{\rho}{a} Q\mu$$

where ρ is the radius of the *end* on which the screw rests.

---

* The student who understands the Integral Calculus will perceive that the above construction is equivalent to integrating the expression $\dfrac{2Q\mu}{\rho^2} r^2 dr$ between the limits of $r = 0$ and $r = \rho$.

[Referring to *Ex.* 384 the equation deduced from the principle of moments will become

$$\text{P}a = rp_1 + rp_2 + rp_3 + \ldots + \tfrac{2}{3}\rho Q \mu]$$

Ex. 394.—It is required to compress a substance with a force of 10,000 lbs.; the screw with which this is done has a diameter of 3 in., and its thread makes 1 turn to the inch; the arm of the lever is 2 ft. long; determine the force P that would be required—(1) if all frictions were neglected; (2) if the friction between the thread and groove were taken into account; (3) if the friction on the end of the screw, which is 1 in. in radius, were also taken into account; the surfaces being iron on iron well greased.
Ans. (1) 66·3 lbs.   (2) 129·6 lbs.   (3) 157·5 lbs.

Ex. 395.—An iron screw 4 in. in diameter communicates motion to a nut; the force is applied at the extremity of a lever 1 ft. long; the inclination of the thread of the screw is 6°; determine the relation between the force applied and the weight raised by the nut, taking into account the frictions between the thread and groove, and the end of the screw whose diameter is 3 in.—the surfaces are cast iron—(1) when well greased, (2) when ungreased.
Ans. (1) $\text{P} = 0\cdot0427\text{Q}$.   (2) $\text{P} = 0\cdot0583\text{Q}$.

Ex. 396.—If the angle of the screw were 12°, the diameter of the screw and of its end 4 in., and the lever by which it is turned 2 ft. long, the surfaces being of cast iron and ungreased, what weight will a force of 1 cwt. overcome? *Ans.* 2730 lbs.

Ex. 397.—Determine the force required in *Ex.* 394 if the surfaces are of ungreased oak. *Ans.* 488 lbs.

[The fibres may be reckoned to rest endwise between the thread and the groove as well as between the end and the movable piece.]

Ex. 398.—Given Q the pressure to be produced by the screw, r the radius of the mean thread, R the length of the arm, h the pitch, μ the coefficient of friction between the thread and the groove, if the friction between the thread and the groove is the only one taken into account, show that the force to be applied at the end of the arm is given by the formula *

$$\frac{r}{\text{R}} \cdot \frac{h + 2\pi\mu r}{2\pi r - h\mu} \text{Q}$$

83. *The endless screw.*—It is not very unusual to make a screw work with a toothed wheel; the arrangement of the pieces when this is done will be sufficiently understood by an inspection of the annexed diagram; the screw A B may

* This is the formula given in General Morin's *Aide-Mémoire*, p. 309.

be mounted in a frame, and turned by a winch; the teeth of the wheel (C) work with the worm of the screw, on turning which the wheel is caused to revolve; as the screw has no forward motion, it will never go out of action with the wheel, and is, on that account, termed an *endless screw*. The reader will find in Mr. Willis's 'Principles of Mechanism'* a discussion of the form that must be given to the teeth in order to secure equable working. When the machine is employed, it commonly happens that the screw drives the wheel; sometimes, however, the screw is driven by the wheel, as in the case of the fly of a musical box. In the former case, if P is the force at the end of the arm which turns the screw, and Q the force exerted by the screw on the wheel in a direction parallel to the axis, it is easily shown that the relation between P and Q is the same as that determined in *Ex.* 384.

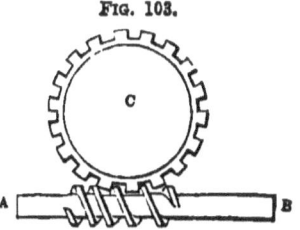

FIG. 103.

Ex. 399.—If a force P acting on the thread of a screw in a direction parallel to its axis is on the point of driving a force Q acting along a tangent to its base, show that

$$Q = P \tan(\alpha - \phi)$$

where α is the inclination of the thread of the screw at the working point, and ϕ the limiting angle of resistance between the driving and driven surfaces.

Ex. 400.—If the action of an endless screw is reciprocal, i.e. if it will act whether wheel or worm is driver, show that the inclination of the thread of the screw must be greater than ϕ and less than its complement.

Ex. 401.—An endless screw consists of a cylinder of cast iron the radius of whose base is 3 in.; the thread makes one turn in 4 in.; what is the greatest extent to which the thread can project if the tooth by which it is driven is of cast iron and is ungreased? *Ans.* 0·98 in.

Ex. 402.—In the last example, if the depth of the thread be 1 in. what is the least pitch with which the machine can work if the surfaces are greased? *Ans.* 2·513 in.

---

* P. 160.

156 PRACTICAL MECHANICS.

84. *Friction of guides*.—One or two instances of the friction of guides have been given already (*Ex.* 373–6); the following case will still further illustrate the subject:—E F is a beam constrained to move in a vertical direction by the four guides A, B, C, D; a projection G H at right angles to E F works with a tooth or cam, K, revolving on a wheel: by the action of the cam the beam is lifted and then allowed to fall by its own weight, thereby serving as a hammer. In the fundamental case the forces act as in the figure: and we treat the beam as a straight line, the guides as points, and represent A C by a, G H by b, H C by x.*

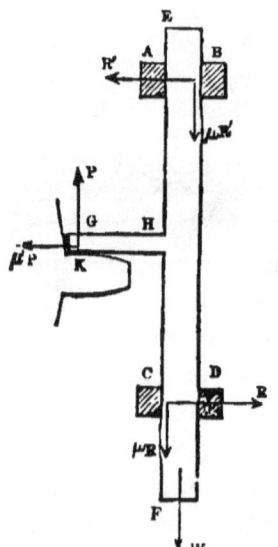

FIG. 104.

Ex. 403.—In the above case show that

$$P \left\{ a - 2b\mu - (a - 2x)\mu\mu' \right\} = aw$$

Ex. 404.—In the above case if $\mu'x > b$ show that the forces will not act exactly as shown in the figure, and that

$$P(1 - \mu\mu') = w.$$

---

* For a fuller discussion of this case, see *Traité de Mécanique appliq. aux Machines*, par J. V. Poncelet, vol. i. pp. 234–238.

CHAPTER VII.

OF THE EQUILIBRIUM OF BODIES RESTING ON AN AXLE, AND OF THE RIGIDITY OF ROPES; WHEEL AND AXLE, PULLEY.

SECTION I.

85. *Fundamental condition of equilibrium in the state bordering on motion, of a body capable of revolving round an axle.*—All the forces acting on the body can be reduced to a single resultant, to which, when the body is at rest, the reaction of the bearing must be equal and opposite; let the annexed figure represent the axle resting on its bearing; let R be the resultant of the forces acting on the body, and let its direction cut the circumference of the bearing at the point P; take O the centre of the bearing and join OP; this line is the normal at the point of contact; the body will therefore be in the state bordering on motion when the angle O P R equals the limiting angle of resistance, the motion being about to ensue in the direction indicated by the arrow-head. This consideration enables us to give a very simple construction, which will apply to all cases in which the forces act on the body along parallel lines. Take O the centre of the bearing (fig. 106), draw a line A O parallel to the directions of the forces; if the body is about to move in the direction indicated by the arrow-head, make the angle A O P equal to the limiting

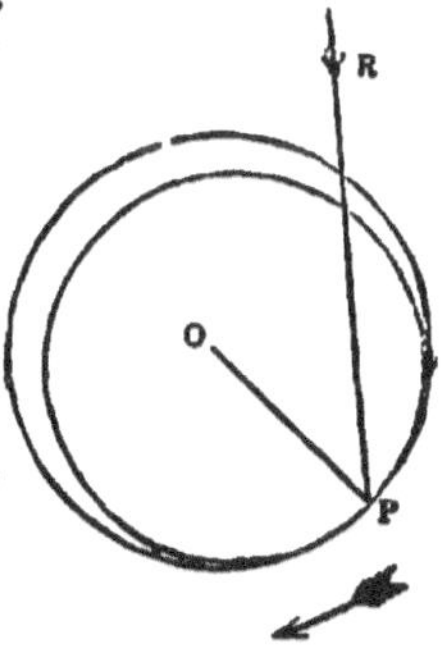

FIG. 105.

angle of resistance; then the resultant force must act along the line R P parallel to O A, since this is the only line drawn parallel to O A which will cut the circumference in a point P such that the angle O P R equals the limiting angle of resistance; hence if we measure moments round P, we shall obtain the required relation between the forces, the sum of those moments being equal to zero by Art. 58. Of course if the motion is about to ensue in a contrary direction, the angle A O P must fall on the other side of O A. It will be remarked that the radii of the axle and its bearing are sensibly equal, so that though in the diagram they are represented as different, that difference never enters the question.

Fig. 106.

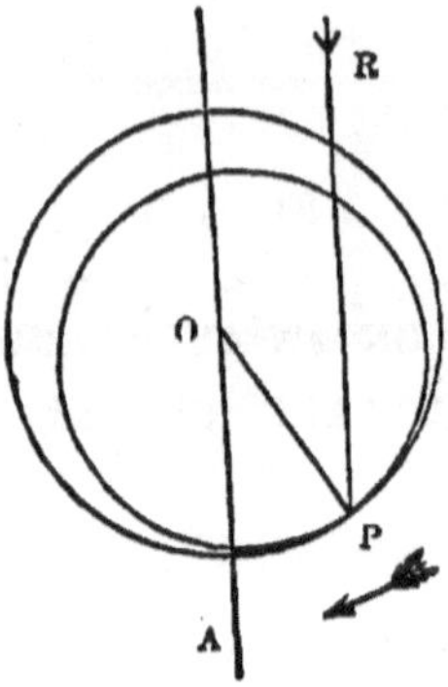

86. *Friction of axles.*—When the body is in the state bordering on motion, the values of the coefficient of friction are the same as those given in the last chapter; the same is also true in cases of motion where no unguent is interposed; in nearly all cases of motion, however, an axle is kept well greased, both to prevent wear and to diminish the resistance; the unguent may be supplied at intervals, as in the case of a common cart-wheel, or continuously, as in the case of the wheel of a railway carriage; as might be expected, a continuous supply of unguent is found to be the most effective means of diminishing the resistance. The following table gives the values of the coefficients of friction, and the limiting angle of resistance for the axles and bearings most commonly used; the coefficients of friction are taken from the experimental determinations of General Morin,[*] from which the limiting angle of resistance has

---

[*] *Notions Fondamentales*, p. 309. To avoid ambiguity, the means of some of Gen. Morin's results have been taken; thus, instead of 0·07 to 0·08, the following table gives 0·075.

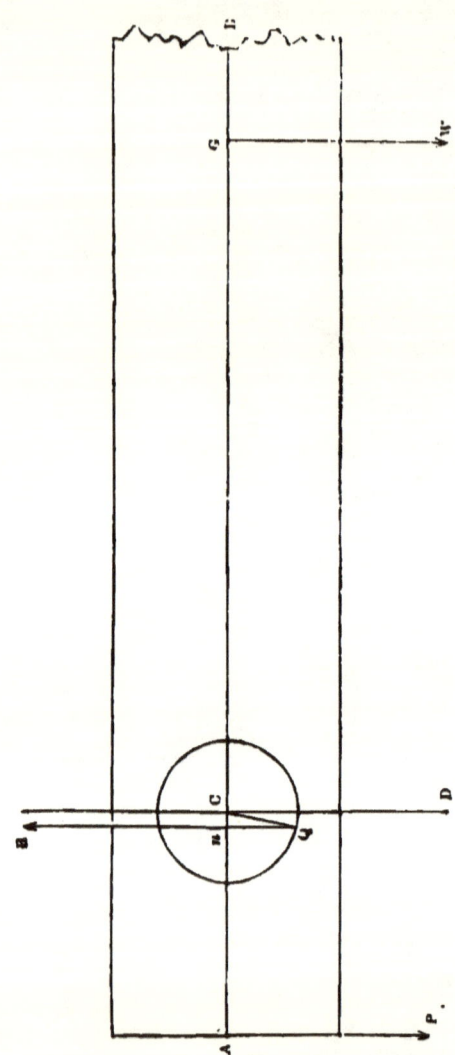

Fig. 9, page 159.

FRICTION OF AXLES. 159

been calculated—those cases have been selected in which the unguent is *most effective* in diminishing friction.

TABLE XII.
FRICTION OF AXLES MOVING ON THEIR BEARINGS.

Axle and Bearings	Unguents	Renewed at intervals		Renewed continuously	
		$\tan \phi$ or $\sin \phi$	ϕ	$\mu \tan \phi$ or $\sin \phi$	ϕ
Cast iron on cast iron	Oil of olives, tallow, or hog's lard	0·075 (mean)	4° 20'	0·054	3° 6'
Wrought iron on cast	Do.	0·075 (mean)	4° 20'	0·054	3° 6'
Wrought iron on brass	Do.	0·075 (mean)	4° 20'	0·054	3° 6'
Wrought iron on lignum-vitæ	Oil, or hog's lard	0·11	6° 20'		
Brass on brass	Do.	0·095 (mean)	5° 30'		
Brass on cast iron	Oil or tallow	. . . .	. .	0·0485 (mean)	2° 47'

Ex. 405.—Let A B (fig. *g*) be a beam movable about a wrought-iron axle which rests on a cast-iron bearing, and whose axis passes at right angles through the axis of the beam;* the centre c of the axle is 12 in. from A, and 30 in. from the centre of gravity of the beam and axle, the radius of the axle being 3 in.; the weight of the whole (i.e. of the beam and axle) is 400 lbs.: find the weight which, when hung at A, will just cause the end A to descend.

Draw the figure to scale; draw through c the vertical line c D, and make the angle D c Q equal to the limiting angle of resistance (10° 45'); draw the

---

* Of course there are in reality two bearings situated symmetrically with reference to the length of the beam, each of which supports half the united

FIG. 107.

pressures P and W; the *plan* of the machine being shown in the accompanying figure.

vertical line Q R cutting A B in n; then this being the direction of the reaction the principle of moments gives us

$$P \times An = W \times nG$$

but since nc is very small, it is desirable to construct the axle on a larger scale; this is done in fig. h, from which we obtain en equal to 0·57 in.; hence we find P equal to 1069·8 lbs.; a result precisely the same as that obtained by calculation.

If A C is represented by p, C G by q, C Q by ρ, and if ϕ is the limiting angle of resistance between the axle and its bearing, we shall have $cn = \rho \sin \phi$, and therefore $An = p - \rho \sin \phi$ and $nG = q + \rho \sin \phi$, whence generally

$$P (p - \rho \sin \phi) = W (q + \rho \sin \phi)$$

In future ρ will be used to denote the radius of any axle that may be under consideration.

Ex. 406.—In the last example determine the value of P which will just prevent the beam from falling when no unguent is used. *Ans.* 936·5 lbs.

Ex. 407.—Determine the magnitude and position of the resultant pressure in *Ex.* 405 if we suppose P = 1020 lbs.; and determine the magnitude of the angle its direction makes with the normal to the point of its application.
Ans. (1) 1420 lbs. (2) $cn = \frac{12}{11}$ in. (3) $c_{Q}n = 3° 13' 47''$.

Ex. 408.—There is a beam of oak A B whose length is 30 ft., depth 2 ft., and thickness 1 ft.; at right angles to its face passes an axle of wrought iron the part of which within the beam is 8 in. square, the projecting part on each side is 6 in. in diameter and 6 in. long (so that its total length is 2 ft.), its axis is situated 10 ft. from the end A, at which end is exerted a force of 5000 lbs.; find the force at B which will just keep the beam from turning and the amount to which that force must be increased if it is on the point of overcoming the force at A; the axle rests on an oaken bearing ungreased.
Ans. (1) 1550 lbs. (2) 1700 lbs.

Ex. 409.—If a string were wrapped round the grindstone described in *Ex.* 16, determine the greatest weight that could be tied to the end of the string without causing motion, supposing the bearing to be of cast iron well greased. *Ans.* 4·8 lbs.

Ex. 410.—If P and Q are two parallel forces acting in contrary directions and keeping a body in equilibrium, and if P, the one more remote from the axle, is on the point of causing motion, show that

$$P (p + \rho \sin \phi) = Q (q + \rho \sin \phi)$$

[If we gradually increase P while Q continues constant, it is plain that their resultant will be made to act at a continually increasing distance from Q. Consequently, in the case supposed in the question, the resultant acts along a line as remote from Q as is consistent with equilibrium.]

87. *Wheel and axle, pulleys.*—The wheel and axle and

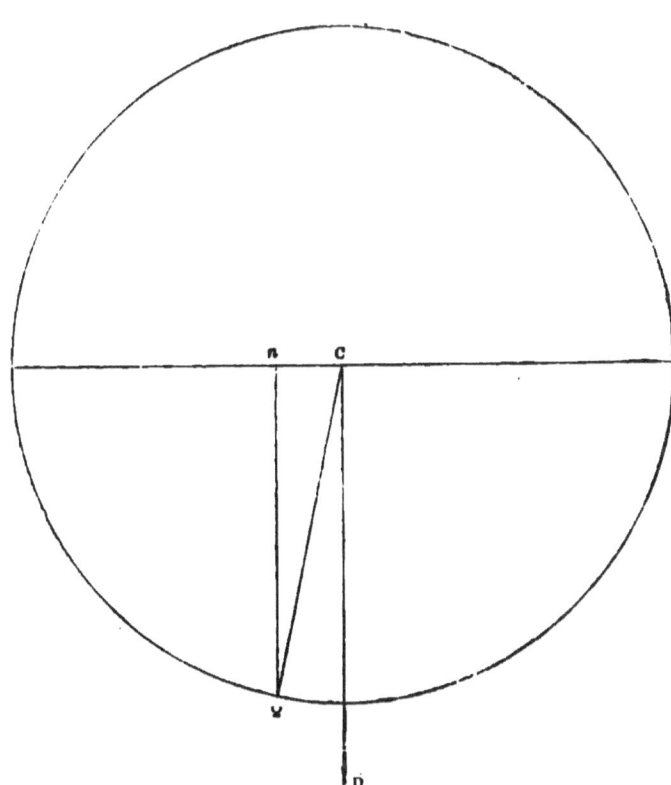

Fig. *h*, page 160.

WHEEL AND AXLE.

the pulley are familiar examples of bodies capable of moving round a fixed axle; they may be sufficiently described as follows:—

(1) *The wheel and axle.*—Let A B represent a cylinder of wood or some other material called the axle, to the end of which is firmly fixed a cylinder of a large diameter E C called the wheel; they rest on a pair of bearings by means of a small cylindrical axis, one end of which is D, the geometrical axes of all these cylinders being coincident; ropes are wrapped in opposite directions round the wheel and axle respectively, to the ends of which weights P and Q are attached; if P is so large as to descend, it will do so by turning the machine; this will wind up Q's rope, and thereby cause that weight to ascend. It is usual to describe the wheel and axle in the above form, in order to give definiteness to the calculation; in practice, however, a winch commonly supplies the place of the wheel.

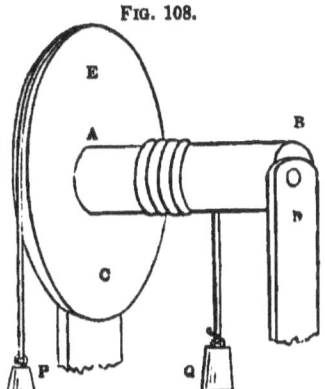

Fig. 108.

(2) *The pulley* is simply a thin cylinder with a groove cut in its circumference, on which a rope can rest: the cylinder is capable of turning round an axis, which is supported by a piece called a block; this well-known machine is represented in the accompanying diagram.

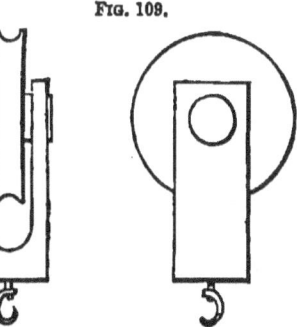

Fig. 109.

When several pulleys are combined into a single machine, they constitute what is called a system of pulleys; the system most commonly used is called the block and tackle; it consists of two blocks containing pulleys (under these circumstances called sheaves) which are either equal in number, or else the upper block contains one more sheave than the lower; the upper block is fixed, while the lower carries the weight; one end of the rope by which the weight is raised is fastened to one of the blocks, and passes in succession round each of the sheaves, as represented in fig. 110; but it must be added that the sheaves in each block are commonly made equal, and arranged one behind the other on a common axis. Another system of pulleys, called the Barton, is sometimes employed; it consists of one fixed and any number of movable pulleys; to the block containing each movable pulley is fastened a rope, which after passing under the next pulley (thereby supporting it) is fastened to a fixed beam. The last of these pulleys carries the weight to be raised; the rope which carries the first movable pulley passes over the fixed pulley; on shortening this rope the pulleys, and with them the weight, are raised; the arrangement is shown in fig. 111; it rarely happens that more than one movable pulley is employed.

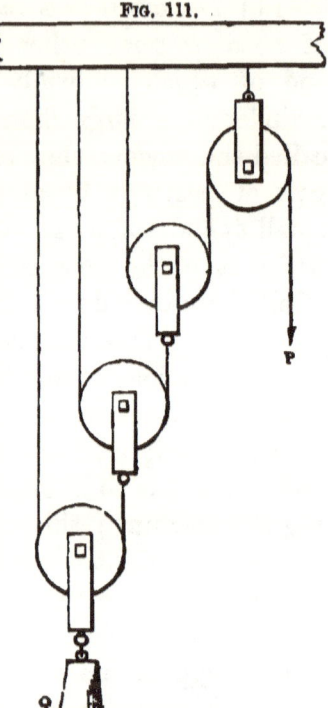

Fig. 111.

It is to be observed that the rigidity of the cords, i.e. their want of perfect flexibility, plays an important part

RIGIDITY OF ROPES. 103

in calculations concerning the mechanical power of the wheel and axle, and of the pulley; we will therefore proceed to explain the method of taking that resistance into account.

88. *Rigidity of ropes.*—Let A B C represent a drum or pulley, movable about an axis C, and let a rope A B D pass over it, to whose ends are applied forces P and Q respectively, the friction of the rope being sufficient to prevent sliding; if one of the forces P overcome the other Q, it must do so by causing the drum to revolve, thereby winding on the rope A B D. Now the portion A B being circular, and B D being straight, the rope must be bent at the point B, and the rope not being perfectly flexible will offer a resistance to being thus bent, and a certain portion of the force P will be expended in overcoming the resistance. It is found that this 'rigidity' of the rope can be taken account of by supposing Q to act along the axis of the rope, i.e. at a distance from C equal to ½ of the sum of the diameters of the rope and drum, and then increasing Q by a certain force; it is found by experiment that this additional force consists of a part depending only on the rope, and another part proportional to Q; it is also found that, when other circumstances are the same, this additional force is greater as the curvature of the axis of the rope is greater, and therefore it can be correctly represented by the formula

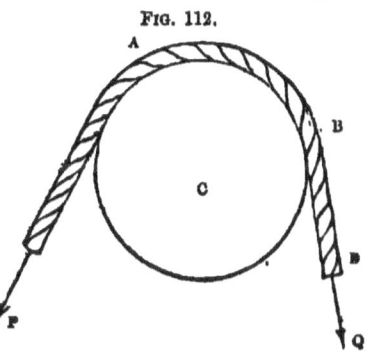

FIG. 112.

$$\frac{A + BQ}{R}$$

where A and B are constants to be determined by experiment, and R is the effective radius of the drum, i.e half the sum of the diameters of rope and drum.

The principal experiments on the rigidity of ropes are due to M. Coulomb,* whose results have been discussed by various writers. M. Morin considers that M. Coulomb's experiments are sufficient for the construction of empirical formulæ only in the cases of new dry ropes and of tarred ropes: from a discussion of the experiments † he obtains values of A and B which, after reduction, give the following values of the above formula:—

(1) For new dry ropes, the resistance due to rigidity in lbs. equals

$$\frac{c^2}{R}\left\{0\cdot062994 + 0\cdot253868c^2 + 0\cdot034910_Q\right\}$$

(2) For tarred ropes, the resistance due to rigidity in lbs. equals

$$\frac{c^2}{R}\left\{0\cdot222380 + 0\cdot185525c^2 + 0\cdot028917_Q\right\}$$

where Q is estimated in lbs., C is the circumference of the rope in inches, and R the effective radius of the drum or pulley in inches. From these formulæ the following table has been calculated:—

TABLE XIII.
RIGIDITY OF ROPES.

Radius of Rope	Circumf. of Rope	New Dry Ropes		Tarred Ropes	
		A	B	A	B
0·16 in.	1 in.	0·32	0·034910	0·41	0·028917
0·24	1·5	1·43	0·078543	1·44	0·065068
0·32	2	4·31	0·139640	3·86	0·115668
0·40	2·5	10·31	0·218183	8·64	0·180731
0·48	3	21·13	0·314190	17·03	0·260253
0·56	3·5	38·87	0·427643	30·56	0·354233
0·64	4	66·00	0·558560	51·05	0·462672
0·72	4·5	105·38	0·706723	80·08	0·585569
0·80	5	160·23	0·872750	121·50	0·722925

* An abstract of Coulomb's Memoirs is given in Young's *Nat. Phil.* vol. ii. p. 171.
† *Notions Fondamentales*, pp. 316–332

RIGIDITY OF ROPES.

Rule.—Multiply B by Q in lbs., add the product to A, divide this sum by the effective radius of the drum or pulley in inches, the quotient is the resistance in lbs.

If the resistance added to Q give Q', the relation between P and Q will be the same as that which obtains between P and Q', acting by means of a perfectly flexible thread on a drum or pulley whose radius equals the effective radius.

It is to be remarked, that the resistance due to rigidity is only called into play when the rope is wound on to a drum; there is no resistance when the rope is wound off.

For example: If the diameter of a pulley is 11 in. and a new dry rope 3 in. in circumference is used to lift a weight of 500 lbs., we have the effective radius of pulley 5·98 or 6 in., and hence

$$\frac{A+BQ}{R} = \frac{21\cdot13 + 0\cdot31419 \times 500}{6} = 30 \text{ lbs.}$$

so that we may consider that a weight of 530 lbs. has to be raised by means of a perfectly flexible string over a pulley 6 in. in radius.

Ex. 411.—To determine the relation between P and Q in the case of the wheel and axle.

In the annexed figure, let C A, the radius of the wheel, be represented by p; CB, the radius of the axle, by q; CD, the radius of the axis, by ρ; the power P and the weight Q act vertically at A and B, and the weight of the machine W acts vertically through C. If P is on the point of preponderating over Q, make W C D equal to ϕ (the limiting angle of resistance between the axis and the bearing), then the reaction of the bearing will act vertically upward through D; and if its direction cuts the line A B in n, we have from the principle of moments

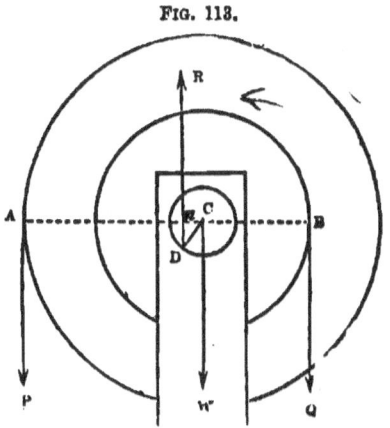

FIG. 113.

$$P.\overline{nA} = Q.\overline{nB} + W.\overline{nC}$$

but $nc = \rho \sin \phi$, therefore $n_A = p - \rho \sin \phi$, and $n_B = q + \rho \sin \phi$; also if we take into account the rigidity of the rope, the effective value of Q is

$$Q + \frac{A + BQ}{q}$$

Hence the required relation between P and Q is

$$P(p - \rho \sin \phi) = \left(Q + \frac{A + BQ}{q}\right)(q + \rho \sin \phi) + W\rho \sin \phi$$

If no account be taken of the rigidity of the rope, the relation between P and Q will be

$$P(p - \rho \sin \phi) = Q(q + \rho \sin \phi) + W\rho \sin \phi$$

Ex. 412.—A wheel and axle weigh 1 cwt., the radius of the wheel is 2 ft., of the axle 6 in., the radius of the axis is 1 in., it is of wrought iron, and rests in a bearing of cast iron well greased; if Q equals 1000 lbs., find the magnitude of P (1) when it will just support, (2) when it is on the point of raising Q—the rope being considered perfectly flexible.

Ans. (1) 244·3 lbs. (2) 255·7 lbs.

Ex. 413.—In the last example, if Q is supported by a new dry rope 3 in. in circumference, determine the value of P when on the point of raising Q.

Ans. 290 lbs.

[The increase of the radius of the axle due to the thickness of the rope must not be overlooked.]

Ex. 414.—If P and Q are two parallel forces, and P is on the point of drawing up Q over a pulley whose effective radius is r, and weight W, show that

$$P(r - \rho \sin \phi) = Q(r + \rho \sin \phi) \pm W\rho \sin \phi$$

where the positive sign is used if P and Q act downward, and the negative sign if they act upward; and that when the rigidity of the rope is taken into account the formula becomes

$$P(r - \rho \sin \phi) = Q\left(1 + \frac{B}{r}\right)(r + \rho \sin \phi) + \frac{A}{r}(r + \rho \sin \phi) \pm W\rho \sin \phi$$

[The proof of the above formulæ exactly resembles that given in *Ex.* 411, except that c_A and c_B are equal.]

89. *Remark* —It appears from the formula of *Ex.* 414 that the part of P expended on the friction caused by the weight of the pulley is small, since it is represented by $W \rho \sin \phi$, in which W is commonly small compared with P and Q, and $\rho \sin \phi$ is always small compared with r; now if we omit the last term the formula will be the same

SYSTEMS OF PULLEYS. 167

whether P and Q act vertically upward or vertically downward, and can be written:
$$P = aQ + b$$
where a and b are written instead of the complicated expressions
$$a = \left(1 + \frac{B}{r}\right) \cdot \frac{r + \rho \sin \phi}{r - \rho \sin \phi} \text{ and } b = \frac{A}{r} \cdot \frac{r + \rho \sin \phi}{r - \rho \sin \phi}$$

In the following questions a and b will have these values, and it will be understood in every question relating to combinations of pulleys that the effect of the weight of the pulley on the friction of the axle is neglected; it must also be remembered that this is not the same thing as neglecting the weight entirely.

Ex. 415.—A pulley 6 in. in radius has an axle of 1 in. in radius of wrought iron, turning on an ungreased bearing of cast iron; a weight of Q lbs. attached to a rope 3 in. in circumference is on the point of being raised over the pulley by a weight of P lbs. attached to the other end of the rope: show that
$$P = 1{\cdot}1117Q + 3{\cdot}4$$

Ex. 416.—If P is on the point of lifting Q by means of a Barton consisting of one fixed and one movable pulley, as shown in the annexed figure, determine the relation between P and Q.

FIG. 114.

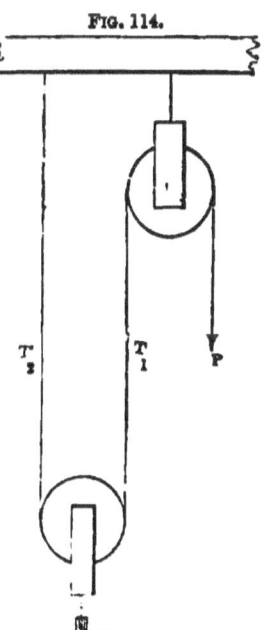

[Let T_1 and T_2 represent the tensions of the portions of the rope against which they are written; then since the rope is the same and the pulleys like one another, we shall have:—
since P is on the point of overcoming T_1, and T_1 on the point of overcoming T_2, and both T_1 and T_2 together lift Q,

$$P = aT_1 + b$$
$$T_1 = aT_2 + b$$
$$Q = T_1 + T_2$$

Therefore $(1 + a) P = a^2 Q + (1 + 2a) b$.]

Ex. 417.—If the pulleys and ropes are of the kind specified in *Ex.* 415, and if the whole weight lifted is 1000 lbs., determine P; also determine P supposing that all passive resistances are neglected. *Ans.* (1) 590 lbs. (2) 500 lbs.

[The weight of 1000 lbs. of course includes the weight of the lower block.]

Ex. 418.—If Q is raised by means of a block and tackle each containing a single sheave, show that the same relation exists between P and Q as that given in *Ex.* 416.

Ex. 419.—If P is on the point of raising Q by means of a block and tackle containing in all n equal sheaves—the parts of the rope being all parallel—find the relation between P and Q.

[See fig. 110. If $t_1, t_2, t_3, \ldots t_n$ are the tensions of the successive portions of the rope, we shall have

$$P = at_1 + b$$
$$t_1 = at_2 + b$$
$$t_2 = at_3 + b$$
$$\vdots \quad \vdots \quad \vdots$$
$$t_{n-1} = at_n + b$$

and $\quad t_1 + t_2 + t_3 + \ldots + t_n = Q$

whence, eliminating $t_1, t_2, t_3, \ldots t_n$, we obtain

$$P = Q \frac{a^n(a-1)}{a^n - 1} + \frac{nb}{a^n - 1}\frac{a^n}{} - \frac{b}{a - 1}]$$

Ex. 420.—Show from the formula in the last example, and also from first principles, that when the passive resistances are neglected $nP = Q$.

Ex. 421.—There is a block and tackle consisting of six sheaves each 3 in. in radius, whose axles are ½ in. in radius, and are of ungreased wrought iron turning on cast iron; the rope used is untarred and is 4 in. in circumference, the total weight raised (i.e. the mass and lower block) is 1000 lbs.: find the force required (1) taking into account the passive resistances, (2) neglecting them. *Ans.* (1) 390 lbs. (2) $166\frac{2}{3}$ lbs.

FIG. 115.

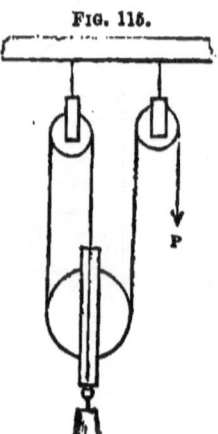

Ex. 422.—When the pulleys are arranged as in the annexed diagram (fig. 115) show that the relation between P and Q is given by the following formula:

FIG. 116.

$$P(1 + a + aa_1) = a^2 a_1 Q + b(1 + a + 2a a_1) + ab_1(1 + a)$$

where a, b refer to the smaller pulleys and a_1, b_1 to the large pulley.

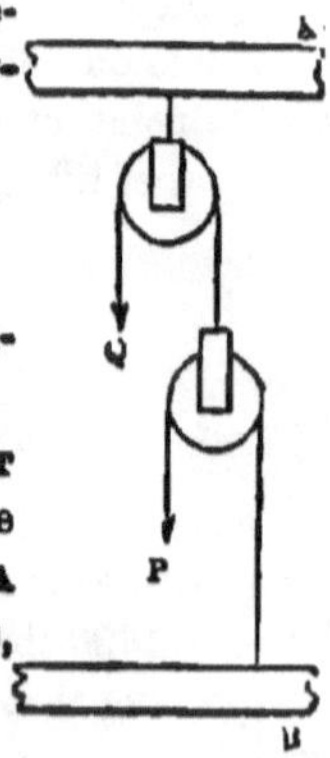

Ex. 423.—If a pair of similar pulleys is arranged as shown in the annexed diagram (fig. 116), where A and B represent immovable beams, show that

$$P = \frac{a^2 Q}{a + 1} + b - \frac{a w}{a + 1}$$

where w is the weight of the movable pulley.

CAPSTAN.

Ex. 424.—In the last example suppose each pulley to be similar to that described in *Ex.* 421, and the movable pulley with its block to weigh 50 lbs.; the rope being dry and 4 in. in circumference, find the force required to raise a weight Q of 1000 lbs. and determine the corresponding value of P when the passive resistances are neglected.

Ans. (1) 658 lbs. (2) 475 lbs.

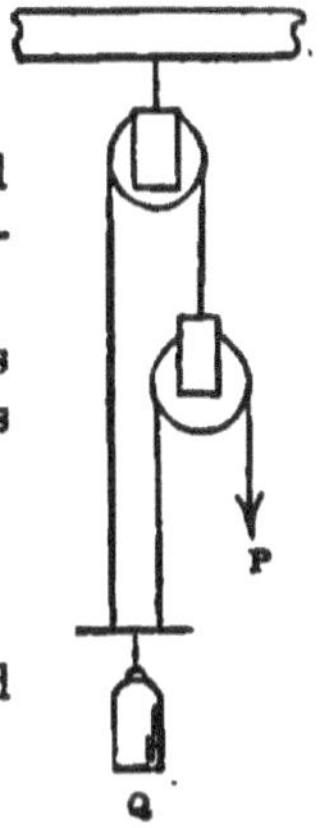

Ex. 425.—If two equal pulleys are employed to raise a weight Q in the manner indicated in fig. 117, show that

$$(2a+1) P = a^2 Q + b(2a+1) - aw$$

FIG. 117.

and determine P when Q weighs 1000 lbs., the pulleys and ropes being the same as in *Ex.* 424; and when passive resistances are neglected. *Ans.* (1) 432 lbs. (2) 317 lbs.

Ex. 426.—In the case of a tackle with three equal sheaves show that the force P which will just support a weight Q is given by the formula

$$P = \frac{(a-1)Q}{a(a^3-1)} + \frac{3b}{a(a^3-1)} - \frac{b}{a-1}$$

and show that when the passive resistances are neglected the equation reduces to $3P = Q$.

90. *The capstan.*—This machine in one of its commonest forms consists of a cylindrical mass of wood, C D, along the axis of which is cut a cylindrical aperture, which receives an axis A B (commonly of metal) on the top of which it rests; in the upper part of the capstan holes are cut, into which are inserted arms, such as E F, by means of which the capstan is turned, thereby winding up the rope G H which carries the weight.

FIG. 118.

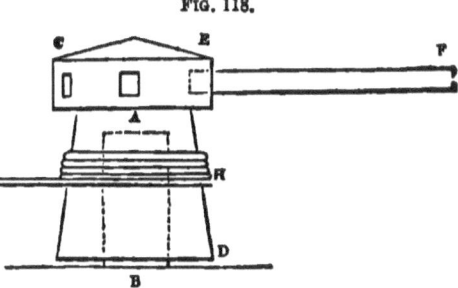

Ex. 427.—A capstan is turned by two equal parallel forces P acting in opposite directions at equal distances a from the geometrical axis of the figure, which are on the point of overcoming a force Q; let b be the radius of the cylinder round which the rope is wrapped, r the radius of the metal axle, μ_1 the coefficient of friction between the top of the axle and the capstan, and μ or $\tan \phi$ that between the side of the axle and the capstan; show that when the friction on the top of the axle is neglected

$$2Pa = (b + r \sin \phi)\left(Q + \frac{A + BQ}{b}\right)$$

and when the friction on the top of the axle is taken into account

$$2Pa = (b + r \sin \phi)\left(Q + \frac{A + BQ}{b}\right) + \tfrac{2}{3}r\mu_1 W$$

where W is the weight of the capstan.

[For friction on top of axle, see Art. 82.]

91. *Equilibrium of two forces acting in given directions on a body capable of turning round an axle.*—Let P and Q be the forces whose directions intersect in A, and

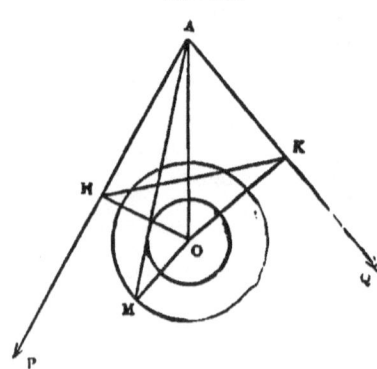

FIG. 119.

let P be on the point of preponderance; let O be the centre and ρ the radius of the axle, and ϕ the limiting angle of resistance between the axle and the bearing; with centre O, and radius $\rho \sin \phi$ describe a circle; and within the angle O A P draw the line A M touching that circle (Eucl. 17–3), join M O, then the angle A M O equals ϕ, and if P and Q are such that their resultant acts along A M, P will be on the point of preponderating over Q, i.e. P will be on the point making the body turn round its axle.

Draw O H, O K at right angles to A P and A Q respectively, join H K, denote O H by p, O K by q, H K by L, P A O by a, Q A O by β, and M A O by θ.

Ex. 428.—In the above case show that

$$P(p \cos \theta - \rho \sin \phi \cos a) = Q(q \cos \theta + \rho \sin \phi \cos \beta)$$

Ex. 429.—Show that the following formula gives a close approximation to the relation between P and Q when p is very much greater than $\rho \sin \phi$

$$Pp = Qq\left(1 + \frac{L\rho \sin \phi}{pq}\right)$$

[Observing that A H O K is a quadrilateral, about which a circle can be described, it is plain that O H K = β and O K H = a, consequently L = $p \cos \beta + q \cos a$.]

Ex. 430.—A weight Q hangs from one end of a rope, which after passing over a pulley (whose weight is neglected) takes a horizontal direction; it is now supported by n equal pulleys, placed at equal distances apart; show that the force P applied to the other end of the rope, which is on the point of lifting Q, is given approximately by the formula

$$P = \left(Q + \frac{A + BQ}{r}\right) \frac{r\sqrt{2} + \rho \sin \phi}{r\sqrt{2} - \rho \sin \phi} + (n w + w) \frac{\rho' \sin \phi'}{r' + \rho'}$$

where r, ρ, ϕ belong to the first pulley, r', ρ', ϕ' to the remaining n pulleys, w is the weight of *one* of the n pulleys, and w the weight of the rope which rests upon them.

92. *The two-wheeled carriage.*—In this case we may consider that the weight of the carriage is equally distributed upon each wheel. Now it will be observed that at each instant the wheel is lifted over a small obstacle A; then if O is the centre of the axle, and B the point of contact with the road, the angle A O B must have a certain magnitude, which we will denote by the letter γ. We will also denote the inclination of the road by a,

Fig. 120.

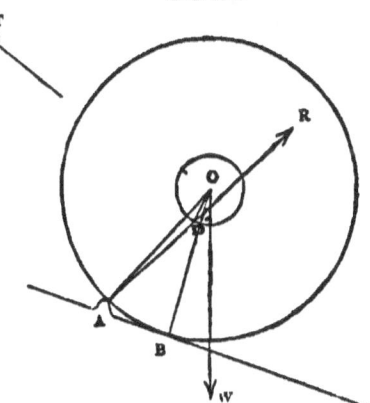

and the angle between the direction of the traction and the road by β. Then the forces concerned are, the traction T, the weight W, and the reaction R, of the point A, which, when T is on the point of moving W, must cut the circumference of the axle in a point D, such that O D R = ϕ; then if we denote the angle O A R by θ, the relation between T and W will be easily obtained by the triangle of forces.

Ex. 431.—When the wheel, as above explained, is on the point of moving, show that

$$T = W \cdot \frac{\sin (a + \gamma + \theta)}{\cos (\beta - \gamma - \theta)}$$

Ex. 432.—If A is the length of the arc A B, r and ρ the radii of the wheel and axle respectively, and if the road and the direction of traction are horizontal, show that

$$rT = W(A + \rho\phi) \text{ very nearly.}$$

Remark.—It appears from the experiments of General Morin that the traction is sensibly proportional to the weight directly and the radius of the wheel inversely, when the roads are paved or hard macadamised, and both the road and direction of traction are horizontal;* consequently it appears that for such roads, under the circumstances assigned in *Ex.* 432, the traction, as found by experiment, equals $\dfrac{kW}{r}$, where k is a constant quantity; but from the example it appears that $k = A + \rho\phi$, and hence the length of the arc A must be very nearly the same for the same road whatever be the radius of the wheel.

---

* Morin, *Notions Fondamentales*, p. 353. The account of the carriage wheel given in the text is taken from Mr. Moseley's *Mechanical Principles of Engineering*, pp. 395, 6, 7. The general results of M. Morin's experiments will be found in the Appendix to Mr. Moseley's work. The reader will find a great deal of condensed information on the subject of carriage wheels in Dr. Young's *Natural Philosophy*, Lecture 18.

CHAPTER VIII.

THE STABILITY OF WALLS.

THE general principles which regulate the relations that exist between the dimensions of a wall and the pressure it can sustain on its summit have been already discussed (Arts. 42, 43); in the present chapter we shall extend the application of the same principles to a few other cases. Several questions intimately connected with the subject of the present chapter are not discussed, as being too difficult for a purely elementary work—such are the conditions of the equilibrium of arches, vaults, domes, the more complicated forms of roofs, &c.

93. *The line of resistance.*—Let A B L M represent any structure divided into horizontal courses by the lines C D, E F, G H.... and let it be subjected to the action of any pressure P along the line Pa; produce Pa to meet C D in a'; if the mass A B C D were without weight the pressure on C D would act on the point a'; but the total pressure on C D is the resultant (R_1) of P and the weight of A B C D; the direction of this resultant must cut C D at some determinate point between a' and D, say at b, and let the direction of R_1 be bb'; now the total pressure on E F will be the resultant

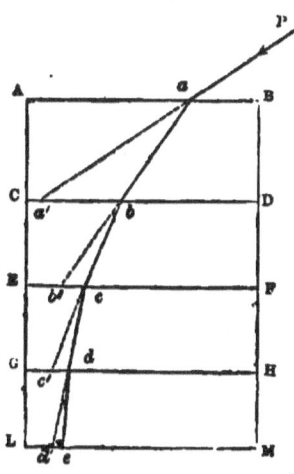

Fig. 121.

(R_2) of R_1, and the weight of C D F E, which will cut E F at a determinate point c, between b' and F; in the same manner, the pressure on the joint G H will act through a determinate point d, and on L M through a point e. Now if we join the points a, b, c, d . . . we shall obtain a polygonal line which cuts each joint in the point through which the direction of the resultant pressure on that joint passes; and if, further, we suppose the number of joints to be indefinitely great, the polygonal line will become a curved line, which is then called the line of resistance. It will be remarked that the directions of the resultants do not coincide with the sides of the polygon $ab, bc,$ . . . . . and therefore the line of resistance determines only the point at which the pressure on each joint acts, not the direction of the pressure at that point.

The line of resistance can be determined without much difficulty in a large number of cases: when this has been done, the condition of equilibrium—so far as the tendency of the structure to turn round any of its joints is concerned—is that this line cut each joint at a point within the structure; and, of course, the stability of a structure about any joint will be greater or less according as the intersection of the line of resistance with the joint is at a greater or less distance within the surface to which it is nearest.

It is plain that since the resultant of the pressures that act on a wall passes through the point of intersection of the line of resistance within its base, the algebraical sum of the moments of the pressures acting on the wall taken with respect to that point must equal zero. It may also be remarked that, in the case of most walls of ordinary shapes, the line of resistance continually approaches the extrados or outward surface; and hence, if the wall possess a certain degree of stability with reference to its lowest joint, it will possess a greater degree of stability with reference to any higher joint. Most of the following questions can, accordingly, be solved without the actual determination of the line of resistance.

EQUILIBRIUM OF WALLS. 175

Ex. 433.—A wall of Portland stone 30 ft. high and 2 ft. thick has to sustain on each foot of its length a pressure equal to the weight of 3 cubic ft. of the stone acting in a direction inclined to the vertical at an angle of 45°. Find the point of a bracket to which this force must be applied that the line of resistance may cut the base 6 in. within the extrados.

FIG. 122.

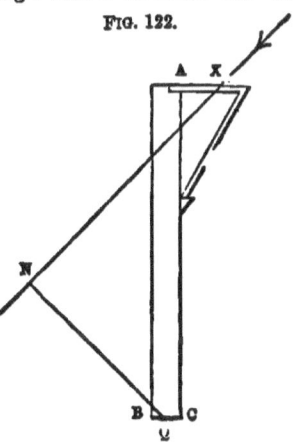

[Let the annexed figure represent a section of the wall; let the force act along the line X N, and let A X equal x; take B Q equal to 6 in.; then the condition of equilibrium is that the moments of the force and of the weight of the wall round Q be equal. Draw Q N perpendicular to X N; it can be easily shown that

$$Q N = A C \cos A X N - Q C \sin A X N - A X \sin A X N$$

$$\text{i.e. } Q N = \frac{28 \cdot 5 - x}{\sqrt{2}}$$

Whence we obtain

$$\frac{28 \cdot 5 - x}{\sqrt{2}} \times 3 = 60 \times \tfrac{1}{2}$$

$$\therefore x = 14 \cdot 36 \text{ ft.}$$

It may be remarked that the determination of a perpendicular resembling Q N occurs in many of the following questions. It may also be added that it is sometimes convenient to resolve the pressure into its horizontal and vertical components at X and obtain the moment of each.]

Ex. 434.—Determine the point of application of the pressure in the last article if the line of resistance cut the base 3 in. within the extrados.

Ans. 7·04 ft.

Ex. 435.—A roof, whose average weight is 20 lbs. per square foot, is 40 ft. in span and has a pitch of 30°, i.e. the rafters make an angle of 30° with the horizon; the walls of the building are of brickwork, and are 50 ft. high and 2 ft. thick; they are supported by triangular buttresses reaching to the top of the wall; the buttresses are 2 ft. wide, and 20 ft. apart from centre to centre. Determine their thickness at the bottom that the line of resistance may fall 6 in. within their extrados: determine also the answer that results from neglecting the weight of the buttress.

Ans. (1) 1·1675 ft. (2) 1·1754 ft.

Ex. 436.—A roof weighing 20 lbs. per square foot has a pitch of 60°; the distance between the walls that support it is 30 ft.; they are of Portland stone and are $2\tfrac{1}{2}$ ft. thick; the pressure of the roof being received on the inner edge of the summit, what is the extreme height to which the walls can be built? *Ans.* The walls can be carried to any height whatever.

Ex. 437.—If the weight of each square foot of a roof is 15 lbs., its pitch $22\frac{1}{2}°$, and the length of the rafters 30 ft., determine—(1) the thrust along the rafters, supposing them to be 4 ft. apart; (2) the tension of the tie-beam if one is introduced; (3) the magnitude and direction of the pressure on each foot of the length of the wall-plate,* if there is no tie-beam; (4) the thickness of the wall, which is of brickwork and 20 ft. high, when the line of resistance cuts the base 2 in. within the extrados, the pressure of the roof being received on the inner edge of the summit; (5) the distance from the axis of the wall at which the pressure of the roof must act if the line of resistance cuts the base of the wall 3 in. within the extrados.

Ans. (1) 2352 lbs.   (2) 2173 lbs.   (3) 705 lbs. at an angle of 50° 21′ 40″ to the vertical.   (4) 3 ft.   (5) 2·7 ft.

Ex. 438.—If w is the weight supported by each rafter of an isosceles roof whose pitch is a, show that the thrust on each rafter is $\dfrac{w}{2 \sin a}$ and the tension of the tie $\dfrac{w}{2 \tan a}$

94. *The pressure produced against a wall by water.*—The following construction can be easily proved from the principles of hydrostatics. Let A B represent a section of the wall made by a vertical plane, C D the surface of the water; draw the vertical line B E; draw B F, at right angles to A B and equal to B E; join C F; then the pressure on any length of the wall will equal the weight of a prism of water whose base is C B F and height the length of the wall; or, in other words, the pressure on each foot of the length of the wall will be the weight of as many cubic feet of water as the triangle B C F contains square

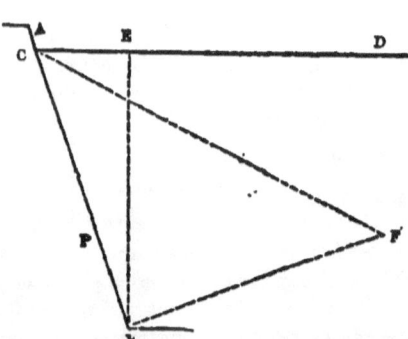

Fig. 123.

---

* The wall-plate is the beam on which the feet of the rafters rest: its office is to distribute the pressure along the wall.

feet; this pressure will act perpendicularly to the face of the wall through a point P, where $BP = \frac{1}{3}BC$.

Ex. 439.—There is a wall supporting the pressure of water against its vertical face; determine the pressure produced by the water on each foot of its length when 20 ft. of its height are covered. *Ans.* 12,500 lbs.

Ex. 440.—In the last case determine the pressure on the lower 10 ft. of the wall. *Ans.* 9375 lbs.

Ex. 441.—An embankment of brickwork has a section whose form is a right-angled triangle A B C; the base B C is 6 ft. long; the height A B is 14 ft.; will the embankment be overthrown when the water reaches to the top, if A B is the face which receives the pressure?

Ans. Yes; the excess of the moment of pressure of water is 9767.

Ex. 442.—In the last case will the embankment be overthrown if A C is the face which receives the pressure?

Ans. Yes; excess of moment of pressure of water 8675.

Ex. 443.—In *Ex.* 441 what horizontal pressure applied at A would keep the embankment steady? *Ans.* 698 lbs.

Ex. 444.—If the section of a river wall of brickwork have the form shown in the accompanying diagram, in which A B = 5 ft., D C = 15 ft., and B C equals 50 ft.; B C being vertical, and the angles B and C right angles, find the height to which the water must rise against B C to overturn it. *Ans.* 37·2 ft.

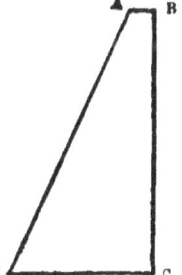

FIG. 124.

Ex. 445.—If in the last example the dimensions were B C equal to 30 ft., A B equal to 3 ft., and D C equal to 10 ft., would the wall be overthrown if the water rose to the summit? *Ans.* Yes.

Ex. 446.—There is the cofferdam sustaining the pressure of 26 ft. of water, supported by props 20 ft. long, 20 ft. apart, one end of each is placed $\frac{2}{5}$rds below the surface of the water and the other end on the ground; determine the thrust on each prop. *Ans.* 468,800 lbs.

Ex. 447.—If the section of an embankment of brickwork were of the form shown in fig. 124, and the dimensions were A B equal to 4 ft., D C equal to 12 ft., and B C equal to 24 ft., would it support the water when it rises to the top and presses on the face A D?

Ans. Yes; excess of moment of weight of wall 5184.

Ex. 448.—If the coefficient of friction between the courses of brickwork in the last example be 0·75, will the wall slide on its lowest section?

Ans. No; defect of horizontal pressure 2628 lbs.

Ex. 449.—In *Ex.* 446 what vertical pressure must by some means be supplied that equilibrium may be possible? *Ans.* 203,100 lbs.

178   PRACTICAL MECHANICS.

Ex. 450.—There is a river wall of Aberdeen granite 15 ft. high and having a rectangular section; the water comes to the distance of one foot from the top of the wall; find its thickness when the line of resistance cuts the base 6 in. within the extrados. *Ans.* 5·34 ft.

Ex. 451.—In the last example if the wall had a section of the form shown in fig. 124, where A B is 1 ft. long, the vertical face of the wall being towards the water; determine the width at the bottom when the line of resistance cuts the base 6 in. within the extrados. If the walls in this example and the last are 200 ft. long, determine the solid contents of each.

Ans. (1) 5·86 ft.   (2) 10,290 and 16,020 cub. ft.

Ex. 452.—In each of the last examples determine the distance from the extrados of the point at which the line of resistance cuts a horizontal joint 8 ft. below the surface of the water.   *Ans.* (1) 1·98 ft.   (2) 1·75 ft.

[The point will, of course, be that round which the moment of the weight of the incumbent portion of the wall equals the moment of the pressure of the water on the eight feet.]

Ex. 453.—A river wall whose section is a right-angled triangle just supports the pressure of water when its surface is on a level with the top of the wall; show that the thickness of the base

$$= \text{height} \times \sqrt{\frac{w}{w_1 + 2w}}$$

if the hypothenuse of the triangle is turned towards the water; but when the perpendicular is turned towards the water the thickness of the base

$$= \text{height} \times \sqrt{\frac{w}{2w_1}}$$

where w is the weight of a cubic foot of water, and w_1 that of a cubic foot of the material of the wall. And show from hence that in the former case the thickness of the base is greater or less than in the latter according as the specific gravity of the wall is greater or less than 2.

Ex. 454.—A wall of brickwork is to be built round a reservoir 20 ft. deep; its slope is inward; it is 1 ft. thick at top; what must be its thickness at the bottom, that when the reservoir is full, the line of resistance may cut the base 6 in. within the extrados?   *Ans.* 10·74 ft.

Ex. 455.—The wall of a reservoir full to the brim is of brickwork and is 20 ft. high and 2 ft. thick; it is supported by props at intervals of 6 ft.; the length of each is 20 ft., and its inclination to the horizon 30°: determine the thrust on each prop, its weight being neglected.   *Ans.* 54,632 lbs.

Ex. 456.—In the last example determine the thickness of the wall that would just support the pressure of the water if the props were removed. If the wall stand on its lowest section without the aid of cement, what must be the coefficient of friction between the surfaces?

Ans. (1) 8·6 ft.   (2) 0·65.

PRESSURE OF EARTH.

Ex. 457.—A reservoir is divided by a brickwork wall 12 ft. high and 2 ft. thick; the water on one side of the wall is 10 ft. deep; what must be the depth on the other side if the wall is just overthrown?   *Ans.* 10·4 ft.

Ex. 458.—A cofferdam sustains the pressure of 26 ft. of water, and is supported at intervals of 10 ft. by props D E and C F; given that B C and B D are respectively 4 ft. and 18 ft. and that D E and C F are respectively 30 ft. and 18 ft.; find the thrust on each prop. And what must be the weight of the struts, and of 10 ft. of the length of the cofferdam, that the whole be not overthrown? The thickness of the cofferdam and the adhesion at B are to be neglected.

Ans. (1) Thrust on D E = 88,020 lbs.; on B C = 144,400 lbs. (2) 84,900 lbs.

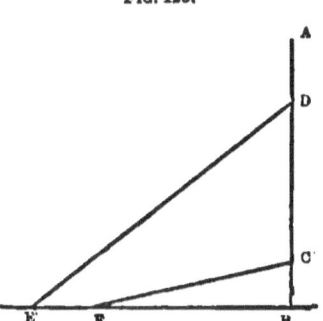

FIG. 125.

95. *The pressure of earth.*—Let A B represent a section of a wall supporting earth, whose surface is A C, it is required to determine the pressure produced on A B by the earth. Now, it must be remembered that two extreme cases may come under consideration: the first arises when the earth is thoroughly penetrated with water, in which case the pressure is the same as would result from hydrostatic pressure; the second arises when the cohesion of the earth is so considerable that it would stand with its face vertical even if the wall were removed. Dismissing these two extreme cases, let us suppose the wall A B removed, the following result will then ensue: the earth being friable will weather and break away until its surface has taken a slope B C, inclined to the horizon at an angle equal to the limiting angle of resistance; when reduced to this state it will have no further tendency to break away, and, unless washed down by rain or removed by some other extrinsic cause, will remain permanently at rest at that slope, which is therefore called

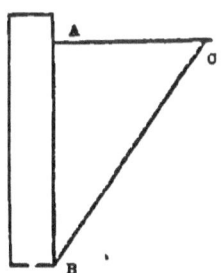

FIG. 126.

its *natural slope*. Hence, in the case we are considering, the wall is required to give a certain degree of support to the wedge of earth A B C; this wedge is generally supported in some degree by the cohesion of its parts with each other and with the earth below B C, so that the wall will be sufficiently strong if it will support the earth, on the supposition that the cohesion is quite destroyed, unless (which is not contemplated) the earth should be saturated with water. The angle of the natural slope of fine dry sand is about 35°; of dry loose shingle about 40°; of common earth, pulverised and dry, about 45°.*

Proposition 20.

If w is the weight of a cubic foot of earth, and ϕ its natural slope, the pressure produced on the vertical face of a retaining wall by earth which does not rise above its summit, and which has a horizontal surface, is the same as that produced by a fluid the weight of a cubic foot of which is $w \tan^2 \left(\dfrac{\pi}{4} - \dfrac{\phi}{2} \right)$.

Let A B be the section of the wall, B A C of the earth; take any portion A X equal to x of the wall, and suppose its length to be 1 foot; draw X Y, making an angle θ with the horizon greater than ϕ; then the weight W of the wedge A X Y equals $\frac{1}{2} w x^2 \cotan \theta$, and acts vertically through a point P, where $XP = \frac{1}{3} XY$; it is supported by the reaction R_1 of X Y and by the reaction R of the wall: the latter reaction is equal and opposite to the pressure produced by the earth on the wall, and its direction is perpendicular to A X: also, since the surface X Y

FIG. 127.

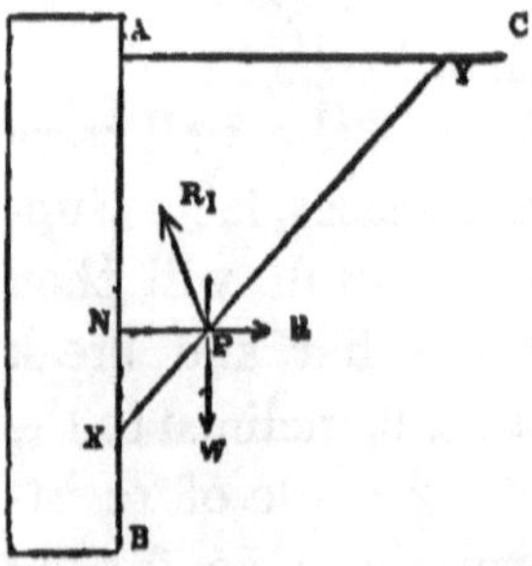

* See Mr. Moseley's *Mechanical Principles of Engineering*, p. 441.

will not exert a greater pressure than is just necessary to support A X Y, the direction of R_1 must be inclined to the normal to X Y at an angle equal to ϕ; also, the directions of R and R_1 must pass through the point P, in which W's direction cuts X Y, so that NX will equal $\frac{1}{3}$ of AX; moreover,

$$R : W :: \sin R_1PW : \sin R_1PR :: \sin(\theta - \phi) : \cos(\theta - \phi)$$
$$\therefore R = W \tan(\theta - \phi) = \tfrac{1}{2}wx^2 \cotan \theta \tan(\theta - \phi)$$

Now, according as θ has different values R will have different values, and if we determine the value of θ for which R is greatest, the wall cannot be called on to supply a greater reaction, and this must therefore equal the pressure which A X actually sustains. But

$$\cot \theta \tan(\theta - \phi) = \frac{\cos \theta \sin(\theta - \phi)}{\sin \theta \cos(\theta - \phi)} = \frac{\sin(2\theta - \phi) - \sin \phi}{\sin(2\theta - \phi) + \sin \phi}$$
$$= 1 - \frac{2 \sin \phi}{\sin(2\theta - \phi) + \sin \phi}$$

which is manifestly greatest when the fractional part of the expression is least, i.e. when $2\theta - \phi$ equals $\dfrac{\pi}{2}$, so that the required value of θ is $\dfrac{\pi}{4} + \dfrac{\phi}{2}$, and, therefore, the required value of the pressure is

$$\tfrac{1}{2}wx^2 \cotan\left(\frac{\pi}{4} + \frac{\phi}{2}\right) \tan\left(\frac{\pi}{4} - \frac{\phi}{2}\right) = \tfrac{1}{2}wx^2 \tan^2\left(\frac{\pi}{4} - \frac{\phi}{2}\right)$$

acting through a point N which is below A by a distance equal to $\frac{2}{3}x$; but this is the same as the pressure that would be produced by a fluid each cubic foot of which weighs $w \tan^2\left(\dfrac{\pi}{4} - \dfrac{\phi}{2}\right)$. Therefore, &c. Q. E. D.

Ex. 459.—A mass of earth the specific gravity of which is 1·7, whose surface is horizontal, presses against a revêtement wall whose top is on the level of the ground and height 20 ft., the natural slope of the earth being 45°; determine the pressure of the earth on each foot of the length of the wall. *Ans.* 3646 lbs.

Ex. 460.—If the wall in the last example is of brickwork and has a rectangular section, determine its thickness to enable it to sustain the pressure of the earth. *Ans.* 4·65 ft.

Ex. 461.—The vertical face of a revêtement wall of brickwork sustains the pressure of 20 ft. of earth, the surface of which is horizontal and 2 ft. below the summit of the wall; the thickness of the wall at top is 1 ft.: what must be its thickness at bottom if it just sustains the earth, the specific gravity of the earth being 2 and its natural slope 45°? Also determine the thickness that would enable the wall to sustain the pressure if the earth were thoroughly permeated with water.*

Ans. (1) 5·47 ft. (2) 9·6 ft.

Ex. 462.—If a pressure P is applied against a wall supported on the opposite side by earth with its surface horizontal; show that when P is on the point of causing the earth to yield, the resistance of the earth is the same as that of a fluid the weight of a cubic foot of which equals (weight of cubic foot of earth) $\times \tan^2\left(\dfrac{\pi}{4}+\dfrac{\phi}{2}\right)$.

[The reasoning in this case is step by step the same as that given in Prop. 20, except that now the wedge of earth is on the point of being forced *up*, so that the direction of R_1 will be on the other side of the perpendicular to x y.]

Ex. 463.—A reservoir wall of brickwork is 4 ft. thick and 15 ft. above the surface of the ground; the foundations are 15 ft. deep; the natural slope of the earth is 45° and it weighs 100 lbs. per cubic foot; when the reservoir is full (so that the water presses against the whole 30 ft. of wall) will the wall stand, supposing the adhesion of the cement perfect?

Ans. Yes; excess of the moment of the greatest pressure that could support the wall over that of the pressure of the water 73,480.

Ex. 464.—If A B C is a section of a rectangular wall, P the pressure ap-

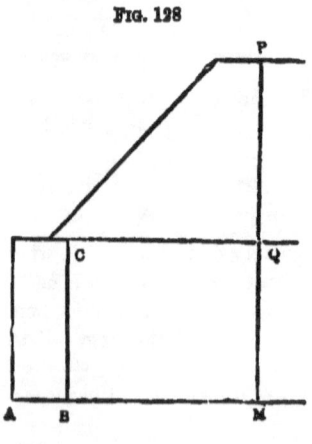

Fig. 128

* It is common for revêtement walls to sustain a surcharge of earth, as shown in the accompanying diagram; an investigation of the pressure in this case will be found in Mr. Moseley's *Mechanical Principles of Engineering*, p. 453. The following practical formula (Morin, *Aide-Mémoire*, p. 417) gives the thickness (x) of a rectangular wall for a given height (H) of the revêtement (Q M) and a surcharge (P Q) whose height is h, viz.

$$x = 0.865\,(\text{H}+h)\sqrt{\dfrac{w}{w_1}}\,\tan\left(\dfrac{\pi}{4}-\dfrac{\phi}{2}\right)$$

w being the weight of a cubic foot of earth and w_1 that of a cubic foot of masonry.

LINE OF RESISTANCE.

plied to every foot of its length at A, the inner edge of its summit; determine the equation to the line of resistance.

[Take any horizontal section of the wall M N; let A N = x, B C = a, then the weight W of A N M = $a x w$, where w is the weight of a cubic foot of the wall; now, if the direction of the resultant cuts M N in R, this will be a point in the line of resistance, and if R N = y we are to determine a relation between x and y. The relation in question can easily be shown to be

$$a w x \left(y - \frac{a}{2}\right) = \mathrm{P}\,(x \sin a - y \cos a)$$

where a is the inclination of P's direction to the vertical.]

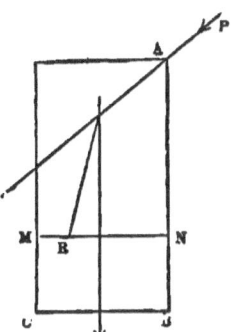
Fig. 129.

*Ex. 465.—In the last example show that the curve is a hyperbola and determine its asymptotes; and show that if the thickness of the wall equals $\sqrt{\dfrac{2\,\mathrm{P}\,\sin a}{w}}$ it may be carried to any height whatever with safety.

*Ex. 466.—If the wall in Ex. 464 has to support the pressure of earth or water reaching to the top of the wall, show that the line of resistance is a parabola with its axis horizontal, and show that in the latter case its focus is in the summit of the wall at a distance from the intrados equal to $\dfrac{a}{2}\left(1 + \dfrac{w}{w_1}\right)$, where w is the weight of a cubic foot of masonry and w_1 of water.

*Ex. 467.—A B C D is the section of a reservoir wall the vertical face of which (B C) is towards the water; the width of the top of the wall (A B) is a; the inclination of A D to the vertical is θ, and s is the specific gravity of the wall; show that when the water reaches to the top of the wall the equation to the line of resistance is—x and y being measured as in Ex. 464—

$$x^2\left(\frac{1}{s} + \tan^2\theta\right) - 3xy \tan\theta + 3ax \tan\theta - 6ay + 3a^2 = 0$$

*Ex. 468.—Show that if the wall in the last example stand, whatever be the depth of the water whose pressure it sustains, then tan θ must be $> \dfrac{1}{\sqrt{2s}}$.

*Ex. 469.—Determine the equation to the line of resistance in a river wall of Aberdeen granite, the thickness of which is 4 ft., and which sustains the pressure of water whose surface is on the level of the top of the wall. *Ans.* $x^2 = 63\,(y-2)$.

*Ex. 470.—Determine from the equation in the last example the height of the wall when the line of resistance intersects the base at a distance of 4 in. within the extrados. *Ans.* 10 2 ft.

CHAPTER IX.

ON THE DEFLECTION AND RUPTURE OF BEAMS BY FORCES APPLIED TRANSVERSELY.*

96. *Neutral surface and neutral line of a beam*.— If we consider a long beam of wood A D supported at its two ends, the effect of its weight will be to bend it into such a shape as that shown in the figure; it is evident that the under surface

FIG. 130.

C D will suffer extension, and the upper surface A B compression: so that there will be a section P Q intermediate to the compressed and extended parts, which will undergo neither compression nor extension; this surface is called the *neutral surface*. Forces may act on the beam in such a manner that the whole of it is either compressed or extended; in such a case the neutral surface will not have a real existence, but there will exist without the body an imaginary surface bearing the same relation to the compressions or extensions as that borne by the actual neutral surface in other cases.

In what follows we shall assume that the forces act in a plane containing the geometrical axis of the beam considered as a prism and at right angles to the axis in its undeflected state; that the cross section of the beam is

---

* This chapter cannot be read with advantage by any student who has not some acquaintance with the Integral Calculus.

symmetrical with respect to the plane containing the forces; and that the deflection is so small as not to change sensibly the moments of the forces. The plane in which the forces act will cut the neutral surface along a line called the *neutral line*. It will appear in sequel that under these circumstances the neutral line coincides with the axis of the beam.

Before going further the student should make himself acquainted with the simpler cases of moments of inertia. They are given in Part II., Chap. V.

97. *The bending moment.*—Let A B be a uniform rod held at rest by three forces P, Q, R acting at right angles to its length; suppose the rod to be so strong as to sustain the action of these forces without being much bent, so that the distances A C, B C are not sensibly altered. We have to

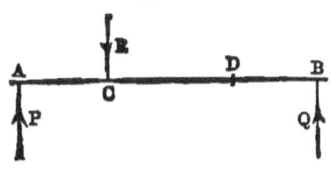

FIG. 131.

do with two sets of forces—(*a*) the external forces P, Q, R; (*b*) the internal forces due to the elasticity of the rod, which are called into play by its extension or compression. The first question to be considered is:—What is the tendency of the forces to break the rod at any assigned point D? Now (Art. 60) P may be replaced by an equal force acting in the same direction at D, and a negative couple A D . P; the like is true of the other forces. Consequently, if we suppose an imaginary plane drawn across the rod at D, as shown in fig. 132, where for convenience the thickness is magnified, we have on the left of D E a force R − P, and a couple whose moment is − A D . P + C D . R; on the right of D E we have the force Q and a couple whose moment is B D . Q. It thus becomes plain:—First, that the forces tend to

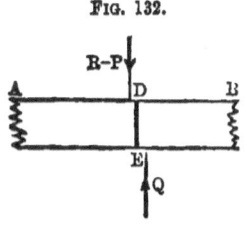

FIG. 132.

cause A E D and B E D to slide in opposite directions along D E, i.e. they produce a *shearing stress* measured by either of the equal forces R—P or Q. Secondly, the forces tend to make the parts A E D and B E D turn in opposite directions, the measure of this tendency being the moment B D . Q or —A D . P + C D . R, which, however, have contrary signs. Either of these moments is called the bending moment at the point D.

In the cases to be considered in the present chapter the shearing stress can be put out of the question; the bending moment alone comes under consideration. We may say, therefore, that the tendency of the forces to break the rod at any point (D) is measured by the sum of the moments taken with respect to D of all the forces on one side of that point, whatever be the number of forces acting on the beam.

Ex. 471.—In fig. 131 suppose the rod to rest on two points under A and B, and let the force R be produced by a weight W hung at C; then if A B is denoted by a, and A C by b, and B D by x, the bending moment at D is $\frac{b\,\mathrm{W}}{a} \cdot x$, if the weight of the rod is neglected. If, however, D is between A and C, the bending moment is $\frac{(a-b)\,\mathrm{W}}{a}(a-x)$, and at C it is $\frac{b\,(a-b)\,\mathrm{W}}{a}$.

Ex. 472.—If the weight of the rod only is considered the bending moment is $\frac{1}{2}wax - \frac{1}{2}wx^2$, where w denotes the weight of a unit of the length of the rod.

Ex. 473.—In *Ex.* 471 if the weight of the rod is taken into account the bending moment is

$$\left(\frac{b\,\mathrm{W}}{a} + \tfrac{1}{2}wa\right)x - \tfrac{1}{2}w\,x^2$$

or

$$\left(\frac{(a-b)\,\mathrm{W}}{a} + \tfrac{1}{2}wa\right)(a-x) - \tfrac{1}{2}w(a-x)^2$$

according as D is between B and C or C and A.

Ex. 474.—In the last example either formula gives for the bending moment at C the value

$$\frac{(a-b)\,b}{a}\{\mathrm{W} + \tfrac{1}{2}wa\}$$

Ex. 475.—In *Ex.* 472 the value of the bending moment is greatest at the middle point, and there equals $\tfrac{1}{8}wa^2$.

BENDING MOMENT. 187

Ex. 476.—In *Ex.* 473 if $\frac{a}{2} > b + \frac{b\,w}{a\,w}$, the bending moment will be greatest at a point between c and the middle of the rod, and the greatest value will be $\frac{w}{2}\left(\frac{a}{2}+\frac{b\,w}{a\,w}\right)^2$; but otherwise its greatest value is at c, and is that given in *Ex.* 474.

Ex. 477.—A rod is supported on many points all on the same horizontal line, so that the only forces acting are weights and the vertical reactions of the fixed points; let A and B be any two consecutive fixed points, B to the right of A; let the distance A B be denoted by l, the load on A B (including the weight of the rod) per unit of length by w; and let P be any point in A B at a distance x from A. Show that the bending moment at P is

$$\text{M} + \text{Q}\,(l-x) + \tfrac{1}{2}w\,(l-x)^2$$

where M is the bending moment at B, and Q the sum of the forces acting at and to the right of B.

[Let there be any number of parallel forces, viz. R acting at B, and $\text{R}_1, \text{R}_2, \text{R}_3 \ldots$ to the right of B at distances $r_1, r_2, r_3, \ldots$ the sum of their moments with respect to B will be

$$\text{M} = r_1\,\text{R}_1 + r_2\,\text{R}_2 + r_3\,\text{R}_3 + \ldots$$

With respect to a point at a distance b to the left of B the sum of their moments will be

$$b\,\text{R} + (b+r_1)\,\text{R}_1 + (b+r_2)\,\text{R}_2 + (b+r_3)\,\text{R}_3 + \ldots$$
$$= b\,(\text{R} + \text{R}_1 + \text{R}_2 + \text{R}_3 + \ldots\ldots) + \text{M.]}$$

98. To obtain a clear view of the reactions by which the rod resists the tendency of the forces to break it, let us suppose it to be divided along D E, and the connection to be re-established as shown in the figure by means of two small pieces of an elastic material D F and E G whose unstretched lengths are equal. The couple acting on A D tends to turn it (as we have already seen) in the same direction as that of the motion of the hands of a watch, and that acting on B F in the opposite direction, so that D F will be compressed and E G stretched. Consequently D F will react with equal forces (R and R_1) against D and F respectively, as shown in the figure; while E G reacts with

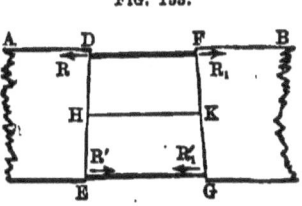

FIG. 133.

equal forces (R' and R'_1) against E and G respectively. The two forces R and R' form a couple with a positive moment, which balances the couple that tends to turn A D E. In like manner the forces R_1 and R'_1 form a couple which balances the couple that tends to turn B F G. Consequently, whether we consider the forces acting on A D E or those acting on B F G, we obtain the relations

$R = R'$ and $R \times D E =$ the bending moment.

Of course D F would be compressed and E G stretched to just the extent needed for calling into play the forces R and R'.

If now we suppose D E G F to be a portion of the beam included between two planes at right angles to its axis in its undeflected state, the only difference will be that the part D H K F above a certain line H K will be compressed, while the other part H E G K will be stretched. The compression will increase gradually from H K to D F and the extension from H K to E G. Thus there will be a distributed force which gradually increases from H to D and acts in R's direction, and in like manner from H to E a distributed force acting in the direction of R'. It is plain that the forces on D E must reduce to a couple with a positive moment, and as it balances the external forces on A D E we must have the moment of this couple equal to the bending moment. The same conclusion would follow if we reasoned on the forces distributed along F G which must balance the couple acting on B F.

Proposition 21.

If a cylindrical or prismatic beam has a cross section symmetrical with respect to a plane containing the axis, and is acted on by forces in that plane at right angles to the axis in its undeflected state, the neutral line will coincide with the axis of the beam (i.e. will pass through

FLEXURE OF BEAMS.

the centre of gravity of each cross section), and the reciprocal of its radius of curvature at any point will equal (the bending moment) $\div \text{E A K}^2$, *where* E *denotes the modulus of elasticity and* A K^2, *the moment of inertia of the cross section taken with reference to an axis passing through the centre of gravity at right angles to the plane of symmetry in which the forces act.*

Fig. 134 corresponds exactly to fig. 133, p. 187; and fig. 135 represents the cross section through D H E at right angles to the plane of symmetry, which is the plane of the paper. Let E D and G F be produced to meet in O. Then as H K is unstretched it is part of the neutral line, and if H K becomes in the limit indefinitely small, H O becomes the radius of curvature (ρ) of the neutral line at the point H. Let H L be denoted by z, and let us consider a lamina L L' of the beam the area of whose cross section $l\,l'$ is denoted by a—the lamina is, of course, supposed to be between two planes very near together, parallel to H K and at right angles to the plane of symmetry. Now, as the uncompressed length of L L' equals H K, it follows from Art. 6 that the reaction of L L' due to its compression equals

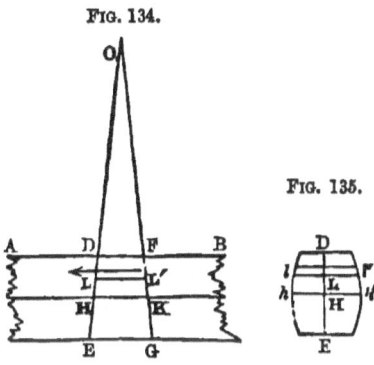

FIG. 134.

FIG. 135.

$$\text{E} a \times \frac{\text{H K} - \text{L L}'}{\text{H K}}$$

But $\quad \rho : \rho - z :: \text{H K} : \text{L L}'$

or $\quad \rho : z :: \text{H K} : \text{H K} - \text{L L}'$

therefore the reaction is $\dfrac{\text{E} a\, z}{\rho}$. This force is distributed uniformly over $l\,l'$, and therefore may be supposed to act at L. If a point is taken below H the lamina corresponding to it is stretched, and therefore the reaction at that point is in the opposite direction to that exerted at L; this, however, is provided for by z being negative in that case, so that the formula just given applies to all points of DE. Hence if $a_1, a_2 \ldots$ be the areas of any other portions of D h E h' taken in the same way as a at distances $z_1, z_2 \ldots$ from H, the resultant of all these reactions will be

$$\frac{E}{\rho}(az + a_1 z_1 + a_2 z_2 + \ldots)$$

Now (Prop. 16) this equals $\frac{E}{\rho} \times $ D h E $h' \times \bar{z}$, where $\bar{z}$ is the distance of the centre of gravity of the cross section from H. As the resultant is a couple, this expression must be zero; but neither E nor the area D h E h' is zero, nor can ρ be infinite (except at particular points), for then there would be no flexure. Consequently $\bar{z}$ must be zero, i.e. H is the centre of gravity of the cross section. Therefore the neutral axis passes through the centre of gravity of every cross section, i.e. it coincides with the axis of the beam.

The sum of the moments of the reactions taken with respect to hh' is

$$\frac{E}{\rho} \times (az^2 + a_1 z_1^2 + a_2 z_2^2 + \ldots)$$

and this must equal the bending moment. Now the quantity within the brackets is the moment of inertia (A K^2) of the cross section taken with reference to hh'. Hence

$$\text{bending moment} = \frac{E A K^2}{\rho}$$

or
$$\frac{1}{\rho} = \frac{\text{bending moment}}{E A K^2}$$

In the ordinary case of a rectangular beam it appears from

Part II., Chap. V., that AK^2 equals $\tfrac{1}{12} b^3 c$, i.e. a twelfth part of the width multiplied by the cube of the depth.

Ex. 478.—Let the beam be held firmly at one end, and a force P applied at the other at right angles to its length; it is required to determine the equation to the neutral line, neglecting the weight of the beam.

[Let LL_1 be the neutral line, LG the position of the beam's axis when unbent, F any point in the neutral line, ρ the radius of curvature at F, x and y the co-ordinates of F, viz. LR and RF; then, since the bending moment at F is $P(a-x)$, we have

FIG. 136.

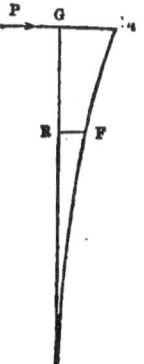

$$\frac{1}{\rho} = \frac{12P(a-x)}{Eb^3c}$$

Now, since the curvature is small, $\dfrac{dy}{dx}$ is small, and therefore $\left(\dfrac{dy}{dx}\right)^2$ can be omitted; consequently

$$\frac{1}{\rho} = \frac{d^2y}{dx^2}$$

therefore $\quad \dfrac{d^2y}{dx^2} = \dfrac{12P}{Eb^3c}(a-x)$

whence $\quad y = \dfrac{12P}{Eb^3c}\left(\dfrac{ax^2}{2} - \dfrac{x^3}{6}\right)$]

Ex. 479.—Show that the deflection of the beam in the last example equals $\dfrac{4P}{Ebc} \cdot \dfrac{a^3}{b^2}$.

Ex. 480.—If in the last example a force is applied to the end of the beam and gradually increased up to P, show that the number of units of work expended in producing deflection equals

$$\frac{2P^2}{Ebc} \cdot \frac{a^3}{b^2}$$

[Compare *Ex.* 149.]

Ex. 481.—The end of a beam of oak is firmly embedded in masonry; the length of the projecting part is 15 ft., its breadth is 3 in., and its depth 6 in.; a force of 2 cwt. is applied perpendicularly at its end; determine the deflection, and the work expended in producing that deflection—the weight of the beam being neglected.

Ans. (1) 5·5 in. (2) 51 ft.-pds. of work.

Ex. 482.—If a beam is held firmly by one end in a horizontal position and is bent simply by its own weight, show that $\dfrac{1}{\rho} = \dfrac{6w(a-x)^2}{Ecb^3}$, where

w is the weight of a unit length of the beam; and that the deflection is $\frac{5}{2} \cdot \frac{wa}{\text{E}bc} \cdot \frac{a^2}{b^2}$.

[These results are true when the beam is loaded uniformly at the rate of w per unit of length.]

Ex. 483.—If the beam in *Ex.* 482 were of elm, were 5 ft. long, 1 ft. broad, and 1 ft. deep, and had to support the pressure of brickwork 14 in. thick and 10 ft. high, determine the depression. *Ans.* 0·15 in.

Ex. 484.—If a horizontal beam A B is supported at its ends and is loaded by a weight w at its middle point, and if ρ is the radius of curvature at a point in the neutral line whose distance from the middle point of the beam is x; show that

$$\frac{1}{\rho} = \frac{3\text{w}(a-2x)}{\text{E}cb^3}$$

and that the deflection at the middle of the beam is $\frac{\text{w}}{4\text{E}bc} \cdot \frac{a^3}{b^2}$.

[If the centre of the beam is taken as the origin of co-ordinates, x being measured horizontally and y vertically, the bending moment is $\frac{1}{2}\text{w}(\frac{1}{2}a-x)$, and the value of y at either end equals the required deflection.]

Ex. 485.—If the beam in the last example were bent by its own weight, which is w per unit of length, show that $\frac{1}{\rho} = \frac{3w\,(a^2-4x^2)}{2\text{E}cb^3}$, and that the depression at the middle point is $\frac{5}{32} \cdot \frac{wa}{\text{E}bc} \cdot \frac{a^3}{b^2}$.

Ex. 486.—A fir batten 3 in. deep, 1½ in. broad, is placed horizontally between two props 5 ft. apart and loaded with a weight of 135 lbs. in the middle; its own weight being neglected, determine the depression; determine also the depression if it were fixed at one end and loaded with the same weight at the other end. *Ans.* (1) $\frac{18}{133}$ in. (2) $\frac{288}{133}$ in.

Ex. 487.—A spar of oak 3·2 in. square is placed horizontally between two props 12·8 ft apart and loaded with 268 lbs. in the middle; determine the deflection, neglecting the weight of the beam. *Ans.* 1·597 in.

Ex. 488.—A piece of elm 2 in. square is placed horizontally between two supports 7 ft. apart, it is loaded in the middle with a weight of 125 lbs.; determine the deflection when its own weight is neglected. *Ans.* 1·65 in.

Ex. 489.—There is a beam of larch 6 in. deep, 4 in. wide, and 12 ft. long, it is supported on a fulcrum whose distance from one end is 4 ft.; the shorter end carries a weight of 2 cwt.; determine the deflection of each arm of the beam, its own weight being neglected. *Ans.* (1) 0·109 in. (2) 0·437 in.

Ex. 490.—A rod, whose weight can be neglected, has a weight tied to each end, and is then placed on a fulcrum so that the weights balance each other; show that the droops at the ends of the rod are inversely proportional to the squares of the weights.

Ex. 491.—Find the force which being applied vertically to the end of the beam in *Ex.* 482 exactly neutralises the droop. Why should not this force equal half the weight of the beam? *Ans.* $3\,wa \div 8$.

Ex. 492.—The ends of a beam rest on horizontal supports, it is deflected by its own weight and a vertical force w acting through its middle point; determine the total deflection, and show that it equals the sum of the separate deflections produced by its own weight and by w, if w act vertically downward, and their difference if w act vertically upward.

Ex. 493.—If A B, A C are the principal rafters of a roof the feet of which are fastened together by a tie-beam B C, the middle point of which is D; if A and D are joined by a 'king-post' which exactly neutralises the bending in the middle of the tie-beam caused by its weight, show that the tension of the king-post equals $\frac{5}{8}$ of the weight of the tie-beam.

Ex. 494.—In *Ex.* 487 determine the deflection when the weight of the spar is taken into account. *Ans.* 1·8 in.

Ex. 495.—A beam of larch supported at each end measures 20 ft. between the points of support, it is 6 in. wide and 10 in. deep, it sustains a wall of brickwork 30 ft. high and 1 ft. thick throughout its whole length; find the deflection. *Ans.* 23·13 in.

Ex. 496.—If the beam in the last example is supported by a column which exactly neutralises the deflection of the middle point, find the pressure on the column. *Ans.* 42,170 lbs.

Ex. 497.—If in the last example the under surface of the beam in its undeflected state is 12 ft. from the ground, the middle point is supported by a column of cast iron 3 inches in diameter, which in its uncompressed state is exactly 12 ft. long; determine the deflection of the beam at its middle point and the pressure on the column. *Ans.* (1) 0·05 in. (2) 42,077 lbs.

[The column being compressible will allow the middle of the beam to descend, whereby the thrust on the column will be diminished: the question to be answered is—At what degree of compression will the tendency of the column to recover its form upward exactly balance the tendency of the beam to deflect downward?]

Ex. 498.—In the last example suppose the measurements to be made at 50° Fahrenheit, at what temperature would there be no deflection at the middle point of the beam? *Ans.* 107° F.

Ex. 499.—If a hollow cylinder (whose weight is neglected) the radii of whose section are r_1 and r be supported horizontally at two points whose distance is a; show that, when it sustains a weight w at its middle point, the radius of curvature of the neutral line at a point distant x from the middle is given by the formula

$$\frac{1}{\rho} = \frac{w\,(a-2x)}{\pi\,\mathrm{E}\,(r_1^{\,4}-r^4)}$$

and the deflection at the middle point by the formula

$$\delta = \frac{w\, a^3}{12\, \pi\, \text{E}\, (r_1{}^4 - r^4)}$$

[The moment of inertia of the space between two concentric circles with respect to a diameter is $\frac{1}{4}\pi (r_1{}^4 - r^4)$; see *Ex.* 760.]

Ex. 500.—If in the last example the cylinder sustains throughout its length a uniform load of w lbs. per unit of length, then

$$\frac{1}{\rho} = \frac{w\,(a^2 - 4x^2)}{2\,\pi\,\text{E}\,(r_1{}^4 - r^4)}$$

and

$$\delta = \frac{5\, w\, a^4}{96\, \pi\, \text{E}\, (r_1{}^4 - r^4)}$$

Ex. 501.—If an iron girder * has a section of the form shown in the annexed diagram, of the following dimensions, $\text{A E} = c_1$, $\text{A B} = b_1$, $\text{C F} = c$, $\text{C D} = b$, the lower end G H being of the same dimensions as the upper, show that when this girder sustains a uniform pressure throughout the whole of its length the deflection at the middle point is given by the formula

FIG. 137.

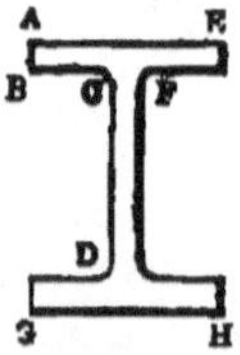

$$\delta = \frac{5\, w\, a^4}{32\{6\,(b + b_1)^2\, b_1\, c_1 + 2 b_1{}^3\, c_1 + b^3 c\}\,\text{E}}$$

Ex. 502.—If there are two beams containing the same amount of materials, of the same length and the same depth, and sustaining the same weight, the one has a rectangular section, the other a section of the form shown in the last example; given that $b = 4$ in., $c = 1$ in., $b_1 = 1$ in., $c_1 = 4$ in., show that the deflection of the rectangular beam will be $\frac{14}{9}$ of the deflection of the other beam.

99. Equation of three moments.

A very interesting application of Prop. 21 is to the determination of the pressures exerted by a beam or a rod on its points of support when there are more than two of them.

For this purpose it is convenient to investigate a relation between the bending moments at any three

---

* In practice the lower flange is commonly made much larger than the upper, since cast iron offers more resistance to pressure than to tension, and of course the greatest economy of materials is effected when the load that would tear the lower flange would also crush the upper. To discuss this question would lead us beyond our present limits.—See Mr. Moseley's *Mechanical Principles*, p. 556; Mr. Rankine's *Applied Mechanics*, p. 319; see also Mr. Fairbairn's *Useful Information*, Append. I.

EQUATION OF THREE MOMENTS. 195

consecutive points of support, which may be done as follows:—

Let A be any one of the points, A_1 and A_2 the points next to it on either side, and let the bending moments at them respectively be M, M_1, M_2; let $A_1 A$ be denoted by l_1, $A A_2$ by l_2, the corresponding weights per unit of length by w_1 and w_2, and the sum of the forces to the right of A_1 (including the reaction of A_1) by Q_1. Then it follows from *Ex.* 477 that

FIG. 138.

$$M = M_1 + Q_1 l_1 + \tfrac{1}{2} w_1 l_1^2 \qquad (1)$$

If we reckon the moments of the weights positive and if $A P$ be denoted by x, the bending moment about P (*Ex.* 477) gives the equation

$$E A K^2 \frac{d^2 y}{dx^2} = M_1 + Q_1 (l_1 - x) + \tfrac{1}{2} w_1 (l_1 - x)^2 \ldots (a)$$

In integrating this equation we must remember that the neutral axis at A need not be horizontal; all that we know is that the curve has some determinate but unknown inclination at that point, which we will denote by a; hence, when $x = 0$ we have $y = 0$ and $\dfrac{d y}{d x} = \tan a$; also when $x = l_1$ we have $y = 0$. On integrating twice, therefore, we obtain

$$E A K^2 y = E A K^2 x \tan a + \tfrac{1}{2} M_1 x^2 + Q_1 (\tfrac{1}{2} l_1 x^2 - \tfrac{1}{6} x^3) \\ + \tfrac{1}{2} w_1 (\tfrac{1}{2} l_1^2 x^2 - \tfrac{1}{3} l_1 x^3 + \tfrac{1}{12} x^4)$$

and therefore

$$0 = 24 \, E A K^2 \tan a + 12 M_1 l_1 + 8 Q_1 l_1^2 + 3 w_1 l_1^3 \qquad (2)$$

If now Q_2 is the sum of the forces to the left of A_2 we have from *Ex.* 477

$$M = M_2 + Q_2 l_2 + \tfrac{1}{2} w_2 l_2^2 \qquad (3)$$

o 2

and if we denote AP_1 by x_1 we shall obtain as before, by considering the bending moment about P_1—

$$E A K^2 \frac{d^2y}{dx^2} = M_2 + Q_2(l_2 - x) + \tfrac{1}{2} w_2 (l_2 - x)^2$$

On integrating this twice we shall obtain, as before—

$$0 = 24 \, E A K^2 \tan \beta + 12 M_2 l_2 + 8 Q_2 l_2^2 + 3 w_2 l_2^3 \quad (4)$$

Now, as there is no sudden change of direction of the curve at A, we have $a + \beta = 180°$, and consequently $\tan a + \tan \beta = 0$. Therefore, by adding (2) and (4), we obtain

$$12 (M_1 l_1 + M_2 l_2) + 8 (Q_1 l_1^2 + Q_2 l_2^2) + 3 (w_1 l_1^3 + w_2 l_2^3) = 0$$

From (1) and (3) we obtain

$$M (l_1 + l_2) = M_1 l_1 + M_2 l_2 + Q_1 l_1^2 + Q l_2^2 + \tfrac{1}{2} (w_1 l_1^3 + w_2 l_2^3)$$

and therefore

$$8 M (l_1 + l_2) + 4 (M_1 l_1 + M_2 l_2) = w_1 l_1^3 + w_2 l_2^3$$

an equation which gives the required relation and is called the *equation of three moments*. It should be observed that the reasoning goes upon the assumptions that the points of support are accurately in a horizontal line, and that the moment of inertia of the cross section of the beam is the same at all points of its length, but the load need not be at the same rate per unit of length on the parts of the beam between different points of support.

By means of this equation the reaction at each point of support can be determined without ambiguity. Suppose there are n points of support 1, 2, 3, 4, ... we can express the bending moment at each point in terms of the weights and the reactions, and we can apply the equation of three moments first to 1, 2, 3, then to 2, 3, 4, then to 3, 4, 5, and so on, thereby obtaining $n-2$ relations between the reactions and known quantities. The two equations of equilibrium between the parallel forces acting on the beam (Prop. 12) give two more equations, and thus we have n equations between the n unknown reactions, which are thereby completely determined. In most particular cases the process can be abridged.

When the reactions have been determined, the circumstances of the flexure of any portion of the neutral line can be determined by means of the equation (a).

Ex. 503.—A beam is supported on five equidistant props (A, B, C, D, E), one being under each end; find the pressures (P, Q, R, S, T) on the points of support.

Here we need five equations, and they could be easily formed as above explained, but we may assume as evident that P is equal to T, and Q to S;

so that $\qquad 2\text{P} + 2\text{Q} + \text{R} = 4aw$

where $4a$ is the length of the beam and w its weight per unit of length. Now if M, M_1, M_2, M_3, M_4 are the bending moments at A, B, C, and D we shall have $\text{M} = 0 = \text{M}_4$, $\text{M}_1 = \frac{1}{2}a^2w - a\text{P} = \text{M}_3$, and $\text{M}_2 = 2a^2w - a\text{Q} - 2a\text{P}$. The equation of three moments must now be applied to M, M_1, and M_2, and then to M_1, M_2, and M_3. In doing this the student must observe the unavoidable change of notation; in Art. 99 M is the moment at the intermediate point, and M_1 and M_2 at the extreme points, so that, as $l_1 = l_2 = a$, the first application gives

$\qquad\qquad 16\ \text{M}_1 a + 4\ (\text{M} + \text{M}_2)\ a = 2wa^3$

or $\qquad\qquad 8\text{M}_1 + 2\text{M}_2 = wa^2$

the second application gives

$\qquad\qquad 16\ \text{M}_2 a + 4\ (\text{M}_1 + \text{M}_3)\ a = 2wa^3$

or $\qquad\qquad 8\text{M}_2 + 4\text{M}_1 = wa^2$

Hence $\qquad\qquad 12\text{P} + 2\text{Q} = 7aw$

and $\qquad\qquad 20\text{P} + 8\text{Q} = 17aw$

Therefore $\qquad 28\text{P} = 11aw,\ 7\text{Q} = 8aw,\ 14\text{R} = 13aw.$

Ex. 504.—A beam is supported on three points, one under each end and one in the middle; find the pressure on each point of support.

[If the pressures are P, Q, P, and the moments are denoted as in Art. 99, we have

$\qquad\qquad 2\text{P} + \text{Q} = 2aw$

$\text{M}_1 = \text{M}_2 = 0$ and $\text{M} = \frac{1}{2}a^2w - a\text{P}$; and then, on applying the equation of three moments, we obtain $8\text{P} = 3aw$, and $4\text{Q} = 5aw$.]

Ex. 505.—In the last example required the equation to the neutral line, the point at which the deflection is greatest, and the amount of the same.

[If the middle point is taken as the origin of co-ordinates, it can be easily shown that the bending moment at a point distant x from the origin is $\tfrac{1}{8}w(a^2-5ax+4x^2)$, whence the equation to the neutral line is easily determined, observing that when $x=0$ we have $y=0$ and $\dfrac{dy}{dx}=0$, and that at the point of greatest deflection we shall also have $\dfrac{dy}{dx}=0$; whence we shall obtain $16x=(15-\sqrt{33})a$ for the position of the required point, and for the amount of greatest deflection (D) we have (nearly) $185\,\text{E}\,\text{A}\,\text{K}^2\,\text{D} = wa^4$.]

Ex. 506.—A beam rests on four points, one under each end and the other two equidistant from the middle point; find the pressures on the points of support.

[If the distances between the props are b, $2a$, and b, and the pressures on them P, Q, Q, P, it is evident that

$$\text{P}+\text{Q}=(a+b)w$$

and it follows from the equation of three moments that

$$8b(3a+b)\text{P}=w(3b^3+12ab^2-8a^3).]$$

Ex. 507.—In the last example explain the result arrived at by making $3b=2a$.

100. *Strength of beams.*—On this subject we may ask either of two questions—(*a*) What is the greatest load applied in a given way that the beam will support with safety? (*b*) What is the load applied in a given way that will break the beam? In either case we must ascertain the point of the beam at which the bending moment is greatest, for it is evident that if the beam is strong enough at this point it is strong enough at all points; and if the load is gradually increased it will break at this point. With regard to the first question, suppose it to be ascertained that the material can be safely stretched $1-n$th part of its natural length. Suppose that fig. 134 shows the section at which the bending moment is greatest, E G is the fibre that is most stretched, and if the load is such that E G is longer than H K by $1-n$th part of H K the load is the greatest consistent with safety.

Now   E G : H K :: E O : O H

therefore   E G − H K : H K :: E H : O H

STRENGTH OF BEAMS. 199

or denoting E H by b_1 and O H by ρ, we have
$$\rho = n b_1$$
and therefore (by Prop. 21)

$n b_1$ (greatest bending moment) $= $ E A K^2

Now if P is the greatest tension per unit of area of cross section that can be applied to the material with safety we have (Art. 6) n P $=$ E;

therefore b_1 (greatest bending moment) $=$ P A K^2

We may reason in the same manner on the greatest pressure (Q) per unit of area of cross section that the material can sustain safely, and we shall obtain

b_2 (greatest bending moment) $=$ Q A K^2

where b_2 denotes H D. Whichever of these equations gives the smallest value of the bending moment will give the greatest value of the load that can be borne safely.

If we suppose the cross section of the beam to be rectangular, and that the greatest pressure per unit of area equals the greatest tension per unit of area that the material can support safely, we shall have, denoting either by Q

greatest bending moment $= \frac{1}{6}$ Q A b

With regard to the second question, let there be two beams of the same material acted on by any transverse forces, and at the sections where the bending moment is greatest suppose the fibres which are most elongated to be equally stretched, i.e. suppose n to have the same value in both cases. Then if ρ' and b'_1 denote in the case of the second beam the quantities corresponding to ρ and b_1 in the first we must have

$$\frac{b_1}{\rho} = \frac{1}{n} = \frac{b'_1}{\rho'}$$

and therefore by Prop. 21

$$\frac{b_1 \text{ (bending moment)}}{A K^2} = \frac{b'_1 \text{ (bending moment)}}{A' K'^2}$$

If either of these beams is on the point of breaking, the other will be on the point of breaking also, for at the weakest part of the beam both are equally stretched. Suppose the second beam to be a foot long, and to have for a cross section a square of one inch; let it be supported at two points, one under each end, and let the force (P) that will just break it when applied to its middle point be found by actual experiment; then, taking all the measurements in inches, the right-hand side of the above equation reduces to 18 P. If we denote this by s we have what is called the modulus of rupture for the material, and we see that when the beam is about to yield at any point by cross breaking

$$\text{the bending moment} = S A \times \frac{K^2}{b_1}$$

It will be observed that the reasoning by which this formula is arrived at assumes that the immediate cause of the rupture is the yielding of the under side of the beam to tension. If the beam gave way by the yielding of the upper side to pressure, precisely similar reasoning could be applied, but b_1 would denote the depth of the neutral axis below the upper surface.

TABLE XIV.*

MODULUS OF RUPTURE.

Substance	Lbs. per Square Inch	Substance	Lbs. per Square Inch
Oak (English)	10,032	Fir (Riga)	7,110
Larch	4,992	Elm	6,078

* From Mr. Moseley's *Mechanical Principles of Engineering*, p. 622.

STRENGTH OF BEAMS. 201

Ex. 508.—A rectangular beam 6 in. deep and 3 in. wide rests horizontally on two points 10 ft. apart; it can be safely subjected to either pressure or tension at the rate of 1,000 lbs. per square inch. What is the greatest weight that can be safely hung from its middle point, its own weight being neglected?

Let P denote the required weight; the greatest bending moment, being at the middle point, will be 30P, the units being pounds and inches; hence

$$3 \times 30\text{P} = 1000 \times 18 \times \tfrac{1}{12} \times 6^2$$

therefore P = 600 lbs.

It will be found that the depression caused by P at the middle point is $\tfrac{2}{5}$ths of an inch, if the modulus of elasticity is 1,000,000 lbs. per square inch.

Ex. 509.—In the case of a rectangular beam, whose weight is neglected, show that the breaking load is as follows:—(a) $\dfrac{sbc}{6} \cdot \dfrac{b}{a}$, when it is held firmly at one end and loaded at the other; (b) $\dfrac{2sbc}{3} \cdot \dfrac{b}{a}$, when supported under the two ends and the load is applied at the middle.

Ex. 510.—In the last example, if the loading is distributed uniformly over its length, show that the breaking load is (a) $\dfrac{sbc}{3} \cdot \dfrac{b}{a}$; (b) $\dfrac{4sbc}{3} \cdot \dfrac{b}{a}$.

Ex. 511.—Given a cylindrical log of wood, show that the strongest rectangular beam that can be cut out of it is one whose sides are in the ratio of 1 : $\sqrt{2}$.

Ex. 512.—A beam of oak is supported in a horizontal position on points 20 ft. apart, it is 3 in. deep and 4 in. wide; determine the weight that can be suspended at a distance of $6\tfrac{2}{3}$ ft. from one point of support without breaking it. What would be the magnitude of the weight if the depth were 4 in. and breadth 3 in.? *Ans.* (1) 1128·6 lbs. (2) 1504·8 lbs.

Ex. 513.—What must be the depth of a beam of Riga fir 4 in. wide and 30 ft. long, that will just sustain a weight of $\tfrac{1}{4}$ a ton at its middle, taking into account its own weight? *Ans* 4·6 in.

CHAPTER X.

VIRTUAL VELOCITIES—MACHINES IN A STATE OF UNIFORM MOTION—TOOTHED WHEELS.

101. *The principle of virtual velocities.*—Let P be a force acting at the point A along the line A P, and let it be represented by A C (Art. 25). Suppose P's point of application to be shifted through an indefinitely small distance to B, draw B n at right angles to A C or C A produced, and let A n be denoted by p, which is commonly reckoned positive when n falls between A and C, and negative when it falls on C A produced, then p is called the virtual velocity of P, and Pp its virtual moment or virtual work.

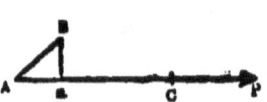

FIG. 139.

The principle of virtual velocities is as follows:—If a system of forces in equilibrium act on any machine which receives any small displacement—consistent with the connection of the parts of the machine—the algebraical sum of the virtual works of the forces will equal zero.

If $P_1, P_2, P_3 \ldots$ are the separate forces, and $p_1, p_2, p_3 \ldots$ their virtual velocities, the principle is expressed algebraically by the following equation, which is commonly called the equation of virtual velocities:

$$P_1 p_1 + P_2 p_2 + P_3 p_3 + \ldots = 0$$

It must be remarked that in the above definition the line A B is considered a small quantity of the first order

(App. Art. 3), and consequently the virtual works $P_1 p_1$, $P_2 p_2$, $P_3 p_3$ ... are in general of the first order; if, however, the virtual velocity of the point of application of any one of the forces be of the second order, the virtual work of that force will vanish in comparison with the virtual works of the other forces and will disappear from the above equation; this will happen in the following cases:—(a) When AB is ultimately at right angles to AC—e.g. when AC is the normal to a curve of which AB is a chord—hence the virtual work of the reaction of a smooth surface equals zero when the body slides along the surface; (b) when the points A and B coincide, e.g. when AC is a portion of a rigid body in the act of turning round the point A, i.e. the virtual work of the reaction of a fixed axis is zero provided the axis can be treated as a line; hence also when an incompressible body rolls without sliding on any surface, rough or smooth, the virtual work of the reaction equals zero.

The principle now enunciated will be seen from the following pages to be one of very great importance in the theory of machines; as the general proof is not by any means easy it will be useful for the student to prove from first principles that it holds good in a few elementary cases.

Ex. 514.—If X and Y are the rectangular components of a force P, show that the virtual work of P equals the sum of the virtual works of X and Y.

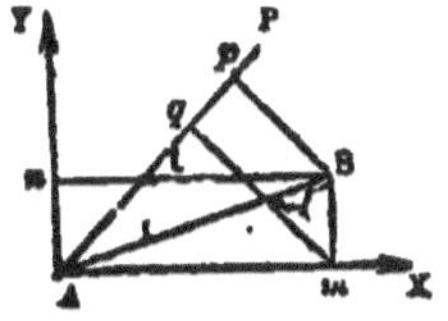

Let A be the point of application of P, and let A be transferred to B; complete the rectangle mn, and draw Bp and mq at right angles to AP; then Ap, Am, An, are the virtual velocities of P, X, and Y, and we have to prove that

FIG. 140.

$$P \cdot Ap = X \cdot Am + Y \cdot An$$

Let XAP be denoted by θ, then it is evident that
$$Ap = Aq + qp = Am \cdot \cos\theta + An \sin\theta$$
therefore
$$P \cdot Ap = Am \cdot P \cos\theta + An \cdot P \sin\theta$$
or
$$P \cdot Ap = X \cdot Am + Y \cdot An \qquad (1)$$

If P had acted in the contrary direction, X, Y, and P would have been in equilibrium; the virtual work of P would be negative; and (1) would become the equation of virtual velocities.

Ex. 515.—In the last example suppose that P balances X and Y, and suppose its point of application to be transferred in a direction at right angles to A P, verify the equation of virtual velocities.

[It must be remembered that in this case P's virtual work equals zero.]

Ex. 516.—Show that the principle of virtual velocities is true in the case of a body in the state bordering on motion up an inclined plane, when a small motion is given to it either up or down the plane.

FIG. 141.

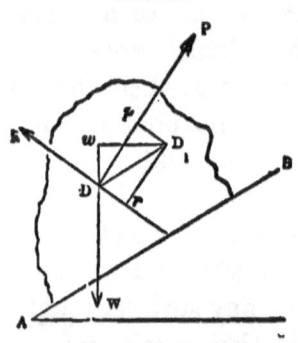

[Draw the figure as in *Ex.* 346, then, if the motion take place up the plane, D will be transferred to a point D_1 along a line $D D_1$ parallel to A B; let fall from D_1 perpendiculars on the directions of the forces, viz. $D_1 w$, $D_1 p$, $D_1 r$, then $D w$, $D p$, $D r$, are the virtual velocities of the forces, and of them $D p$ is positive and the others negative; the equation of virtual velocities therefore becomes

$$P . D p = W . D w + R . D r$$

and this the student is required to prove.]

Ex. 517.—Verify the principle of virtual velocities in the last case, assuming that the plane (and with it the body) is so moved that D describes a straight line at right angles to D R.

Ex. 518.—Verify the principle of virtual velocities in the case of two forces in equilibrium on a straight bar capable of turning round a fixed point.

[Let P and Q be the forces which balance on the rod A B round the fixed point C; suppose the rod to turn through a small angle and to come into the position A′ B′; draw A′ m at right angles to A P and B′ n at right angles to B Q, then A m is the virtual velocity of P and B n of Q, the latter being negative; also the virtual work of the reaction of C is zero (Art. 101); the equation to be proved is therefore

FIG. 142.

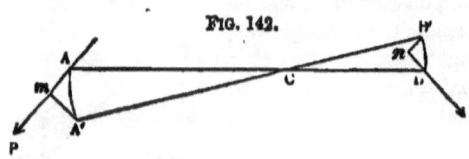

$$P . A m = Q . B n$$

The student must remember that A A′ m and B B′ n are ultimately right-angled triangles.]

VIRTUAL VELOCITIES. 205

Ex. 519.—Verify the principle in the case of two parallel forces P and Q which keep a beam at rest round a rough axle of finite dimensions (as in *Ex.* 405), the motion being given to the beam round the axle.

[Using the notation of *Ex.* 405 and calling θ the small angle through which the beam is turned, the virtual works are severally $P p \theta$, $W q \theta$, and $R \rho \theta \sin \phi$.]

Ex. 520.—In the last example how would it be possible to move the system so that the reaction R should disappear from the equation of virtual velocities? [Round the point Q, fig. *g*.]

Ex. 521.—In *Ex.* 519 show that when the axle is smooth the reaction will disappear from the equation of virtual velocities.

102. *Proof of the principle of virtual velocities.*—The following proof applies to the case of any system of forces acting on a single rigid body and in one plane, in which the displacement is supposed to be made: it can be easily extended so as to include every case of forces that act on any machine.

LEMMA.—*Let* A *and* X *be any two points in a given line, let the line be transferred to any consecutive position* O Y, *so that* A *comes to* B *and* X *to* Y; *then if* B Y *equals* A X, *and if* B *n and* Y *m are drawn at right angles to* A X, *the line* A *n will ultimately equal* X *m.*

FIG. 143.

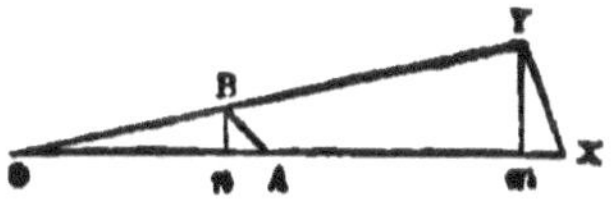

For $n\,m$ equals B Y cos O, i.e. it ultimately differs from B Y, and therefore from A X, by a small quantity of the second order; take away the common part A *m*, then A *n* and X *m* ultimately differ by a small quantity of the second order, but they are themselves of the first order, and therefore are ultimately equal. (See App. I., Art. 3.)

N.B.—If A X be transferred to B Y in such a manner that either A *n* or X *m* is of an order higher than the first, then will the other also be of an order higher than the first; e.g. if A O is a small quantity of the first order, and B A O a finite angle, A B and A *n* are both of the second order; likewise A X Y is ultimately a right angle, and consequently X *m* is also of the second order.

Cor.—Hence if a force act along a certain line, and if two points in the line be rigidly connected, its virtual velocity will be the same at whichever point we suppose it to act; also if there be two equal and opposite forces, their virtual works will be equal and have contrary signs, whether we suppose them to act at the same point or each at one of two rigidly connected points, e.g. Suppose P to act along A x (fig. 143); if it act at A its virtual work is P . A n, if it act at x its virtual work is P . x m; consequently in either case its virtual work is the same. If A n is of the second or some higher order, x m is not of the first order, and in either case the virtual work is zero.

We can now proceed with the general demonstration required, and this is given in the three following steps :—

(a) If a system of parallel forces acting in a given plane have a resultant, and if the points on which the forces and their resultant are supposed to act be rigidly connected, then the sum of the virtual works of the forces will equal the virtual work of the resultant.

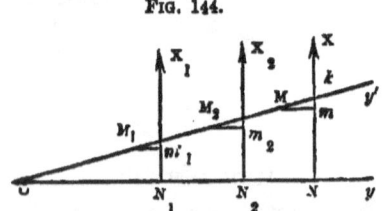

Fig. 144.

Let x_1, x_2, ... be the forces, x their resultant, draw a line (o y) at right angles to their directions, and cutting them in n_1, n_2, . . . n, and suppose these points to be rigidly connected with those at which the forces are supposed to be applied, then the virtual works of the forces in the required case are severally equal to their virtual works if supposed to act at n_1, n_2 . . . . n. Now, suppose these points to receive any small displacement consistent with their rigid connection, and suppose them to be transferred to m_1, m_2 . . . m, these points will be in a straight line (o y') and their mutual distances will be

VIRTUAL VELOCITIES. 207

the same as before; the two lines will (generally) intersect in some point O. Draw $M_1 m_1$, $M_2 m_2$, ... $M m$, at right angles to the directions of the forces, then their virtual velocities are respectively $N_1 m_1$, $N_2 m_2$, ... $N m$. Let the angle $y O y'$ be denoted by θ, and $O N_1$, $O N_2$ .... $O N$, by y_1, y_2, ... y, then it is plain that ultimately *

$$N_1 m_1 = y_1 \theta,\ N_2 m_2 = y_2 \theta, \ldots \ldots N m = y \theta$$

But by Prop. 12 we have

$$X_1 y_1 + X_2 y_2 + \ldots \ldots = X y$$

and therefore

$$X_1 y_1 \theta + X_2 y_2 \theta + \ldots \ldots = X y \theta$$

i.e. the sum of the virtual works of the forces equals the virtual work of their resultant in the case specified.

(b) Next, let us consider the case of any system of forces P_1, P_2, P_3 .... acting in one plane on points rigidly connected.

Resolve the forces in directions respectively parallel to two rectangular axes, then P_1 will be equivalent to its two components X_1, Y_1, and similarly P_2 to X_2, Y_2, P_3 to X_3, Y_3, &c., and the original system is divided into two systems of parallel forces, viz. X_1, X_2, X_3 ... and Y_1, Y_2, Y_3 ... ; let X be the resultant of the former system and Y of the latter, and let their directions intersect at a certain point A, then the direction of their resultant (R) will pass through A, and R will be the resultant of P_1, P_2, P_3 ..... Suppose A to be rigidly connected with the other points, and suppose X, Y, and R to act at A. Now, if the points of application of the forces receive any displacement whatsoever, the virtual work of R equals the sum of the virtual works of X and Y (*Ex.* 514), i.e. (by *a*) equals the sum of

* For let $o y'$ cut $N x$ in k, we shall have $N m = y \tan \theta - M m \tan \theta$, but $M m$ and $\tan \theta$ are small quantities of the first order, so that their product is of the second order, and can therefore be neglected, i.e. $N m$ ultimately equals $y \tan \theta$ or $y \theta$.

the virtual works of $x_1, x_2, x_3 \ldots$ and of $y_1, y_2, y_3 \ldots$; but (*Ex.* 514) the virtual work of P_1 equals the sum of the virtual works of x_1 and y_1, and similarly of $P_2, P_3, \ldots$; hence the virtual work of R equals the sum of the virtual works of $P_1, P_2, P_3 \ldots$; or

$$R\,r = P_1\,p_1 + P_2\,p_2 + P_3\,p_3, \perp \ldots$$

(*c*) If $P, P_1, P_2, P_3, \ldots$ are forces in equilibrium acting in one plane at points of a rigid body, and if that body receive any small displacement, the sum of the virtual works of the forces will equal zero.

For let R be the resultant of $P_1, P_2, P_3, \ldots$ and let it act on the body at any one point in its direction, then (by *b*)

$$P_1\,p_1 + P_2\,p_2 + P_3\,p_3 + \ldots \qquad = R\,r$$

But R is equal and opposite to P, since the given forces are in equilibrium, and hence, since R and P act on rigidly connected points, we have by the corollary to the lemma

$$P\,p + R\,r = 0$$

and therefore, by addition,

$$P\,p + P_1\,p_1 + P_2\,p_2 + P_3\,p_3 + \ldots \qquad = 0$$

Q. E. D.

103. *The work done by a force.*—If the student turn to Art. 11 he will see that the definition therein given might be stated more generally as follows:—*a unit of work* is the work done by a force of one unit when its point of application moves through a unit of distance in the direction of the force; it follows that, when the point of application of a force of P units moves through s units of distance in the direction of the force, P s units of work are done. When the units are pounds and feet it is convenient to call the unit of work a *foot-pound*. We have now to consider the extension of the definition which must be made to meet the case of a force

DEFINITION OF WORK.

whose point of application moves in any manner whatsoever. The required extension will be readily made by observing that *if the point of application of a force receives any small displacement, the virtual work of the force is the work done by the force during the displacement.* The justice of this statement can be illustrated (or proved) by the consideration of the following simple case:—

Let W be a weight attached to the end E of a perfectly flexible and inextensible string without weight, passing over a smooth point C; let W be balanced by a force P acting at A along C A, then will P equal W; now, suppose A to be transferred through a small distance to B, draw B n at right angles to C A, produced, then will W be raised from E to D, and A n is ultimately equal to D E. Now, the work expended in raising W is W × D E, i.e. it ultimately equals P × A n, the virtual work of P.

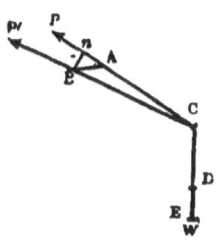
FIG. 145.

Next, let us suppose that the point of application of P is transferred successively to points A, A′, A″, A‴, . . . the successive directions of that force being A P, A′P′, A″P″, . . .; let fall on them the perpendiculars A′N, A″N′, A‴N″, . . . and let AN, A′N′, A″N″ . . . be denoted by $p, p', p'', \ldots$ then the work done by P when its point of application is transferred from A to A′ is its virtual work Pp, and the work done during the successive transfers will be P′p', P″p'', P‴p''', ... and the whole work done will be Pp+P′p'+P″p''+. . . . . whether the successive values of P be the same or not. As, however, this is somewhat general, it will be well to particularise two important cases.

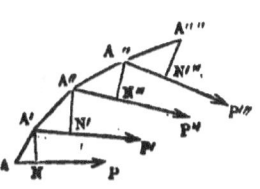
FIG. 146.

(*a*) Let the force continue constant, then if the lines

P

along which it successively acts are parallel, and if the point of application of the force moves in any line straight or curved, the work done will equal the product of the force and the projection of the line on the direction of the force; e.g. take the case of a crank whose arm is a, the extremity of which describes a circle whose diameter is $2a$, let P be the driving force acting along the connecting rod, which we may suppose to be so long as to be virtually parallel throughout its motion, then the work done by P in one revolution is $4\text{P}a$.

(b) Let the force continue constant, then if the direction of the force always touches the curve described by its point of application, the work done will equal the product of the force and the length of the curve, e.g. Suppose a winch whose arm is a to be turned by a force P acting at right angles to the arm, then the work done by P in one turn of the winch will be $2\pi a\text{P}$.

It must be remarked that the virtual work of a force may be either positive or negative, and hence the work done by a force may be either positive or negative; in the latter case, however, it is perhaps better to speak of the work as being expended on or done against the force.

It is scarcely necessary to remark that a force will do no work, in the cases in which its successive virtual works are zero (Art. 101). Another case may also be specified:—A rigid body may be conceived as consisting of a number of points connected by their mutual attractions which act along the lines joining them, and which are so great that the points undergo no relative displacement from the action of the external forces; under these circumstances the sum of the virtual works of each pair of mutual attractions will equal zero (Art. 102 *Cor.*), and therefore the work done by the whole system of internal forces must equal zero. If, however, the body is either compressed or extended, the work done by or expended on the internal forces can be no longer neglected (*Ex.* 149).

MACHINES IN UNIFORM MOTION. 211

104. *Machines in a state of uniform motion.*—Suppose any machine to be acted on by forces P, P_1, P_2, P_3, ... in equilibrium, and suppose the machine to be slightly moved, then if $p, p_1, p_2, p_3, \ldots\ldots$ are the virtual velocities of the forces respectively, we shall have

$$P p + P_1 p_1 + P_2 p_2 + P_3 p_3 + \ldots \quad = 0 \quad (1)$$

In the new position of the machine, suppose the forces, without undergoing any change of magnitude, to be in equilibrium, and suppose the machine to receive a second displacement, then if $p', p_1', p_2', p_3', \ldots$ are the virtual velocities of the forces we shall have

$$P p' + P_1 p_1' + P_2 p_2' + P_3 p_3' + \ldots \quad = 0 \quad (2)$$

Suppose that in this second position the forces are in equilibrium and that the machine receives a third displacement, then if $p'', p_1'', p_2'', p_3'' \ldots$ are their virtual velocities we shall have

$$P p'' + P_1 p_1'' + P_2 p_2'' + P_3 p_3'' + \ldots \quad = 0 \quad (3)$$

and so on for any number of displacements. Hence by addition

$$P(p + p' + p'' + \ldots) + P_1(p_1 + p_1' + p_1'' + \ldots) + P_2(p_2 + p_2' + p_2'' + \ldots) + P_3(p_3 + p_3' + p_3'' + \ldots) + \ldots = 0 \text{ (A)}$$

Now, if we suppose p, p', p'', &c., to be positive, $P(p+p'+p''+\ldots)$ is the work done by P; if $p, p', p'', \ldots$ are negative, $P(p+p'+p''+\ldots)$ is the work expended on P; in the former case P would be called a *power*, in the latter a *resistance*, hence the equation (A) contains the following fundamental theorem, viz. *If a machine be in motion and if at each instant of the motion the powers and resistances form a system of forces in equilibrium, the sum of the units of work done by the powers will equal the sum of the units of work expended on the resistances.*

Now, it will be remarked that if the machine be in

P 2

motion all change of its motion must be due to an excess of the powers over the resistances or of the resistances over the powers; hence, in the case supposed, there can be no change in the motion of the machine at any instant; such a machine moves uniformly,* and hence the theorem above proved justifies the assertion made in Art. 14, viz. that the number of units of work done by the agent equals the number expended on prejudicial resistances, together with the number expended usefully.

105. *The modulus of a machine.*—Let us assume that the machine enables a certain force or *power* P to overcome a second force or *weight* Q, then the relation between P and Q can generally be expressed by means of an equation of the form

$$P = A Q + B \qquad (1)$$

where A and B are numbers depending on the form of the machine, and on the passive resistances (compare Art. 89). Now, by considerations depending on the form of the machine, there will be some fixed relation between the distance (s_1) described by P's point of application and (s_2) the distance described by Q's point of application, let then

$$s_1 = n\, s_2 \dagger \qquad (2)$$

By multiplying (1) and (2) together we obtain

$$P\, s_1 = n\, A\, Q\, s_2 + B\, s_1 \qquad (3)$$

But $P s_1$ is the work (U_1) done by P, and $Q s_2$ is the work (U_2) expended on Q, hence

$$U_1 = n\, A\, U_2 + B\, s_1 \qquad (4)$$

* If the machine has a motion of translation, like a railway train, its motion is said to be uniform when its velocity undergoes no change; if the machine moves round a fixed axis like a fly-wheel, its motion is uniform if its angular velocity undergoes no change; if it has both motions combined, like the wheels of a carriage, it moves uniformly if neither velocity undergoes any change.

† It can be easily shown in regard to any machine the parts of which move without passive resistances that $n\, P = Q$.

MODULUS OF A MACHINE.

If the machine moves with a uniform motion, the equation (4) gives the number of units of work (U_1) actually done by the *power* while (U_2) is expended on the *weight*. If P and Q are not in equilibrium during the motion, U_1 is still the number that must be *expended* on the weight and resistances; if P does a greater number of units than U_1 the surplus will be accumulated in the machine, the motion of which will be accelerated; if P does a less number than U_1 the difference must be withdrawn from the work previously accumulated, and the motion of the machine will be retarded. The subject of accumulated work will be treated further on.

Ex. 522.—If a body be dragged along an inclined plane show that the units of work expended will equal the number that would be expended in dragging it along the base, supposed equally rough, and in lifting it up the perpendicular height.

Let A B C be the inclined plane, M the body whose weight is Q, P the force, which acting along the plane would be on the point of dragging M up the plane, if M were at rest, then

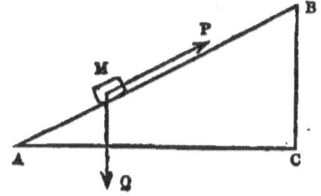

FIG. 147.

$$P \cos \phi = Q \sin (a + \phi)$$
or $$P = Q (\sin a + \mu \cos a)$$

where a denotes the angle B A C, and μ or $\tan \phi$ the coefficient of friction between M and A B. Now, if M is in motion along A B, under the action of P and Q it will move uniformly, and the work done by P will equal the work expended on Q; but the work done by P is P × A B, therefore the work expended on Q equals

$$Q \times AB (\sin a + \mu \cos a)$$
or $$Q \times (BC + \mu \times AC)$$

But μ Q × A C is the work required to drag M along A C, if μ is the coefficient of friction between M and A C, and Q × B C is the work that must be expended in lifting Q from C to B, therefore the number of units of work is as stated. By an exactly similar process it may be shown that the number of units of work required to drag a body *down* a rough inclined plane equals the number required to drag it along the base supposed equally rough *diminished* by the number required to lift the body through the height of the plane.

Ex. 523.—If a train weighs 80 tons and the friction is 7 lbs. per ton, determine the number of foot-pounds of work that must be expended in drawing it for 4 miles up an incline of 1 in 200; and determine the horse-power of the engine that will do this in 10 minutes with a uniform velocity.

Ans. (1) 30,750,720 ft.-pds. (2) $93\frac{2}{11}$ H.-P.

Ex. 524.—In the last example over what distance on a horizontal plane would the same engine have drawn the train in the same time?

Ans. $10\frac{2}{3}$ miles.

Ex. 525.—How long would it take the engine in *Ex.* 523 to draw the same train with a uniform velocity over a space of 4 miles up an incline of 1 in 100?

Ans. $16\frac{9}{13}$ min.

Ex. 526.—A train is drawn with a uniform velocity up an incline 3 miles long of 1 in 250, on which the resistances are 7 lbs. per ton; determine the distance on a horizontal plane over which the same train could be drawn with a uniform velocity by the same expenditure of work.

Ans. $6\frac{21}{25}$ miles.

Ex. 527.—In *Ex.* 346 if the body is in the state of uniform motion up the plane, show that the relation between v_1 the work done by P, and v_2 the work expended on W, is given by the equation

$$v_1 \sin \alpha \cos (\beta - \phi) = v_2 \cos \beta \sin (\alpha + \phi)$$

[The relation between the forces P and W is

$$P \cos (\beta - \phi) = W \sin (\alpha + \phi)$$

Now, if s_1 is the distance through which P's point of application moves measured in the direction of that force

$$s_1 = l \cos \beta$$

and if s_2 is the distance through which W's point of application moves when similarly measured

$$s_2 = l \sin \alpha$$

where l is the length of the plane, hence

$$s_1 \sin \alpha = s_2 \cos \beta$$

whence the relation between v_1 and v_2 is at once found.]

Ex. 528.—If a pivot sustaining a pressure Q is made to revolve once, show that the number of units of work expended on the friction of the end equals $\frac{2}{3} \pi \mu \rho$ Q. [See Art. 82.]

Ex. 529.—In the case of a single fixed pulley the number of units of work expended in raising a weight Q through a height q is given by the formula

$$v = a \, Q \, q + b \, q$$

where a and b have the values assigned in Art. 89.

Ex. 530.—In the case of a tackle of n sheaves show that the number of units of work expended in raising a weight Q through a height q is given by the formula

$$v = Q q \cdot \frac{n a^n (a-1)}{a^n - 1} + \left(\frac{n b a^n}{a^n - 1} - \frac{b}{a-1} \right) n q$$

[See *Ex.* 419.]

MODULUS OF A MACHINE. 215

Ex. 531.—In *Ex.* 421 determine the number of foot-pounds of work expended on the passive and on the useful resistances when the weight of 1000 lbs. is raised through 50 ft.   *Ans.* (1) 67,000.   (2) 50,000.

Ex. 532.—'It is said that in a pair of blocks with five pulleys in each two-thirds of the force are lost by the friction and rigidity of the ropes.' * Determine the degree of truth in this statement when each sheave is 4 in. in radius, and turns on an axle $\frac{1}{4}$ of an inch in radius, the axle being of wrought iron and the bearing of cast iron, and the rope 4 in. in circumference; the weight to be raised being 1000 lbs.

Ans. $\dfrac{\text{Work expended on passive resistances}}{\text{Work done}} = \dfrac{19}{29}$ nearly.

Ex. 533.—In the capstan *Ex.* 427 show that the work that must be done by the forces in order to move the weight Q through a height g is given by the formula

$$U = \left(1 + \frac{r\sin\phi}{b}\right)\left(1 + \frac{B}{b}\right) Q g + \frac{QA}{b}\left(1 + \frac{r\sin\phi}{b}\right) + \frac{2\mu_1 rw}{3} \cdot \frac{Q}{b}$$

Ex. 534.—A rope passes over a single fixed pulley in such a manner that its two parts are at right angles to each other; the one end carries a weight Q; the radius of the pulley is r and of the axle ρ, the angle β such that $\sin\beta = \dfrac{\rho \sin\phi}{r\sqrt{2}}$; then, the weight of the pulley being neglected, show that if P is the force that will just raise Q, we have

$$P = \left(Q + \frac{A + BQ}{}\right) \tan(45° + \beta)$$

Ex. 535.—In the last example show that the relation between P and Q may be very nearly represented by the formula

$$P = Q\left(1 + \frac{B}{r} + \frac{\rho\sqrt{2}}{r}\sin\phi\right) + \frac{A}{r}\left(1 + \frac{\rho\sqrt{2}}{r}\sin\phi\right)$$

Ex. 536.—A weight of 500 lbs. has to be raised from a depth of 50 fathoms; it is fastened to a rope which passes over a fixed pulley in such a manner that the parts of the rope are at right angles to each other; the rope is wound up by means of a capstan which is turned by two equal parallel forces acting at the end of equal arms; the rope is 3 in. in circumference, the pulley 6 in. in effective radius, its axle half an inch in radius, and of wrought iron turning upon cast; the capstan weighs 4 cwt., its axle is 4 in. in radius, oak moving on wrought iron, the effective radius of the capstan 15 in.; determine the number of foot-pounds of work that must be done in order to raise the weight (not weight and rope), and the number expended on passive resistances.   *Ans.* (1) 204,356.   (2) 54,356.

Ex. 537.—There is a fixed pulley 20 inches in radius (r) moving on an axle 1 in. (ρ) in radius ($\sin\phi = 0.15$); a weight of 500 lbs. is raised from a depth of 300 feet (l) by means of a rope 3 in. in circumference which passes

* Dr. Young's *Lectures*, vol. i. p. 206.

over it; the end of the rope falls as the weight rises; determine the error that results from neglecting the weight of the rope in calculating the foot-pounds of work required to raise the weight—the united length of the two hanging parts of the rope being reckoned at 300 ft.

$$\text{Ans. Error} = \frac{0\cdot 405\, l^2\, \rho \sin \phi}{r} = 274.$$

[Compare *Ex.* 141 and 158.]

Ex. 538.—In the last example determine the error that would result from neglecting the weight of the rope if the end were *not* allowed to fall.

Ans. Error 19,000.

Ex. 539.—If a weight Q is raised through a height q by means of a screw, show that if the same notation is employed as in *Ex.* 393 the number of units of work expended is given by the formula

$$\text{U} = \text{Q}\, q\, \left\{ \tan (a + \phi) + \tfrac{2}{3} \cdot \frac{\rho}{r}\, \mu \right\} \cot a$$

where all frictions are neglected except those between thread and groove and on the end of the screw.

Ex. 540.—An iron screw 4 in. in diameter communicates motion to an iron nut, the screw thread is inclined to its base at an angle of 18°, the diameter of the end of the screw is 2 in.; all the surfaces are of cast iron; determine the number of foot-pounds of work that must be expended in raising a weight of 3 tons through a height of 2 ft. by means of this screw.

Ans. 23,358.

Ex. 541.—Determine through what height a man working with this screw could raise a weight of 1 ton in a day; and what would be the best length of the arm of the screw on which he works—pushing horizontally; determine also the part of his work which is expended in overcoming friction.

Ans. (1) 384 ft.   (2) $7\tfrac{1}{4}$ ft.   (3) $\tfrac{3}{7}$.

106. *The end to be attained by cutting teeth on wheels.*—The problem to be solved is this:—Given an axle A, moving with a uniform angular motion round its geometrical axis, it is required to connect it in such a manner with a parallel axis B, as to communicate to it a uniform angular motion which shall have a given ratio to the former. Suppose the axle A to revolve m times in one minute, and it is required to make the axle B revolve n times in one minute; join the centres A and B, divide A B into $m + n$ equal parts, and take A C equal to n of these parts, and therefore B C will contain m of them, so that

$$A C : C B :: n : m$$

TOOTHED WHEELS. 217

with centres A and B, the radii A C, B C respectively, describe circles touching at C; if these circles are fixed each to its own axle, and revolve with them, and if their circumferences are rough, so that they roll on each other, the problem is solved; for take on the circumferences

Fig. 148.

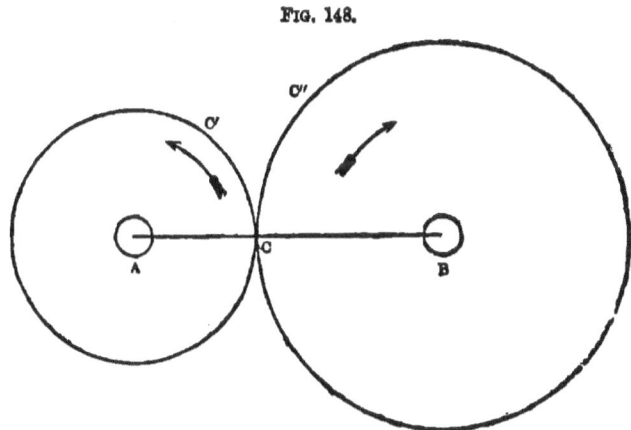

respectively points C' and C'' which were in contact at C, then must the arc C C' equal the arc C C'', since the several points of the arcs have been successively in contact each with each, and this is true whatever be the lengths of those arcs. Now, in one minute the point C' describes an arc whose length is $2\pi A C . m$, and therefore C'' describes an arc whose length is $2\pi A C . m$, i.e. an arc whose length is $2\pi B C . n$, since $A C . m = B C . n$; but $2\pi B C . n$ is n times the circumference of the circle whose radius is B C, and therefore the axle B makes n turns while A makes m turns, i.e. B moves in the required manner.

It is evident that the angular motions will have the same ratio whatever be the time, and therefore when the time is very short; hence if the angular motion of the axle A varies from instant to instant, that of the axle B will also vary, but the ratio of the angular motions will remain constant.

It is also plain that the directions of the angular motions will be contrary, as indicated by the arrow heads.

It may be remarked that the wheel A C is called the driver, and B C the follower.

Ex. 542.—If in the last article a single wheel moving on a parallel axle with its centre in the line A B were interposed between A C and B C, it would cause the follower to revolve in the *same* direction as the driver, and would not produce any change in the ratio of their angular motions, the radii A C and B C being unchanged.

107. *Practical objection to the above solution.*—It is evident that the above solution fails if the surfaces of the wheels rub smooth, so that the motion becomes partly one of sliding and partly one of rolling contact; and also that it will fail if the centres A and B are slightly displaced, since then the contact ceases: one method, in common use, of obviating this objection is to pass a strong band of leather tightly over the wheels; this method is commonly used when the centres A and B are so considerable a distance apart that the wheels would be inconveniently large if in immediate contact; the most effectual means, and the only one with which we are here concerned, is to cut teeth on the circumferences of the wheels; when this is properly done the uniform revolution of the wheel A can be made to communicate a uniform revolution to the wheel B. The problem we have to solve is therefore twofold:—

(1) To determine the form that must be given to the teeth of wheels, in order that any uniform motion of the driver round its axis shall communicate to the follower a *uniform* motion round its axis.

(2) As this cannot be done without causing the teeth of the one wheel to *slide* over those of the other, it is required to determine what amount of work is lost by the friction of the teeth when work is transmitted from one axle to the other.

The limits of the present work will not allow us to do

more than give one solution of the former question, and an approximate solution of the second. Readers who desire further information on this very important subject will be able to obtain it by reference to Mr. Willis's *Principles of Mechanism*, and to Mr. Moseley's *Mechanical Principles of Engineering*: * the former work treats only of the question of *form*; the latter also contains a very full discussion of the question of *force*.

108. *Definition and properties of the epicycloid.*—If a circle carrying on its circumference a pencil-point be made to roll on the outside of the circumference of a fixed circle, the point will trace out a curve called an *epicycloid*: the fixed circle is called the *base*; the moving circle is called the *generating circle*. Thus if Q is a point on the generating circle A D Q, and A P C is the base or fixed circle, then if Q were in contact with A P C at P, the point Q will trace out the epicycloid P Q.

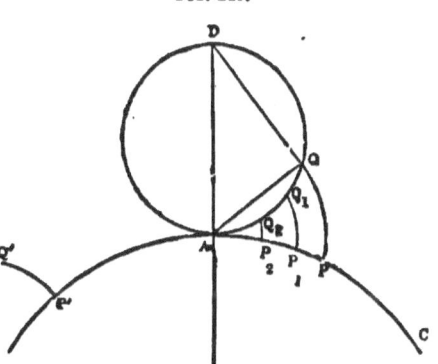

FIG. 149.

(*a*) It is evident that the length of the arc A Q equals that of the arc A P.

(*b*) It is evident that the point Q is at the instant moving in a circle of which the centre is A, and radius A Q, so that the line A Q is the normal to the epicycloid at the point Q, and if D Q be joined that line is a tangent to the curve at Q.

(*c*) It is evident that the form and dimensions of the curve are independent of the particular point Q occupies

* A very clear elementary discussion of the forms of the teeth of wheels will be found in Mr. Goodeve's *Elements of Mechanism*.

on the generating circle, so that if we take a succession of points Q, Q_1, Q_2 ... on the generating circle, and describe with them a succession of epicycloids Q P, Q_1 P_1, Q_2 P_2 ... they will all be exactly like one another, and if P' Q' be any epicycloid described on the same base with the same generating circle as the others, it too will be exactly like the rest: if we now suppose all the former to remain fixed, and the circle P' A C to revolve round its centre, carrying P' Q' with it, then when P' comes to P_2, the curve P' Q' will exactly cover P_2 Q_2, and in like manner it will successively cover P_1 Q_1 and P Q.

Proposition 22.

An epicycloidal tooth can be made to work correctly with a straight tooth.

Let PQ be the tooth described on the base AP, the centre of which is O_1, by a circle whose diameter is A O; suppose the base to revolve round O_1 and let the tooth assume successively the positions $p_1 q_1$, $p_2 q_2$, $p_3 q_3$ .... cutting the circle ADO in points q_1, q_2, q_3, then since the straight lines Oq_1, Oq_2, Oq_3 .... touch the epicycloid in the points q_1, q_2, q_3 .... it is plain that a straight line whose length is O A, and which is movable

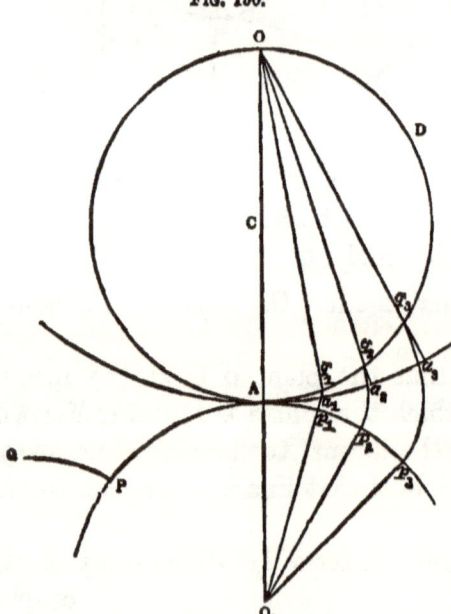

FIG. 150.

round O, will, if driven by the tooth, come successively

RULE FOR FORM OF TEETH. 221

into the positions $O\,a_1, O\,a_2, O\,a_3, \ldots$ passing through the points $q_1, q_2, q_3 \ldots$ respectively. Now, if we suppose the angles $A\,O_1\,p_1, p_1\,O_1\,p_2 \ldots$ to be equal, the arcs $A\,p_1$, $p_1\,p_2, p_2\,p_3 \ldots$ are equal, and therefore (Art. 108 (a)) the arcs $A\,q_1, q_1\,q_2, q_2\,q_3 \ldots$ are equal, and the angles they subtend at the centre O will be equal, and their halves will be also equal, i.e. the angles $A\,O\,a_1, a_1\,O\,a_2, a_2\,O\,a_3$, are equal; so that if the circle $P\,A\,O_1$ move with a uniform angular motion, it will communicate a uniform angular motion to a straight line $A\,O$ movable about the point O, i.e. the straight line works truly with the epicycloidal tooth.

Ex. 543.—If with centre o and radius $O\,A$ a circle be described, show that if this circle work with $A\,P$ by friction, any one of its radii will have the same angular velocity as if it had been driven by the tooth $P\,Q$.

109. *Practical rule for the form of teeth.**—Let O, O_1 be the centres of the two toothed wheels; draw the line of centres $O\,O_1$; when the point of contact of any two teeth is on the line of centres let it be at A; with centres O and O_1 and radii $O\,A$ and $O_1\,A$ respectively describe circles, $a\,A\,a'$, $b\,A\,b'$; these are called the *pitch circles* of the respective wheels, i.e. the two circles which rolling by friction would move with the same angular motions as the wheels. Now, if there are to be m teeth in the wheel O, there must be m_1 in the wheel O_1, where m_1 is given by the proportion $O\,A : O_1\,A :: m : m_1$.

FIG. 151.

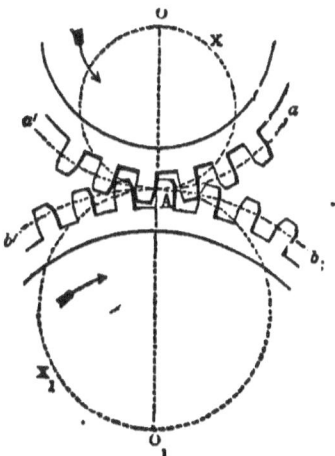

Divide the circumference of $a\,A\,a'$ into m equal parts,

---

* This rule, though not the *best*, is—or, at all events, used to be—very generally employed in practice. (See Willis, p. 106).

of which parts let AA_1 be one; the chord of this arc is called the *pitch* of the wheel; divide it into two (nearly)

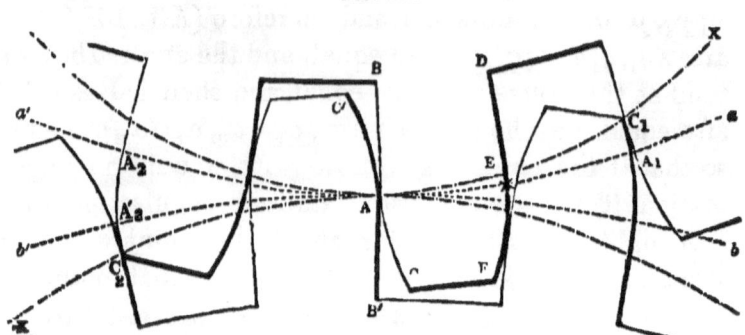

Fig. 152.

equal parts, of these A E (the smaller) is the breadth of a tooth, and EA_1 the space between two teeth; then the flanks B A, D E of a tooth (i.e. the parts of its outline within the pitch circle) are straight lines converging to the centre O; and the faces of the tooth A C, E F (i.e. the parts of its outline on the outside of the pitched circle) are portions of epicycloids described on the pitch circle as a base by a generating circle whose diameter equals the radius of the pitch circle of the wheel with which it is to work, viz. O_1 A. The teeth of the wheel O_1 are cut upon the same principle; the circumference of the pitch circle $b A b'$ is divided into m_1 equal parts, and each is divided into a tooth and a space; the flanks of the teeth converge to O_1, the faces are epicycloids described on the pitch circle as a base by a generating circle whose diameter equals the radius O A. That the two wheels thus constructed will work truly, follows immediately from Prop. 22; thus, if the wheel O revolve uniformly, the tooth B A C driving the tooth $B' A C'$, the epicycloid A C will cause the straight line A B', and therefore the wheel O_1, to revolve uniformly: on the other hand, if the wheel O_1 moving with a uniform motion drive O, the epicycloid A C' will cause the straight line A B, and therefore the wheel O, to revolve uniformly. This is of course true whether the wheels move in the directions indicated by the arrow heads in fig. 151, or in directions opposite to them. In order to prevent the *locking* of the teeth, it is usual to make A E less than A_1 E by $\frac{1}{11}$ of the pitch A A_1; and to cut the space A B′ deeper than the perpendicular length of the tooth A C in such a manner that the distance from C to the centre is less than the distance from B′ to the same centre by $\frac{1}{10}$ of the pitch A A_1; if, however, the workmanship is very good, the differences can in both cases be made smaller.

The rule for determining the length of the teeth commonly adopted by millwrights is to make the length of the tooth beyond the pitch circle (i.e. A C or A C′) equal to $\frac{3}{10}$ of the pitch.* This rule is, however, a very bad one; the following, though not perhaps the best, is very much better:—Suppose O to be the driver, and suppose a pair of teeth to be in contact on the line of centres, the face of the next tooth should be so long that its extreme point c_2 should just be on the circumference of the generating circle A x_1, as shown in the figure; the length of the tooth of the follower is determined by a similar rule; the extreme point of the following tooth c_1 should (under the same circumstances) be on the circumference of the generating circle A X O. The reason of this rule is as follows:—It may be considered that when the wheels are in motion the pair will bear the whole or nearly the whole stress which at any instant will be the next to go out of contact; so that, the above construction being employed, the one pair of teeth is just going out of contact when the next pair comes to the line of centres, and consequently the working stress is not thrown upon any pair of teeth until

---

* Willis's *Principles of Mechanism*, p. 98. The rule which follows is given both by Mr. Moseley, *Mechanical Principles*, p. 267, and by Gen. Morin, *Aide-Mémoire*, p. 280.

it c mes to the line of centres; but it appears that practically the friction between a pair of teeth is very much more destructive when they are in contact before the line of centres than when in contact behind the line of centres; by following, therefore, the rule above given, the friction between any pair of teeth is diminished. (Compare *Ex.* 566.)

In practice the teeth of a wheel are all cut from a pattern; in constructing a pattern the epicycloidal curve may be drawn from the actual rolling of a circle of the proper size; or an approximation may be obtained by means of circular arcs. Rules proper for this purpose will be found in Mr. Willis's Treatise above referred to.

Ex. 544.—To determine the radius of the pitch circle of a wheel which shall contain n teeth of given pitch a. *Ans.* $r = \dfrac{a}{2 \sin \dfrac{180°}{n}}$

Ex. 545.—If a wheel of m teeth drive another of n teeth; then if the driver make p revolutions per minute, the follower will make $\dfrac{mp}{n}$ revolutions per minute.

Ex. 546.—There are three parallel axes, A, B, C; A makes p revolutions per minute, it carries a wheel of m_1 teeth which works with a wheel of n_1 teeth on B; B also carries another wheel of m_2 teeth which works with a wheel of n_2 teeth on C; show that C makes $\dfrac{m_1}{n_1} \cdot \dfrac{m_2}{n_2} \cdot p$ revolutions per minute.*

Ex. 547.—A winding engine is worked in the following manner:—A steam engine causes a crank to make 30 revolutions per minute; the axle of the crank has on it a wheel containing 36 teeth, which works with a wheel containing 108 teeth; the latter wheel is on the same axle as the drum, which is 5 ft. in radius; determine the number of feet per minute described by the load. *Ans.* 314 ft.

---

* The above arrangement is to be found in most *cranes*; if the student is not acquainted with the arrangement of a train of wheels he will do well to examine a good crane, such as is to be seen at most railway stations: the train of wheels in a clock is also a good example but cannot commonly be studied without taking the clock to pieces.

TOOTHED WHEELS. 225

110. *The hunting cog.*—If wheels have to do heavy work, and the precise ratio between the velocities is not of great importance, an additional tooth—called a *hunting cog*—is introduced into one of the wheels, so that the same pair of teeth may seldom work together; by this means they are kept from wearing unequally. For instance, if in the last example we denote the teeth of the driver by the successive numbers 1, 2, 3, . . . 36, and the teeth of the follower by the successive numbers 1, 2, 3, . . . 108; then in every revolution 1 will work with 1, 37, and 73; 2 will work with 2, 38, and 74; and 36 will work with 36, 72, and 108. If now we introduce a hunting cog into the driving-wheel, so that it contains 37 teeth, then on the first revolution 1 will work with 1, 38, and 75; in the next revolution with 4, 41, and 78; in the third with 7, 44, and 81, and not until the 38th revolution will it work with 1 again.

Ex. 548.—If in the last example a 'hunting cog' were introduced into the driver so that it contains 37 teeth, determine the number of feet per minute the load will now travel. *Ans.* 323 ft.

Ex. 549.—If in *Ex.* 546 there are $k+1$ axles and the drivers contain m teeth, and the followers contain n teeth a-piece, show that the number of revolutions made by the last axle will be $p\left(\dfrac{m}{n}\right)^k$

Ex. 550.—If in the last example it is required to multiply the number of revolutions 200 times, how many axles must we use—(1) if we take $m=2n$; (2) if we take $m=4n$; (3) if we take $m=6n$, and determine the number of teeth employed, in each case using the nearest whole numbers?
Ans. Axes (1) 8. (2) 4. (3) 3.
Teeth (1) 24n. (2) 20n. (3) 21n.

Ex. 551.—If each driver has m teeth, and each follower n teeth, and if M is the total number of teeth in the train, and if the last axle makes q revolutions while the first axle makes one revolution, show that

$$q = \left(\dfrac{m}{n}\right)^{\tfrac{\text{M}}{m+n}}$$

Q

* *Ex.* 552.—In the last example show that for given values of M and n we shall obtain the greatest value of q by making $m = 3·59 . n$ nearly.*

[It is easily shown that $\log_\epsilon \left(\dfrac{m}{n}\right) = 1 + \dfrac{n}{m}$, whence the result stated.]

Ex. 553.—In the case of a pair of wheels with epicycloidal teeth show that the distance through which the surfaces of each pair of teeth slide one upon the other while in contact and after passing the line of centres is approximately represented by the formula $\dfrac{2\pi r}{n}\left(\dfrac{\pi}{n} + \dfrac{\pi}{n_1}\right)$ or $\dfrac{2\pi r_1}{n_1}\left(\dfrac{\pi}{n} + \dfrac{\pi}{n_1}\right)$ where r and r_1 are the radii of the driver and follower respectively, and n and n_1 the number of teeth in those wheels respectively.

[The motion of one tooth on the other is partly a sliding and partly a rolling motion. Now, if we refer to fig. 152, it is evident that the pair of teeth just going out of contact touch at c_2; it is also evident that the two points A_2 and A'_2 were in contact at A, so that the space through which the surfaces have slidden over each other is $A_2 A'_2$, which is very nearly equal to the sum of the versed sines of the *arcs* $A A_2$ and $A A'_2$, i.e. to r vers $\dfrac{2\pi}{n}$ + r_1 vers $\dfrac{2\pi}{n_1}$; whence the value assigned in the question.]

Ex. 554.—A weight P balances a weight Q under the following circumstances: P is tied to a rope which is wrapped round an axle whose radius is p; Q is tied to a rope which is wrapped round an axle whose radius is q; to the former is attached a concentric rough wheel, whose radius is r, to the latter in like manner a concentric rough wheel, whose radius is r_1; these two wheels are in contact on the line of centres so that $r + r_1$ equals $O O_1$; show that if we neglect the magnitude of the axes and the rigidity of the cords, we shall have

$$P = Q \frac{q}{p} \cdot \frac{r}{r_1}$$

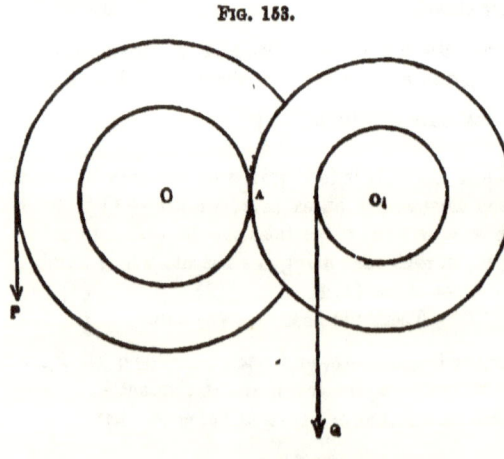

FIG. 153.

[The arrangement described in the above example is represented in the annexed diagram; it

---

* It would appear from this that the best proportion between the number of teeth in driver and follower for multiplying velocity is 1 : 4. This result is due to Dr. Young, *Lectures*, vol. ii. p. 56. Mr. Willis remarks that the rule is not of much practical value (*Principles*, p. 218).

FRICTION BETWEEN TEETH OF WHEELS. 227

is evident that the rough wheels act on each other by means of a mutual action through the point A.]

Ex. 555.—In the last example if we suppose the separate wheel and axles to turn round axes whose radii are ρ and ρ_1 respectively and the limiting angles of resistance between them and their bearings to be ϕ and ϕ_1, show that when P is on the point of overcoming Q we have the following relation (neglecting the rigidity of cords, and the weights of the wheel and axles):—

$$\text{P}(p-\rho \sin \phi)(r_1+\rho_1 \sin \phi_1) = \text{Q}(q+\rho_1 \sin \phi_1)(r+\rho \sin \phi)$$

Ex. 556.—If in the last example, besides the frictions on the axes, we take into account the weights w and w_1 of the wheel and axles, determine the relation between P and Q.

Ex. 557.—If in the last example we neglect powers and products of $\dfrac{\rho \sin \phi}{p}$, $\dfrac{\rho \sin \phi}{r}$, $\dfrac{\rho_1 \sin \phi_1}{q}$, $\dfrac{\rho_1 \sin \phi_1}{r_1}$, show that the number of foot-pounds of work that must be done in order to raise a weight of Q lbs. through s ft. is given by the formula

$$\text{U} = \text{Q}s \left\{ 1 + \left(\frac{1}{p}+\frac{1}{r}\right)\rho \sin \phi + \left(\frac{1}{q}-\frac{1}{r_1}\right)\rho_1 \sin \phi_1 \right\}$$
$$+ \frac{r_1 s}{q}\left\{ w \frac{\rho \sin \phi}{r} + w_1 \frac{\rho_1 \sin \phi_1}{r_1} \right\} *$$

Ex. 558.—In the last example if we suppose the rough wheels to be replaced by a pair of toothed wheels whose pitch circles have the same radii as the wheels; then if the wheel O contains n teeth, and the wheel O_1 contains n_1 teeth, show that when Q is raised through a distance s the work lost by the friction of the teeth is approximately represented by the formula $\mu \text{Q} s \left(\dfrac{\pi}{n}+\dfrac{\pi}{n_1}\right)$, where μ is the coefficient of friction between the teeth.

[If the wheel $O_1 A$ revolves through an angle $\dfrac{2\pi}{n_1}$ the distance through which the surfaces of the driving and driven teeth slide is $\dfrac{2\pi r_1}{n_1}\left(\dfrac{\pi}{n}+\dfrac{\pi}{n_1}\right)$ and therefore, supposing R, *the mutual pressure, to continue constant* during the contact of the teeth, the number of units of work expended on friction equals $\mu \text{R} \dfrac{2\pi r_1}{n_1}\left(\dfrac{\pi}{n}+\dfrac{\pi}{n_1}\right)$. Now, approximately, $\text{R}r_1 = \text{Q}q$, and therefore the work expended on one pair of teeth equals $\mu \text{Q} \dfrac{2\pi q}{n_1}\left(\dfrac{\pi}{n}+\dfrac{\pi}{n_1}\right)$; but $\dfrac{2\pi q}{n_1}$ is the distance through which Q is raised during

---

* If P instead of being a weight, were a force acting vertically upward it is easily shown that the third term of this equation is

$$\text{Q}s\left(\frac{1}{q}+\frac{1}{r_1}\right)\rho_1 \sin \phi.$$

Q 2

the action of one pair of teeth, and the same being true of every pair of teeth, we obtain the result stated in the question. Of course, the addition of the expression contained in the present question to that obtained in the last is the correct approximate formula for the work expended in raising a weight through the intervention of a pair of toothed wheels.]

Ex. 559.—A force P acting at the end of an arm OA, two feet long, causes the toothed wheel OB to make 10 turns per minute; this wheel working with the wheel O_1 B turns the drum O_1 C and raises the weight Q; given that P does at the point A 330,000 foot-pounds of work per minute, determine approximately the weight Q that will be raised by the drum, having given the radius of OB to be 1 ft., O_1 B to be 3 ft., the number of teeth in OB to be 40, and the radius of the drum 5 ft.; the teeth, axles, and bearing are all of cast iron without unguents; the radii of the axles are 3 in., the weight of the axles and appendages of O is 3600 lbs., and that of O_1 is 5400 lbs. *Ans.* 2752 lbs.

[See Note to *Ex.* 557.]

Ex. 560.—Show that in a train of p pairs of wheels and pinions* the work lost by friction between the teeth is given by the formula

$$\mu Q S \pi \left\{ \frac{1}{n_1} + \frac{1}{n_2} + \frac{1}{n_3} + \ldots + \frac{1}{n_{2p}} \right\}$$

where $n_1, n_2, n_3 \ldots n_{2p}$ are the number of teeth in the successive wheels and pinions.

Ex. 561.—There is a train of p equal pairs of wheels and pinions; the numbers of teeth are such that the last axle revolves m times faster than the first; show that if U is the number of units of useful work yielded, the work lost by the friction between the teeth is represented by the formula

$$\frac{\mu U \pi p}{n} (1 + m^{\frac{1}{p}})$$

where n is the number of teeth in each wheel.

Ex. 562.—It is required to make the last axle move m times faster than the first, show that the loss of work is least when p, the number of pairs of wheels and pinions, is given by the formula

$$m^{-\frac{1}{p}} + \log_e m^{-\frac{1}{p}} + 1 = 0$$

Ex. 563.—If in the last example it is required to multiply the velocity 100 times, show that the proper number of pairs of wheels and pinions is 3 or 4, i.e. show that the equation in the last example gives a value of p between 3 and 4; and determine the number of teeth employed in each case if the first pinion have 20 teeth, using the nearest whole numbers.
Ans. (1) 339. (2) 333.

---

* When a small wheel drives a large one the former is frequently called a pinion and the latter a wheel.

FRICTION BETWEEN TEETH OF WHEELS.

Ex. 564.—If in the pair of wheels already described (Art. 109) all but a single pair of teeth be cut away, so that the remaining teeth act on each other while the wheel o moves through an angle $\frac{2\pi}{n}$ before coming to the line of centres, and also while it moves through an equal angle after having passed the line of centres, and if we suppose P and Q to act on the pitch circles of their respective wheels, show that when the point of contact is in such a position that the wheel o has to revolve through an angle θ before the point of contact comes to the line of centres we have

$$P\{r_1 - (r+r_1)\tan\theta\tan\phi\} = Qr,$$

and that when the point of contact is so situated that the driver has revolved through an angle θ from the line of centres we have

$$Pr = \{r + (r+r_1)\tan\frac{r\theta}{r_1}\tan\phi\}$$

[If in the accompanying figure x is the point of contact of the teeth before they come to the line of centres, that point x will be on the circum-

FIG. 154.

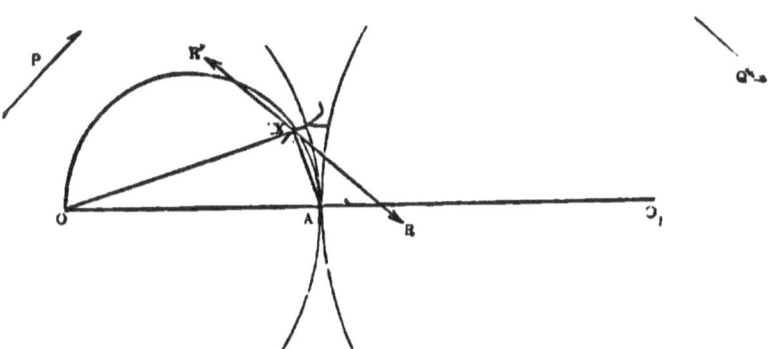

ference of a circle whose diameter is OA; if then we draw a line RR such that the angle RXA equals ϕ, this will be the line of the mutual action of the teeth; remembering that the angle AOX equals θ it is easily shown that the perpendiculars on RR' from O and O_1 are respectively equal to

$$r\cos\theta\cos\phi$$

and $\qquad (r+r_1)\cos(\theta+\phi) - r\cos\theta\cos\phi$

whence the first equation is obtained; the second is obtained in a similar manner, by determining the relation between P and Q when the follower has revolved through an angle θ' which will be found to be

$$Pr = Q\{r + (r+r_1)\tan\theta'\tan\phi\}$$

whence we obtain the answer.]

Ex. 565.—If A B be any diameter of a circle A P B; if C be any point taken in the prolongation of A B (so that B is between A and C), and if A P, B P, C P be joined, show that

$$BC = AC \tan PAB \tan BPC$$

and hence explain the action of the forces which produces the result that follows from the first equation in *Ex.* 564, viz. that when $r_1 = (r + r_1) \tan \theta \tan \phi$ the force P must be infinitely large to bring Q into the state bordering on motion.

Ex. 566.—If the driver be not greater than the follower, show from the equations of *Ex.* 564, that for a given value of Q the value of P is greater when the driving tooth is in a given position *before* it comes to the line of centres than when it is in a corresponding position after having passed the line of centres.

[If m be written for the ratio of r to r_1 (so that m cannot be greater than unity) the equations in *Ex.* 564 can be written thus:—

$$P\{1 - (1+m)\mu \tan \theta\} = Q$$

and
$$P' = Q\left\{1 + \left(1 + \frac{1}{m}\right)\mu \tan m\theta\right\}$$

consequently

$$P - P' = Q\left\{(1+m)\mu \tan \theta - \left(1 + \frac{1}{m}\right)\mu \tan m\theta + \text{positive terms}\right\}$$

and this, on expanding in powers of θ, is found to equal

$$\mu Q\left[(1+m)\theta\left\{\tfrac{1}{3}(1-m^2)\theta^2 + \tfrac{2}{15}(1-m^4)\theta^4 + \ldots\right\} + \text{positive terms}\right]$$

PART II.

DYNAMICS.

CHAPTER I.

INTRODUCTORY.*

111. *Velocity.*—Before considering *force* as the cause of *change* of *velocity*, it will be necessary to define accurately the means of estimating velocities numerically.

Def.—A body moves uniformly or with a uniform velocity when it passes over equal distances in equal times.

The units of distance and time commonly employed are feet and seconds: † and whenever a body is said to be moving with any particular velocity, e.g. 5 or 6, this will always mean with a velocity of 5 or 6 ft. per second.

Def.—When a body moves with a variable velocity, that velocity is measured at any instant by the number of units of distance it would pass over in a unit of time *if it continued to move uniformly from that instant.*

It will be seen from the definition that variable velocity is measured in a manner that exactly falls in with the

---

* The student is particularly recommended to make himself thoroughly master of this chapter before proceeding further.

† To prevent mistake, it may be stated that the time referred to is *mean solar time*.

ordinary way of speaking: thus, when we say that a train is moving at the rate of 40 miles an hour, we mean that if it were to keep on moving uniformly for an hour, it would pass over 40 miles. Again, if we were to drop a small heavy body, we should find that at the end of a second it is moving at the rate of about 32 feet per second, or, as it is commonly stated, it acquires in a second a velocity 32, meaning that if it were to move uniformly from the end of that second it would pass over 32 feet in each successive second.

112. *Relation between uniform velocity, time, and distance.*—In the case of a body moving with a uniform velocity, it is evident that the number of feet (s) passed over in t seconds must be t times the number of feet passed over in one second (v),

$$\therefore s = v t.$$

The distance s can, of course, be represented geometrically by the area of a rectangle whose sides severally represent on the same scale the velocity and the time.

Ex. 567.—A body moves uniformly over $2\frac{1}{2}$ miles in half an hour; determine its velocity. *Ans.* $7\frac{1}{3}$.

Ex. 568.—A body moves at the rate of 12 miles an hour; determine its velocity. *Ans.* $17\frac{3}{5}$.

Ex. 569.—The equatorial diameter of the earth is 41,847,000 ft., and the earth makes one revolution in 86,164 seconds; determine the velocity of a point on the earth's equator. *Ans.* 1526.

Ex. 570.—A body moves with a velocity of 12; how many miles will it pass over in one hour? What would be its velocity if we used yards and minutes as units instead of feet and seconds? *Ans.* (1) $8\frac{2}{11}$. (2) 240.

113. *The velocity acquired by falling bodies.*—It appears as the result of the most careful experiments that at any given point on the earth's surface, a body falling freely in vacuo acquires at the end of every second a

FALLING BODIES. 283

certain constant additional velocity:* this velocity is slightly different at different places, but is always the same at the same place, and never differs greatly from 32; so that if at any instant the falling body has a velocity v, it will have at the end of the next second a velocity v + 32. This additional velocity is the *accelerative effect*, or, as it is sometimes called, the *accelerating force*, or simply the *acceleration*, of gravity, and is denoted by the letter g;— in all the following examples it will be assumed that g equals 32, unless the contrary is specified.

From what has been said it is plain that if a body is let fall, it acquires a velocity g at the end of the first second, $2g$ at the end of the second second, $3g$ at the end of the third second, and so on: consequently, if v is the velocity acquired at the end of t seconds, we shall have

$$v = g t.$$

By the same reasoning it appears that if the body is thrown *downward* with a velocity V, and if v is its velocity after falling t seconds, then

$$v = V + g t$$

Moreover, when a body is thrown *upward* so as to move in a direction opposite to that in which gravity acts, it appears that it loses in every second a velocity g; consequently in that case

$$v = V - g t$$

Ex. 571.—A body falls for 7 seconds; with what velocity is it moving at the end of that time? *Ans.* 224.

Ex. 572.—If a body is let fall, how long will it take to acquire a velocity of 200 ft. per second? *Ans.* $6\frac{1}{4}$ sec.

Ex. 573.—A body is projected downward with a velocity of 80 ft. per second; determine the velocity it will have at the end of 5 seconds, and the number of seconds that must elapse before its velocity equals twice its initial velocity. *Ans.* (1) 240. (2) $2\frac{1}{2}$ sec.

---

* It may be remarked, that the difference between the velocities with which a feather and a bullet descend is entirely due to the resistance of the air.

Ex. 574.—A body is thrown downward with a velocity of 160 ft. per second; determine its velocity at the end of 4 seconds, and the number of seconds in which a body that is merely dropped would acquire that velocity. *Ans.* (1) 288. (2) 9 sec.

Ex. 575.—A body A is projected downward with a velocity of 160 ft. per second; at the same instant another body B is projected upward with an equal velocity; determine how much faster A will be moving than B at the end of 4 seconds. *Ans.* 9 times.

Ex. 576.—A body is thrown upwards with a velocity of 96 ft. per second; with what velocity will it be moving at the end of 4 seconds?

[The formula gives -32, i.e. it will be moving *downward* with a velocity of 32 ft. per second.]

Ex. 577.—In the last case how long will it take the body to reach the highest point?

[It will be at the highest point when $v=0$, i.e. after 3 seconds.]

Ex. 578.—A body is at any instant moving *upward* with a given velocity v; show that it will be moving downwards with an equal velocity after $\dfrac{2v}{g}$ seconds; and that it will reach its highest point after $\dfrac{v}{g}$ seconds.

Ex. 579.—A body is thrown up with a velocity mg; after how long will it be descending with a velocity ng? *Ans.* $m+n$ sec.

114. *The distance described in a given time by a falling body.*—It admits of proof that if a body is allowed to fall freely from rest for t seconds the number of feet (s) which it will pass over is given by the formula

$$s = \tfrac{1}{2} g t^2$$

If, however, it is *thrown downward* with a velocity v, we shall have

$$s = v t + \tfrac{1}{2} g t^2$$

and if *upward* with a velocity v, it will, at the end of t seconds, be s feet above the point of projection, where

$$s = v t - \tfrac{1}{2} g t^2$$

Ex. 580.—How many feet will be described in 4 seconds by a body that moves freely from rest under the action of gravity? *Ans.* 256 ft.

Ex. 581.—Through how many miles would a body falling freely from rest descend in one minute? *Ans.* $10\tfrac{10}{11}$ mi.

Ex. 582.—A body is projected downward with a velocity of 20 ft. per second; how far will it fall in $1\tfrac{1}{2}$ second? *Ans.* 66 ft.

FALLING BODIES.

Ex. 583.—A body is projected upward with a velocity of 100 ft. per second; how high will it have ascended in 3 seconds? *Ans*. 156 ft.

Ex. 584.—Show that the greatest value of $vt-\tfrac{1}{2}gt^2$ is found by making $t=\dfrac{v}{g}$. [Compare this result with *Ex*. 578.]

Ex. 585.—If a body be projected *upward* with a velocity of 96 ft. per second, where will it be at the end of 7 seconds, and what will be the whole distance it will have described?

Ans. (1) 112 ft. below the point of projection. (2) 400 ft.

Ex. 586.—A body is projected upward with a velocity of 100 ft. per second; determine where the body will be, with what velocity, and in what direction, the body will be moving at the end of 4 seconds.

Ans. (1) 144 ft. above the point of projection. (2) 28 ft. per sec. downward.

Ex. 587.—A body is projected upward with a velocity v; show that it will return to the point of projection after $\dfrac{2v}{g}$ secs.

[Compare this result with *Ex*. 578.]

Ex. 588.—A body falls for a time t, and has a velocity v at the beginning, and v at the end of that time: show that it describes the same distance as another body describes in the same time with a uniform velocity $\tfrac{1}{2}(v+v)$.

115. *Relation between velocity acquired and distance passed over by a falling body*.—The above relations between the velocity (v) which the body has at the end of a time (t) and between the distance (s) which it describes in the same time (t) enable us to determine the relation between v and s; thus, if the body is simply let fall we have

$$v = gt$$
and $$s = \tfrac{1}{2}gt^2$$
whence $$v^2 = 2gs$$

an equation which gives the velocity acquired in falling from rest through s feet. In like manner if we take the equations

$$v = V + gt$$
and $$s = Vt + \tfrac{1}{2}gt^2$$
we see that $$2gs = 2Vgt + g^2 t^2$$

and
$$\text{V}^2 + 2gs = \text{V}^2 + 2\text{V}gt + g^2 t^2$$
$$= (\text{V} + gt)^2$$
therefore $\quad v^2 = \text{V}^2 + 2gs$

This equation gives the velocity (v) which the body has after falling through s feet from the point at which it was moving downward with a velocity V. Similarly we can show that
$$v^2 = \text{V}^2 - 2gs$$
in which v is the velocity which it has when it is s feet above the point at which it was moving upwards with a velocity V; whether the direction of the velocity v is upward or downward does not appear from the equation, and must be determined by other considerations. When a body is moving with a given velocity (V), a certain height (H) can always be found such that if a body fell down it freely from rest it would acquire the given velocity; under these circumstances V is said to be the velocity due to the height H. These quantities are, of course, connected by the equation
$$\text{V}^2 = 2g\text{H}.$$

Ex. 589.—If a body is thrown upward with a velocity V, show that it will ascend through $\dfrac{\text{V}^2}{2g}$ feet.

Ex. 590.—If a body is thrown upward with a velocity of 200 ft. per second, find its greatest height. *Ans.* 625 ft.

Ex. 591.—If a body falls freely through 150 ft., find the velocity it acquires. *Ans.* 98.

Ex. 592.—A body is projected vertically upward with a velocity of 200 ft. per second; how long will it take to reach the top of a tower 200 ft. high, and with what velocity will it reach that point?
Ans. (1) 1·1 sec. (2) 164·9.

Ex. 593.—Let A be the highest point of a vertical line A B: at the same instant one body is dropped from A and another thrown up from B, they meet at the middle point of A B; find the initial velocity of the second body. *Ans.* $\sqrt{g \times \text{A B}}$.

Ex. 594.—A stone (A) is let fall from a certain point; one second after another stone (B) is let fall from a point 100 ft. lower down; in how many seconds from the beginning of its motion will A overtake B, and what distance will it have described? *Ans.* (1) $3\frac{5}{8}$ sec. (2) $210\frac{1}{4}$ ft.

Ex. 595.—A stone (A) is let fall from the top of a tower 350 ft. high; at the same instant a second stone (B) is let fall from a window 50 ft. below the top; how long before A will B strike the ground? *Ans.* 0·35 sec.

Ex. 596.—A stone (A) is projected vertically upwards with a velocity of 96 ft. per second; after 4 seconds another stone (B) is let fall from the same point; how long will B move before it is overtaken by A, and at what point will this happen? *Ans.* (1) 4 sec. (2) 256 ft. below the point of projection.

Ex. 597.—In the last example if only 3 seconds had elapsed when B was let fall, would A ever have overtaken it? *Ans.* No.

Ex. 598.—The point A is 128 ft. above B; a body is thrown upward from A with a velocity of 64 ft. per second, and at the same instant another is thrown upward from B with a velocity of 96 ft. per second; show that after 4 seconds they will both be at A; moving downward with velocities 64 and 32 respectively.

Ex. 599.—Determine the heights to which velocities of 20, 59, and 760 ft. per second are respectively due.

Ans. (1) $6\frac{1}{4}$ ft. (2) $54\frac{25}{64}$ ft. (3) 9025 ft.

116. *Other cases of uniformly accelerated motion.*— The velocity of a body is said to be uniformly accelerated when it is increased by equal amounts in equal intervals of time. Thus, taking feet and seconds as the units of distance and time, if the velocity of a body is, during any second of its motion, changed from v ft. a second at the beginning of the second to $v+20$ ft. a second at the end of the second, the acceleration is said to be 20 in feet and seconds. In like manner if the acceleration is f in feet and seconds, this means that if at any instant the body is moving at the rate of v feet a second, its velocity will become at the end of a second $v+f$ feet a second. The velocity is said to be uniformly retarded when it is diminished by equal amounts in equal intervals of time; thus, if at any instant the velocity is v feet a second, and at the end of a second it becomes $v-f$ feet a second, the velocity is said to undergo a uniform retardation f. It is usual to reckon retardation as a negative acceleration. If we write f for g in the formulæ of the preceding articles they will apply to the rectilineal motion of bodies whose velocities undergo any uniform acceleration or retardation. Thus, if V is the initial velocity, f the acceleration, t the time at the end of which the body has a velocity v, and is at a distance s from the starting point, we shall have

$$v = V + ft$$
$$s = Vt + \tfrac{1}{2}ft^2$$
$$v^2 = V^2 + 2fs$$

Ex. 600.—At the distance of the moon the accelerative effect of gravity is reduced to about $\tfrac{1}{3600}$; if a body fell freely from this distance for one hour, with what velocity per minute would it then be falling? and in how many seconds would a body falling in the neighbourhood of the earth's surface acquire the same velocity?

Ans. (1) 1928⅔. (2) 1 sec. very nearly.

Ex. 601.—If a body were to begin to fall to the earth from the distance of the moon, how many yards would it fall through in half an hour?

Ans. 4821 yards.

Ex. 602.—In the last example if a body were thrown upward with a velocity of 4 miles an hour, how long would it take to return to the point of projection? *Ans*. 1314 sec.

117. *The acceleration of the motion of a given body produced by a given force.*—In most cases the moving body is acted on by several forces, which to a certain extent neutralise each other, and its motion is caused by their resultant. Suppose that a body is placed on a smooth horizontal plane and moved by a force (P) acting horizontally; the forces acting are the weight of the body, the reaction of the plane, and the force P; of these the two former neutralise each other, and the latter produces the motion. So long as the force producing motion remains unchanged, it will uniformly accelerate (or retard) the motion. The amount of the acceleration is determined by the fundamental principle *—*If any given body is acted on*

* The evidence for this principle, as for all the other fundamental principles of dynamics, is experiment, though it is very difficult to devise experiments which shall exhibit them in a state of isolation: Galileo, who discovered most of them, possessed a rare sagacity in detecting the *parts* of

RELATION BETWEEN FORCE AND MOTION. 239

successively by two forces, the accelerations due to the action of the forces are in the same ratio as the forces. Now, if the weight of a body be W lbs., we know that the sensible attraction the earth exerts on it at London is a force of W lbs.—the term *pound* being used to denote the unit of force, as in Art. 23. Also, if this body fall freely in vacuo in London it has been ascertained (see Table XV., p. 250) that its velocity is increased in each second by a velocity of 32·1912 ft. per second; this is, therefore, the acceleration of that body's velocity when acted on by a force of W lbs. Suppose the same body to be acted on by a force of P lbs. and the corresponding acceleration to be f;—suppose, as in the above instance, that the body is placed on a smooth horizontal plane, and urged along a straight line on the plane by the force P;—then in each successive second of its motion its velocity will be increased by a velocity of f feet per second, where f is given by the proportion

$$W : P :: 32·1912 : f$$

In the following examples 32 will be used as an approximate value of 32·1912.

It follows from the remark already made (Art. 116) that the formulæ previously given for falling bodies will be true in the present case when f has been substituted for g. Thus we shall have

$$v = ft \qquad s = \tfrac{1}{2} f t^2 \qquad v^2 = 2fs \quad \&c.$$

Ex. 603.—A body weighing 30 lbs. slides along a smooth horizontal plane under a constant force of 15 lbs.; determine—(1) the additional velocity it acquires in every second; (2) the velocity it will have at the end of 5 seconds; (3) the distance it will pass over in 5 seconds.

Ans. (1) 16. (2) 80. (3) 200 ft.

---

a phenomenon which were due to disturbing causes, and thus was enabled to get at the fundamental principles. The experimental verification of these principles is nearly always *indirect*, and consists in comparing actual cases of motion (e.g. that of planets, of pendulums, &c.) with the secondary principles which have been derived from them.

240   PRACTICAL MECHANICS.

Ex. 604.—A mass weighing w lbs. is urged along a rough horizontal plane by a force of P lbs. acting in a direction parallel to the plane: the coefficient of friction is μ; if the body's velocity is increased in every second by f, show that

$$f = \frac{P - \mu W}{W} g$$

where g denotes 32·1912 or (approximately) 32.

Ex. 605.—A weight of 100 lbs. is moved along a horizontal plane by a constant force of 20 lbs.; the coefficient of friction is 0·17; determine—(1) the distance it will describe in 10 seconds; (2) the time in which it will describe 200 ft.   *Ans.* (1) 48 ft.   (2) 20·4 sec.

Ex. 606.—A train weighing 50 tons is impelled along a horizontal road by a constant force of 550 lbs.; the friction is 8 lbs. per ton; what velocity will it have after moving from rest for ten minutes, and what distance will it describe in that time? *

Ans. (1) 17½ miles per hour.   (2) 7714 ft.

Ex. 607.—If in the last example the steam were cut off at the end of the 10 minutes, how many seconds will elapse before the train stops, and how far will it go?   *Ans.* (1) 225 sec.   (2) 2893 ft.

Ex. 608.—A train is observed to move at the rate of 30 miles per hour, the steam is cut off, and it then runs on a horizontal plane for 10,000 ft.; find how many lbs. per ton the resistances amount to supposing them independent of the velocity.

[It is easily shown that $f = 0·0968$; then the resistance (P) in lbs. per ton (W) is found to equal 6·776 lbs.]

Ex. 609.—A sphere lies on the deck of a steamer and is observed to roll back 20 inches; if the resistance to rolling is the $\frac{1}{20}$th part of its weight, determine the change in the velocity of the steamer.

Ans. 2·309 ft. per sec.

118. *The motion of connected bodies.*—The meaning of the term *reaction* has been already explained in Art. 28, where the law is stated that when a body (A) acts on another body (B) the action is mutual; whatever force A exerts on B, B exerts an equal opposite force on A. It

---

* If the resistances which oppose the motion of the train were constant, it would be possible to attain any velocity, however great; in reality the resistance of the air always imposes a limit to the velocity that can be attained by a train moved by a force that exceeds the frictions by any given amount; thus Mr. Scott Russell's formula for the resistance contains a term involving the square of the velocity of the train (Rankine, p. 620).

must be understood that this law is perfectly general, and is true whether the bodies are at rest or in motion. Suppose that two bodies, whose weights are P and Q, are connected by a very fine weightless thread, supported by a smooth point on which it hangs. If P is the heavier of the two, it will descend, and in doing so it will draw Q up. If the question is to determine the velocity acquired and the distance described by the bodies in a given time, we may proceed thus :—The system whose weight is P + Q is moved by the excess of the weight of P over that of Q; hence if f is the accelerative effect of that force we have

$$P+Q : P-Q :: g : f$$

and as this proportion gives f the question can be answered easily. But if the question is to find the force which causes P's motion or Q's motion, we must proceed as follows :—If P acts on Q with a force T, Q will react on P with an equal opposite force, and, as both these forces are transmitted along the thread, T is the tension of the thread. Now the velocities of the bodies are always equal, therefore the acceleration of P's velocity must equal that of Q's velocity. But P moves downward under a force of P — T pounds, and Q upward under a force of T — Q pounds. Therefore if f is the required acceleration we have

$$\frac{f}{g} = \frac{P-T}{P} \quad \text{and} \quad \frac{f}{g} = \frac{T-Q}{Q}$$

whence we obtain

$$f = \frac{P-Q}{P+Q} \cdot g \quad \text{and} \quad T = \frac{2PQ}{P+Q}$$

The value of f is the same as that previously determined, and T is the value in lbs. of the force with which P acts on Q, and of that with which Q reacts on P.

Ex. 610.—If in the case explained in the last article P and Q weigh $12\frac{1}{2}$ lbs. and $11\frac{1}{4}$ lbs. respectively; find (1) the acceleration of P's and Q's velocity, (2) the distance they would describe in 5 sec. from a state of rest, (3) the tension of the thread. *Ans.* (1) $1\frac{1}{3}$. (2) $16\frac{2}{3}$ ft. (3) $11\frac{17}{43}$ lbs.

Ex. 611.—A weight Q is tied to a string, and rests on a rough horizontal table; to the other end of the string is tied a weight P which hangs vertically over the edge of the table; if the weight of the string and its friction against the edge of the table are neglected, show that when P falls it accelerates Q's velocity in every second by f, where

$$f = \frac{P - \mu Q}{P + Q} \cdot g$$

[The student will remark that in this case a weight $P + Q$ is moved by a force $P - \mu Q$.]

Ex. 612.—A mass of cast iron weighing 100 lbs. is drawn along a horizontal plane of cast iron by means of a cord which is parallel to the plane, and to the end of which a weight of 20 lbs. is attached (as in *Ex.* 611); determine—(1) the acceleration; (2) how far it will move in 4 seconds.
Ans. (1) $1\frac{1}{3}$. (2) $10\frac{2}{3}$ ft.

Ex. 613.—If in the last example the mass had described 5 ft. in $1\frac{1}{2}$ seconds, what must have been the coefficient of friction? *Ans.* $\frac{1}{30}$.

Ex. 614.—If in *Ex.* 611 Q weighs 1 lb. and P weighs 1 oz.; if moreover the length of the string is 12 ft. and P is placed at the edge of the table which is 3 ft. above the ground, find—(1) how long P will take to reach the ground; (2) how long it will take Q to arrive at the edge of the table, the friction between Q and the table being neglected.
Ans. (1) 1·78 sec. (2) 4·46 sec.

Ex. 615.—In the last example suppose P and Q each to weigh one pound; determine the coefficient of friction between Q and the table if that body just reaches the edge. *Ans.* $\frac{1}{2}$.

Ex. 616.—In *Ex.* 611 show that the tension of the string equals $\frac{(1 + \mu) PQ}{P + Q}$

Ex. 617.—In *Ex.* 612 find the tension of the cord. *Ans.* $19\frac{1}{4}$ lbs.

Ex. 618.—A plane is observed to be descending with a uniform acceleration of 8; a body weighing w lbs. rests on the plane: show that the mutual pressure between w and the plane is $\frac{3}{4}$w.

Ex. 619.—Three bodies P, Q, R, weighing 100 lbs. apiece, are connected by threads and placed one after another on a smooth horizontal plane; they are set in motion by a weight of 20 lbs. which is connected by a thread to P and hangs over the edge of the plane; find the tensions of the threads.
Ans. $6\frac{1}{4}$ lbs., $12\frac{1}{2}$ lbs., $18\frac{3}{4}$ lbs.

Ex. 620.—A chain hangs over a point; if we suppose the chain perfectly flexible, the point perfectly smooth, and the hanging parts of unequal length, it will not stay at rest, but will run off the point; show that during

ACCUMULATED WORK. 243

the motion the tension of the chain at its middle point equals the weight of the shorter of the two hanging parts.

119. *The work accumulated in a moving body, or its kinetic energy.*—A moving body has, in virtue of its velocity, the power of doing work against a resistance. For instance, if a train is in motion, and the steam is cut off, it will run for a considerable distance before coming to rest; and all the while it is moving it is doing work against the friction and other resistances. This power which a moving body has of doing work may be called the work accumulated in the body, but it is more commonly called its *vis viva*, or kinetic energy. Let the body weigh W lbs. and have a velocity of V feet a second; suppose its motion to be opposed by a constant force of P lbs.; the work originally accumulated in the body when its velocity was V will be gradually expended in overcoming the force, and will be exhausted when it comes to rest. Let the body move through h feet before being brought to rest by the force; it will then have done Ph foot-pounds of work against the resistance, and this must have been the work accumulated in it, or its kinetic energy, at the instant its velocity was V. Now if f is the retardation of the velocity due to P we shall have

$$V^2 - 2fh = 0$$

but we know (Art. 117) that $\dfrac{f}{g} = \dfrac{P}{W}$

therefore $\qquad P h = \dfrac{W V^2}{2g}$

where g stands for 32·1912 or approximately 32. If then W is given in pounds, and V in feet per second, $\dfrac{W V^2}{2g}$ is the work accumulated on the body, or its kinetic energy in foot-pounds.

Ex. 621.—A body whose weight is 10 lbs. moves with a velocity of 16 ft. per second, it has to overcome a constant resistance of half a pound; determine the number of feet it will describe before stopping.

[There are 40 foot-pounds of work accumulated in the body; now, if x be the number of feet required, $\frac{1}{2}x$ is the number of foot-pounds of work done, whence x equals 80 ft.]

Ex. 622.—In a similar manner obtain the answers to the *Ex.* 607, 608, and 609.

Ex. 623.—If in two bodies moving with the same velocities there are accumulated v foot-pounds of work, and if their weights are P and Q, show that the number of foot-pounds accumulated in P is $\frac{PU}{P+Q}$.

Ex. 624.—P weighing 10 lbs. is attached by a fine thread to Q weighing 40 lbs.; Q is placed on a rough table over the edge of which P hangs; P falls through 5 ft. and then comes to the ground; Q moves on for 8 ft. more and comes to rest on the table. What is the coefficient of friction between Q and the table? *Ans.* $\frac{1}{15}$.

Ex. 625.—A railway truck weighing with its contents 10 tons—resistances being 8 lbs. per ton—is drawn from rest by a horse; after going 300 ft. it is observed to be moving at the rate of 5 ft. per second; determine the number of foot-pounds that has been done by the horse.

Ans. 32,750.

Ex. 626.—A train weighs 100 tons—resistances are 8 lbs. per ton—determine the smallest number of foot-pounds expended in a run of 100 miles on a level road.* *Ans.* 422,400,000.

Ex. 627.—In the last example if the train stops 10 times and the driver in each case gets the speed up to 30 miles an hour, and to save time turns off the steam and puts on the break at 1000 ft. before each station, determine the total loss of work; and the proportion it bears to the total number of foot-pounds that need be expended.

Ans. (1) 59,760,000. (2) nearly $\frac{1}{7}$.

Ex. 628.—A shot weighing 6 lbs. leaves the mouth of a gun with a velocity of 1000 ft. per second: determine the number of foot-pounds accumulated in it, and the mean pressure exerted by the exploded powder behind it if the length of the bore is 5 ft.

Ans. (1) 93,750. (2) 18,750 lbs.

Ex. 629.—If the shot in the last example penetrates 24 in. into a piece of sound oak, determine the mean resistance offered by the wood.

Ans. 46,875 lbs.

---

* It is supposed that at the end of the journey the steam is turned off at such a point that the train just runs into the station without putting on the break.

120. *Velocity acquired by a body in sliding down a smooth curve.*—Let h be the *vertical* height of a point A above another point B, the points being anywhere situated; let us suppose them joined by any *smooth* line, whether *straight* or *curved*; then if a body is supposed to slide in vacuo from A to B along the curve, and if V is the velocity it has at A, and v the velocity it has at B, it can be proved that

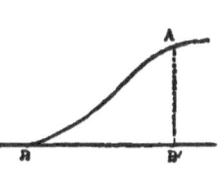

Fig. 155.

$$v^2 = V^2 + 2gh$$

Now, if a point B' were to be taken vertically under A, and in the same horizontal line as B, a body that is thrown down from A with a velocity V will have at B' the same velocity, viz. v—i.e. the change in the velocity of the body between A and the horizontal line B B' is irrespective of the path it describes. In explanation of this remarkable fact it may be observed, that at every instant the reaction is perpendicular to the direction in which the body is moving, and therefore cannot accelerate the velocity. The same formula is true of a body suspended by thread, and oscillating; for the tension of the thread will act at each point perpendicularly to the direction of the body's motion, and will neither accelerate nor retard the velocity.

Ex. 630.—A stone is tied to the end of a string 10 ft. long and describes a vertical circle of which the string is the radius; if at the highest point it is moving at the rate of 25 ft. per second, find its velocity after describing angles of 90°, 180°, and 270° respectively from the highest point.

Ans. (1) 35·6. (2) 43·6. (3) 35·6.

Ex. 631.—Show that if a body oscillate in any arc of a circle, the arc of ascent would always equal the arc of descent if there were no passive resistances.

Ex. 632.—A body is tied to the end of a string 12 ft. long, the other end of which is fastened to a point A; at a distance of 4 ft. vertically below A is a peg B; the body descends through an angle of 30° when the string comes to the peg B; find the angle through which the body will rise.

Ans. 36° 58'.

Ex. 633.—Suppose a body to move in a vertical circle whose radius is r and lowest point A; let V be the velocity it has at a point P and v that which it has at Q; let the chords AP and AQ be denoted by C and c respectively; show that
$$v^2 = V^2 + \frac{g}{r}(C^2 - c^2)$$

Ex. 634.—If a body slide down an arc of a vertical circle to the lowest point of the circle, show that the velocity at that point is proportional to the chord of the arc described.

121. *Centrifugal force.*—If a stone is tied to the end of a string and whirled round, there arises a very peculiar case of the action of forces, and one which requires careful consideration. Suppose the string to be r feet long, the stone to weigh W lbs., and to move with a velocity of V feet per second; now, the tendency of the stone at each instant is to move off in the direction of a tangent to the circle it describes, therefore there must be exerted on it at each instant a certain force (P) acting along the radius and towards the centre sufficient to deflect it from the tangent and to keep it in the circle; this force is given in pounds by the formula

$$P = \frac{W}{g} \cdot \frac{V^2}{r}$$

where g denotes 32·1912 or approximately 32. In the case supposed this force is supplied by the hand which pulls the stone towards the centre; the stone consequently exerts against the hand a reaction equal and opposite to P. This reaction of the stone against the hand is its centrifugal force. The string is, therefore, stretched by two equal forces, viz. P exerted by the hand and the centrifugal force of the body. It must be added, that when any heavy body moves in a circle under the action of any forces whatever, the sum of the resolved parts of the forces in pounds along the radius must at each instant equal $\frac{W}{g} \cdot \frac{V^2}{r}$ or the body will not continue to move in the circle.

CENTRIFUGAL FORCE. 247

We have already seen (Art. 117) that if a body whose weight is W lbs. is acted on by a force of P lbs., it would acquire at the end of every second an additional velocity f equal to $\dfrac{P}{W}g$; in the present case therefore

$$f = \frac{v^2}{r}$$

Ex. 635.—A weight of 1 lb. is fastened to the end of a string 3 ft. long and made to perform 50 revolutions in 1 min. with a uniform velocity; the revolutions take place in a horizontal plane: determine the tension of the string. *Ans.* 2·57 lbs.

Ex. 636.—In *Ex.* 630 determine the tension of the string at the highest and at the other points, supposing the body to weigh 10 lbs.
Ans. (1) 9·53 lbs. (2) 39·53 lbs. (3) 69·53 lbs. (4) 39·53 lbs.

Ex. 637.—If a body moves in a vertical circle the radius of which is 5 ft., determine the velocity at the highest point that the body may just keep in the circle. *Ans.* 12·65.

[Let T be the tension of the string, then $T + W = \dfrac{W}{g} \cdot \dfrac{v^2}{r}$ and the body will just keep in the circle if $T = 0$. If $\dfrac{W}{g} \cdot \dfrac{v^2}{r}$ were less than W the body would fall within the circle; if it were greater than W there would be a certain tension on the string.]

Ex. 638.—In the last example show that the tension of the string at the lowest point will equal 6 times the weight of the body; and that when the body has described a quadrant from the highest point the tension is 3 times the weight of the body.

Ex. 639.—Show that the centrifugal force (f, Art. 121) at the equator equals 0·11129 or the $\frac{1}{289}$ part of what the acceleration due to gravity would be if the earth were at rest.
[See *Ex.* 569 and Table XV.]

Ex. 640.—How many revolutions would the earth have to make in 24 hours, if bodies would *just* stay on her surface at the equator? *Ans.* 17.

Ex. 641.—Given that the moon makes one revolution round the earth in about 2,360,000 seconds, and nearly in a circle whose radius is 59·964 times the earth's equatorial radius, show that the accelerative effect of gravity on the moon must equal $\dfrac{1}{112 \cdot 48}$ reckoning in feet and seconds: what inference can be deduced from this as to the law of the decrease of the earth's attraction?

248   PRACTICAL MECHANICS.

Ex. 642.—A body moves in a circle whose radius is r, the force tending towards the centre necessary to keep it in the circle is P; if U is the work accumulated in the body show that
$$2\,U = P\,r$$

122. *Time of oscillation of a simple pendulum.*—If a small bullet is suspended by a very fine thread, and caused to oscillate in any small arc (e.g. not exceeding 2° or 3° on each side of the lowest point), then the time of each oscillation * is given approximately by the formula

$$t = \pi \sqrt{\frac{l}{g}}$$

where t is the required time in seconds, and l the distance in feet from the point of suspension to the centre of the bullet. It may be remarked that the above formula would be rigorously true if the bullet were reduced to a point, the thread perfectly flexible and without weight, and the arc of vibration indefinitely small: a pendulum possessing these properties (which is of course an abstraction) is called a *simple pendulum,* and the above formula is said to give the time of a small oscillation of a simple pendulum.

Ex. 643.—If $g = 32\cdot2$, determine the number of oscillations made in one hour by a pendulum 3 ft. long. *Ans.* 3754·2.

Ex. 644.—It is found that at a certain place a pendulum 39·138 inches long oscillates in one second; determine the accelerative effect of gravity at that place. *Ans.* 32·1897.

Ex. 645.—Find the time of 100 oscillations of a pendulum 11 ft. long at a place where the length of the seconds pendulum is 39·047 in.
 Ans. 183·9 sec.

Ex. 646.—If L is the length of a seconds pendulum show that
$$g = \pi^2 L$$

Ex. 647.—If L is the length of a seconds pendulum at any place, and l the length of a pendulum that oscillates in n seconds at the same place, show that
$$l = n^2 L$$

---

* I.e. the time of moving from the highest point on one side to the highest on the other.

Ex. 648.—A pendulum at the average temperature oscillates in one second; it is found that its length is L; after a certain time it is found to lose 50 secs. a day; determine the increase of its length.

Ans. 0·00116L.

123. *Centre of oscillation and of percussion.*—Let
A B represent any body capable of oscillating about an axis passing through S perpendicular to the plane of the paper, which plane contains the centre of gravity G: let the body be made to oscillate round the axis, and let the time of its small oscillations be noted; determine the length l of the simple pendulum which would make a small oscillation in the same time; in S G produced take O, such that S O equals l; then the point O is called the *centre of oscillation* of the body corresponding to the *centre of suspension* S. If A B has a definite geometrical form S O can be determined by calculation, as will be shown hereafter; but in any case it can be determined by observation as explained.

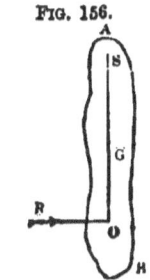

FIG. 156.

In the plane of the paper draw O R at right angles to S O; it admits of proof that if A B * were struck a blow of any magnitude along the line O R, there would be no impulse communicated to the axis; the point O is therefore also called the *centre of percussion.*

Ex. 649.—A mass of oak is suspended freely by a horizontal axis; it is observed to make 43 oscillations in one minute; at what distance below the point of support must a shot be fired into it so that there may be no impulse on the point of support? *Ans.* 6·313 ft.

Ex. 650.—A tilt hammer when allowed to oscillate about its axis is observed to make 35 small oscillations per minute; at what distance from its axis must the point be at which it strikes the object on the anvil in order that no impulse may be communicated to the axis?

Ans. 9·528 ft.

---

* The body A B is supposed to be symmetrical with reference to the plane of the paper.

Table XV.
THE VALUE OF THE ACCELERATIVE EFFECT OF GRAVITY AT DIFFERENT PLACES.

Observer	Place	Latitude	Length of Seconds Pendulum in Inches	Accelerative Effect of Gravity; Feet and Seconds
Sabine	Spitzbergen	N. 79°50′	39·21469	32·2528
Sabine	Hammerfest	70°40′	39·19475	32·2363
Svanberg	Stockholm	59°21′	39·16541	32·2122
Bessel	Königsberg	54°42′	39·15072	32·2002
Sabine	Greenwich	51°29′	39·13983	32·1912
Borda, Biot, and Sabine	Paris	48°50′	39·12851	32·1819
Biot	Bordeaux	44°50′	39·11296	32·1691
Sabine	New York	40°43′	39·10120	32·1594
Freycinet	Sandwich Islands	20°52′	39·04690	32·1148
Sabine	Trinidad	10°39′	39·01888	32·0913
Freycinet	Rawak	S. 0° 2′	39·01433	32·0880
Sabine and Duperrey	Ascension	7°55′	39·02363	32·0956
Freycinet and Duperrey	Isle of France	20°10′	39·04684	32·1151
Brisbane and Rumker	Paramatta	33°49′	39·07452	32·1375
Freycinet and Duperrey	Isles Malouines	51°35′	39·13781	32·1895

124. *Variations in the force of gravity at different places of the earth's surface.*—When experimental determinations of the force of gravity are made with great care, it is found to have different values at different places; these differences are due to two principal causes. (1) The spheroidal form of the earth, in consequence of which the attraction of the earth at different places is not the same. (2) The diurnal rotation of the earth, which causes the sensible or apparent force of gravity to be less than the actual attraction, as part of the latter is employed in keeping bodies on the surface of the earth. Besides these general causes, variations are produced in the determinations made at particular places by differences in their level, and differences in the density

VARIATION OF GRAVITY. 251

of the strata in their immediate neighbourhood. The apparent force of gravity at any place is determined by ascertaining the length L of a simple pendulum which beats seconds at that place, and then the accelerative effect of gravity is determined by the formula (*Ex.* 646)

$$g = \pi^2 \text{L}$$

The preceding table gives the length of the seconds pendulum at different places, according to Sir G. Airy,* and the values of g which can be deduced from them.

* *Figure of the Earth*, p. 229

CHAPTER II.

ON UNIFORMLY ACCELERATED MOTION.

125. *Change in the numerical value of an acceleration due to a change in the units of distance and time.*—In the preceding introductory chapter we have used feet and seconds as the units of distance and time, and we shall continue to do so except in cases where the contrary is specified. In fact, if we take the accelerative effect of gravity as 32·1912 or as 32 approximately—and this is almost universally done—we have tacitly taken feet and seconds as the units of distance and time. It would be quite possible, however, to use any other units we please, as yards and minutes, miles and hours, mètres and sidereal minutes, &c. The question may therefore be asked—Suppose an acceleration to be f when feet and seconds are the units, what will be the numerical value of the acceleration when any other units, as yards and minutes, are used? To answer this question we may reason thus:—Since f is the velocity in feet per second by which the velocity of the body is increased in each second, we have to find in yards per minute the velocity by which the velocity of the body is increased in each minute. Now a velocity of f feet per second is one of $f \times 60 \div 3$ yards per minute; and as this is the increase in each second, the increase in each minute must be $60 \times f \times 60 \div 3$ or $f \times 60^2 \div 3$; this is therefore the required value of the acceleration when the units are yards and minutes. Similar reasoning will apply to other cases.

UNIFORMLY ACCELERATED MOTION. 253

Ex. 651.—What velocity would' be acquired by a body that fell freely for one minute? *Ans.* 1920 ft. per sec.

Ex. 652.—If a body moves with a velocity of 1920 ft. per second, what is its velocity estimated in yards per minute? *Ans.* 38,400.

[Now, it must be remembered that at the end of each *second* a falling body acquires an additional velocity of 32 *feet* per *second*; it appears from the last two examples that the same body would acquire in each *minute* an additional velocity of 38,400 *yards* per *minute*; but the former number represents the accelerative effect of gravity in feet and seconds, and therefore the latter represents the accelerative effect of gravity in yards and minutes.]

Ex. 653.—If the unit of distance were a fathom, and the unit of time 15 sec., what would be the numerical value of g? *Ans.* 1200.

Ex. 654.—Given that the accelerative effect of gravity at the distance of the moon equals $\frac{1}{112}$ in feet and seconds, find its value in miles and hours. *Ans.* 21·91.

Ex. 655.—A certain acceleration has a numerical value f when certain units are employed; show that its value will be $\frac{m^2 f}{n}$ when the new unit of time contains m, and the new unit of distance n, of the old units respectively.

Proposition 23.

If A B C D *be any area bounded by a line straight or curved* C D, *and by straight lines* A B, A D, B C, *of which the two latter are at right angles to* A B; *and if the line* D C *be such that for any point* P *the ordinate* P N *represents the velocity with which a body moves at the end of a time t, that is represented on the same scale by* A N, *then the area of the curve will represent the distance described by the body in the time* A B.

FIG. 157

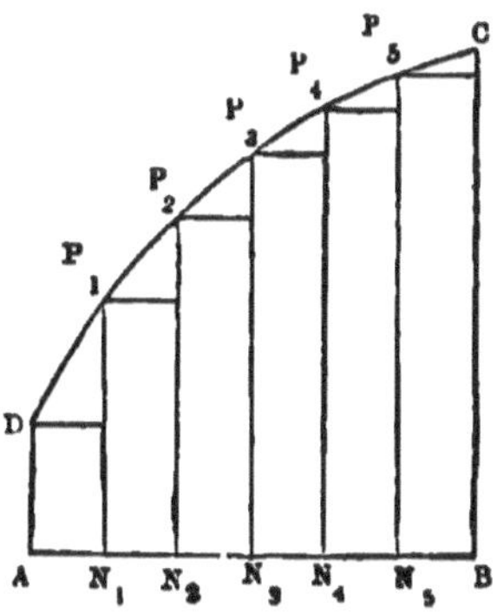

Divide A B into any number of equal parts in $N_1, N_2, N_3, \ldots$ draw the ordinates $P_1 N_1, P_2 N_2, P_3 N_3 \ldots$ and complete the rectangles $D N_1, P_1 N_2, P_2 N_3, \ldots$

Now, if we suppose the body to move during each interval

of time with the velocity it has at the commencement of that interval, it will (Art. 112) describe a distance represented by the sum of the rectangular areas DN_1, P_1N_2, P_2N_3, ...; and this will be true, however great the number of intervals may be, and therefore when the velocity changes continuously, the space described will be correctly represented by the limit of the sum of those areas, i.e. by the curvilinear area A B C D.

Proposition 24.

If a body begins to move along a straight line with a velocity V, *and its velocity undergoes a uniform acceleration f, the distance* (s) *described by it in a time t is given by the formula*

$$s = \mathrm{V} t + \tfrac{1}{2} f t^2$$

Let A B represent the time t on scale; at right angles to A B draw A D and B C representing on the same scale V, the velocity at the beginning of the time, and $\mathrm{V} + ft$ the velocity at the end of the time (Art. 113); join D C, then the area A B C D will represent s the distance described.

FIG. 158.

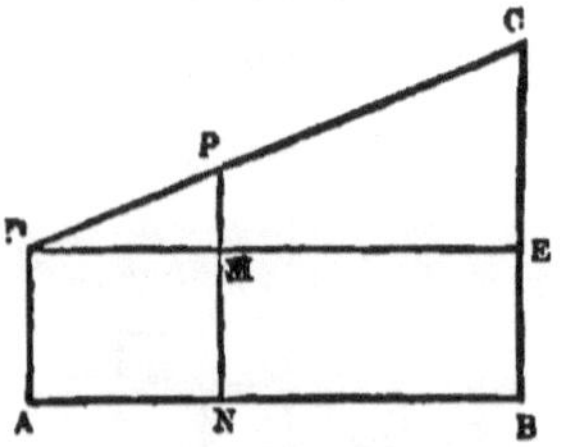

Draw D E parallel to A B; in D C take any point P, and draw P N parallel to D A, cutting D E in M. Now, since D A (which represents V) is equal to B E, we shall have C E equal to $f.t$, or $f \times$ A B. But by similar triangles

$$PM : CE :: DM : DE :: AN : AB$$

and C E equals $f \times$ A B; therefore P M equals $f \times$ A N, therefore P N equals $A D + f \times A N$, i.e. (Art. 113) P N represents the velocity with which the body is moving at the end of the time represented by A N; hence the area A B C D represents the required distance s (Prop. 23).

Now area $A B C D = \frac{1}{2} A B (A D + B C)$

therefore $\qquad s = \frac{1}{2} t (v + v + f t)$

or $\qquad s = v t + \frac{1}{2} f t^2$

If the velocity is uniformly retarded, a precisely similar process leads to the equation

$$s = v t - \frac{1}{2} f t^2$$

and if the body begin to move from rest, we obtain

$$s = \frac{1}{2} f t^2$$

Obs.—On account of the great importance of the formula proved in the proposition it may be well to give briefly another proof. Suppose the time t to be divided into any number (n) of parts each equal to τ, so that t equals $n \tau$. The velocities at the beginning of these successive intervals are v, $v + \tau f$, $v + 2\tau f$, $v + 3\tau f$, ... $v + (n-1)\tau f$. If we suppose the body to move throughout each interval with the velocity it has at the beginning of that interval the distances respectively described will be $v \tau$, $v\tau + \tau^2 f$, $v\tau + 2\tau^2 f$, $v\tau + 3\tau^2 f$, ... $v\tau + (n-1)\tau^2 f$; and the whole distance described on this supposition (s_1) will be their sum, viz.

$$n v \tau + \tau^2 f \{1 + 2 + 3 + \ldots + (n-1)\}$$

therefore $\qquad s_1 = v n \tau + f \tau^2 \dfrac{n(n-1)}{2}$

$\qquad\qquad = v n \tau + \frac{1}{2} f n^2 \tau^2 \left(1 - \dfrac{1}{n}\right)$

$\qquad\qquad = v t + \frac{1}{2} f t^2 \left(1 - \dfrac{1}{n}\right)$

Now s is the limit to which s_1 approximates when n is increased; hence, as $\dfrac{1}{n}$ becomes ultimately zero,

$$s = v t + \frac{1}{2} f t^2$$

Ex. 656.—If a body is thrown up with any velocity, and if t_1 and t_2 are the times during which it is respectively above and below the middle point of its path, show that

$$t_1 : t_2 :: 1 : \sqrt{2} - 1$$

Ex. 657.—If a body falls under the action of any uniformly accelerating force f, show that the distances described in successive seconds form an arithmetical series of which the first term is $\frac{f}{2}$ and the common difference f.

Ex. 658.—In the nth second of its motion a falling body describes h feet: determine its initial velocity.   *Ans.* $v = h - \frac{1}{2}g(2n-1)$.

Ex. 659.—If a body is let fall and describes a certain distance, and this distance is divided into n equal parts, show that the time of describing the first part is to that of describing the last as 1 is to $\sqrt{n} - \sqrt{n-1}$.

Ex. 660.—A falling body describes in the nth second m times the distance described in the $(n-1)$th second; find the whole distance described in the n seconds.

Ex. 661.—There is a chasm with water at the bottom: on dropping a stone down it the splash is heard n seconds after the stone leaves the hand; show that the distance of the surface of the water below the hand is given by the formula ($g = 32.2$)

$$s = 1130\,(35 + n - \sqrt{1225 + 70n})$$

[The velocity of sound may be taken at 1130 ft. per second; it will be observed also that $1130 \div 16.1 = 70$ very nearly; now let x be the time the stone takes to fall, $n-x$ is the time the sound takes to rise, and if s is the required depth we have

$$s = \tfrac{1}{2}gx^2 = 16.1x^2$$

and $\qquad s = 1130\,(n-x)$

therefore $\qquad x^2 = 70\,(n-x)$

whence s is easily found.]

Ex. 662.—When n is but a few seconds, show that the formula in the last example can be written

$$s = \frac{565n^2}{35 + n}$$

Ex. 663.—Determine the values of s from the formulæ of *Ex.* 661 and 662 when n equals 3, 4, and 5 seconds respectively.

Ans. (1) 134·3 ft. and 133·8 ft.   (2) 232·4 ft. and 231·8 ft.   (3) 354·5 ft. and 353·1 ft.

Ex. 664.—A body during the 2nd, 5th, and 7th second of its motion describes respectively 16, 24, and 46 feet. Is this consistent with uniform acceleration?

Ex. 665.—A body moves from rest under the action of a certain force; at the end of 5 sec. the force ceases to act, in the next 4 sec. the body describes 180 ft. Find the accelerative effect of the force.   *Ans.* 9

126. *Composition of velocities.*—Suppose a body to be at any instant at the point A, and suppose it to be moving with such a velocity as would in a certain time carry it to B along the line A B; suppose that at that instant another velocity were communicated to it such as would in the same time carry the body along the line A C to C, if it had moved with that velocity only; complete the parallelogram A B C D and join A D, then at the end of the given time the body will arrive at D, having moved along the line A D. That this is so appears at once from the well-known fact, that when a ship is in a state of steady motion, a man can walk across her deck with precisely the same facility as if she were at rest; thus if he were to walk across the deck when the ship is at rest he would go from A to C; but if we suppose the ship to have such a velocity as will in the same time carry the point A to B, he will come to the point D; and if the velocities have been uniform he will have moved along the line A D. Now, let v and u be the two velocities, then

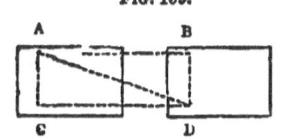

Fig. 159.

$$A B : A C :: u : v$$

and if V is the velocity compounded of them

$$A D : A B :: V : u$$

So that if A B and A C represent the given velocities in magnitude and direction, A D will represent the velocity compounded of them in magnitude and direction. Hence the rule for the composition of velocities is the same, *mutatis mutandis*, as that for the composition of forces.

If the velocities vary from instant to instant, the rule will give the magnitude and direction of the component velocity at any instant; this is the case which commonly happens, for example, when a body is thrown in any

s

direction transverse to the action of gravity. The method of determining the motion of the body may be described in general terms as follows:—Conceive the time to be divided into a great number of intervals, and suppose the velocity that is actually communicated by gravity during each interval to be communicated at once,* then, by the composition of velocities, we can determine the motion during each interval, and therefore during the whole time; the actual motion is the limit to which the motion, thus determined, approaches when the number of intervals is increased.

Proposition 25.

A body is thrown in vacuo with a given velocity (v) *in a direction making any angle with the horizon, to determine its position at the end of any given time* (t).

Let the body be projected along the line A N; take A N equal to vt, and divide t into n parts, each equal τ; then if A N is divided into the same number of equal parts in N_1, N_2, N_3 . . . each part will equal $v\tau$. Now, the effect of gravity is to increase the velocity of a falling body by $g\tau$ in a time τ; we may therefore conceive the body to move through each interval with the velocity it has at the beginning of that interval, and at the end the velocity to be compounded with $g\tau$.

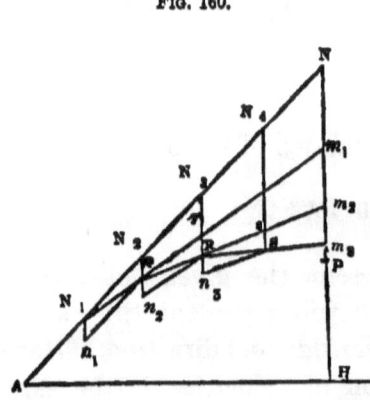

Fig. 160.

During the first interval the body will move over the distance AN_1; draw $N_1 n_1$ vertical and equal to $g\tau \times \tau$; complete

---

* It is immaterial whether we conceive it to be communicated at the beginning or at the end of the interval.

MOTION OF A PROJECTILE. 259

the parallelogram $n_1 N_2$; then since the sides $N_1 n_1$ and $N_1 N_2$ are proportional to the velocities $g\tau$ and v, the body will, during the next interval, move along the line $N_1 Q$, and at the end of the interval will arrive at a point Q vertically under N_2; the actual velocity with which the body has moved being, of course, equal to $N_1 Q \div \tau$. At the point Q we have to compound this velocity with $g\tau$; to do this we must produce $N_1 Q$ to r, making Qr equal to $N_1 Q$; take $Q n_2$ equal to $g\tau \times \tau$, and complete the parallelogram, then the sides of this figure are proportional to the component velocities, and therefore the diagonal is in the same proportion to the velocity compounded of them; at the end of the third interval therefore the body will be at R vertically under N_3; the same construction will apply to any number of intervals, and the required point P will be vertically under N. To determine NP; produce $N_1 Q$ to m_1, QR to m_2, RS to m_3, &c., then will NP equal the limit of the sum of $Nm_1, m_1 m_2, m_2 m_3$, &c.; but by similar triangles $N m_1$ is the same multiple of $N_2 Q$ that $N_1 N$ is of $N_1 N_2$, therefore $N m_1$ equals $(n-1) g \tau^2$, similarly $m_1 m_2$ equals $(n-2) g \tau^2$, $m_2 m_3$ equals $(n-3) g \tau^2$, &c., and therefore their sum equals

$$(n-1) g \tau^2 + (n-2) g \tau^2 + (n-3) g \tau^2 + \ldots + 2 g \tau^2 + g \tau^2$$

$$= g \tau^2 \{(n-1) + (n-2) + \ldots + 2 + 1\}$$

$$= g \frac{t^2}{n^2} \cdot \frac{n(n-1)}{2} = \tfrac{1}{2} g t^2 \left(1 - \frac{1}{n}\right)$$

Now, however great the number of intervals, Q, R, S, &c., will remain vertically under N_2, N_3, N_4, &c., so that in the limit P will remain vertically under N. Also the limit of $\tfrac{1}{2} g t^2 \left(1 - \dfrac{1}{n}\right)$ is $\tfrac{1}{2} g t^2$; therefore the true position of the

s 2

body will be found by measuring downward from N a distance equal to $\frac{1}{2}g t^2$.*

Ex. 666.—A body moves along a smooth horizontal plane with a velocity of 3 ft. per second; at the end of 2 seconds a velocity of 8 ft. per second is impressed on it in a direction at right angles to its motion; after how long will its distance from the starting-point be 20 ft.? *Ans.* 4 sec.

Ex. 667.—A body is projected in vacuo with a given velocity in a given direction; determine its range on the horizontal plane passing through the point of projection and the time of flight.

Let A be the point of projection, A N the direction of projection, A B the horizontal plane through A, let the velocity of projection be denoted by V, and the angle N A B by a. Let the body strike the plane at B after T seconds; we have to determine A B and T; draw B N at right angles to A B, then (Prop. 25)

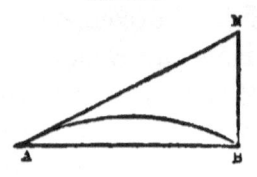

FIG. 161.

$$A N = V T$$

and $\qquad N B = \frac{1}{2} g T^2$
but $\qquad N B = A N \sin a$
therefore $\qquad \frac{1}{2} g T^2 = V T \sin a$
or $\qquad T = \dfrac{2 V}{g} \sin a$
Again $\qquad A B = A N \cos a = V T \cos a$
therefore $\qquad A B = \dfrac{2 V^2}{g} \sin a \cos a$
or $\qquad A B = \dfrac{V^2 \sin 2a}{g}$

Ex. 668.—A body is projected with a velocity of 100 ft. per second in a direction making an angle of 37° with the horizon: determine the time of flight and range on a horizontal plane. *Ans.* 3·76 sec. and 300·4 ft.

Ex. 669.—If a body is thrown with a given velocity the horizontal range is greatest when it is projected at an angle of 45°; and for angles of projection one as much less as the other is greater than 45° the horizontal ranges are the same.

---

* This result is, of course, true for any constant force acting along parallel lines, the accelerative effect of which is f. Hence, if a body has at a certain point a given velocity (V) along a given line A N, and is acted on by such a force, its position at the end of t seconds is given by the construction—take A N equal to V t, draw N P parallel to the direction of the force, take N P = $\frac{1}{2} f t^2$; P is the required position of the body.

MOTION OF A PROJECTILE. 261

Ex. 670.—Show that the least velocity with which a body can be projected to have a horizontal range R is $4\sqrt{2R}$ ft. per second.

Ex. 671.—Determine the angle of elevation and velocity of projection that will enable a body to strike the ground after 10 seconds at a distance of 5000 ft. from the point of projection.
Ans. (1) 17° 45'. (2) 525 ft. per sec.

Ex. 672.—A body is projected with a velocity V in a direction making an angle *a* with the horizon; if R is its range on a plane passing through the point of projection and inclined at an angle *θ* to the horizon, and T the time of flight, determine R and T.

FIG. 162.

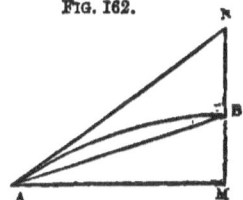

Let A be the point of projection; draw A M a horizontal line through A, A B the inclined plane, A N the direction of projection; let the projectile strike the plane at B, then we have

A N = V T and N B = $\frac{1}{2} g$ T²

but   A N : N B :: sin A B N : sin N A B

or   V T : $\frac{1}{2} g$ T² :: cos *θ* : sin (*a* − *θ*)

therefore $T = \frac{2V}{g} \cdot \frac{\sin(a-\theta)}{\cos\theta}$

Again   A N : A B :: sin A B N : sin A N

or   V T : R :: cos *θ* : cos *a*

therefore $R = \frac{2V^2}{g} \cdot \frac{\sin(a-\theta)\cos a}{\cos^2\theta}$

Ex. 673.—Determine the time of flight and range on a plane inclined at an angle of 10° upward from the horizon in the case of a body projected as in *Ex.* 668.   *Ans.* 2·88 sec. and 233·6 ft.

Ex. 674.—A body is projected with a velocity of 120 ft. per second in a direction making an angle of 28° 45' with the horizon ; determine the time of flight and range on a plane passing through the point of projection—(1) when it is horizontal ; (2) when inclined upwards from the horizon at an angle of 12° ; (3) when inclined downward at the same angle.
Ans. (1) 3·61 sec. and 379·5 ft. (2) 2·21 sec. and 237·7 ft.
(3) 5 sec. and 538·3 ft.

Ex. 675.—A body is thrown horizontally with a velocity of 50 ft. per second from the top of a tower 100 ft. high; find after how long it will strike the ground, and at what distance from the foot of the tower.
Ans. (1) 2·5 sec. (2) 125 ft.

Ex. 676.—If any number of bodies are thrown horizontally from the top of a tower, they will all strike the ground at the same instant whatever be the velocities of projection.

Ex. 677.—There is a hill whose inclination to the horizon is 30°; a projectile is thrown from a point on it at an angle inclined to the horizon at 45°; show that if it were projected down the plane its range would be nearly $3\frac{3}{4}$ times what the range would be if it were thrown up the plane.

Ex. 678.—In the last example suppose the slope of the hill to be due north and south, and the azimuth of the plane of projection to be A; show that the sum of the two ranges obtained by throwing the body towards the ascending and descending parts of the hill equals

$$\frac{2v^2}{g}\sqrt{1 + \frac{\cos^2 A}{3}}$$

[The azimuth is the bearing of a point from the south measured on a *horizontal plane.*]

Ex. 679.—If there are two inclined planes and the angle between them is bisected by the horizontal plane, and if the ranges of the projectile on the three planes are R_1, R_2, and R respectively, show that

$$R_1 + R_2 : R :: 2 : \cos \text{inclination}.$$

Ex. 680.—If in *Ex.* 672 the body is so projected as to obtain the greatest range with a given velocity, show that the direction of projection must bisect the angle between the vertical and the plane.

[It must be remembered that $2 \sin(a-\theta) \cos a = \sin(2a-\theta) - \sin\theta$.]

Ex. 681.—Referring to the figure in Prop. 25, if A H and H P are denoted by x and y respectively, show that

$$x = v t \cos a$$
$$y = v t \sin a - \tfrac{1}{2} g t^2$$

Ex. 682.—Show that the highest point the projectile can reach is $\frac{v^2}{2g} \sin^2 a$ feet above the middle point of the horizontal range.

Ex. 683.—Two bodies of unequal weight are thrown from the same point in different directions with different velocities; find the position of their centre of gravity after t seconds.

Proposition 26.

The curve described by a projectile in vacuo is a parabola whose directrix is horizontal, and at a height above the point of projection equal to that to which the velocity of projection is due.

Let P be the point, and PM the direction of projection;

MOTION OF A PROJECTILE. 263

let P Q be the path of the projectile, and Q its position at the end of t seconds; draw the vertical lines D P N and M Q, also draw Q N parallel to P M, then

FIG. 163.

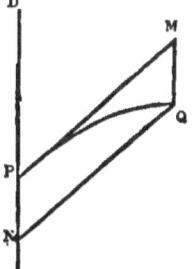

$$P N = Q M = \tfrac{1}{2} g t^2$$
$$Q N = P M = V t$$
$$\therefore Q N^2 = \frac{2 V^2}{g} \cdot P N$$

Now, this relation between Q N and P N is the same, wherever on the curve we may take Q; but if a parabola were drawn through P touching P M, with its diameter vertical and its directrix passing through a point D so taken that $4 P D$ equals $\dfrac{2 V^2}{g}$ we should have for any point of it

$$Q N^2 = \frac{2 V^2}{g} \cdot P N$$

i.e. it would coincide with the curve described by the projectile: hence that curve is a parabola whose directrix is horizontal and passes through the point D; but it will be remarked that

$$V^2 = 2 g \cdot D P$$

So that D P is the height to which the velocity of projection is due. (See Art. 115.)

Cor.—The velocity of the projectile at any point is that due to the height of the directrix above that point.

Let P Q P' be the path of the projectile, of which D D' is the directrix; at the point Q let the body be moving with a velocity v in the direction Q T; now, it is plain that if another body were thrown from Q in the direction Q T with an equal velocity v, it would move in exactly the same

FIG. 164.

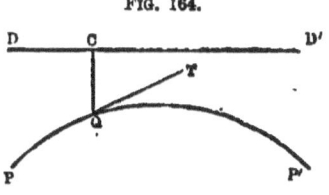

manner as the projectile, i.e. it would describe the curve Q P'; but if a body is thrown from Q so as to describe that curve, it must be thrown with the velocity due to the height Q C, i.e. $v^2 = 2g \cdot CQ$.

Ex. 684.—In fig. 161, show that if B N is divided into any number of equal parts, and the points of section joined to A, the curve will be divided into parts that are described in equal times.

Ex. 685.—Show that the velocity v of the projectile at any time t is given by the formula
$$v^2 = \mathrm{v}^2 - 2\mathrm{v}\,gt \sin \alpha + g^2 t^2$$

Ex. 686.—There is a wall b feet high, a body is thrown from a point a feet on one side of it so as just to clear the wall and to fall c feet on the other side: show that
$$\tan \text{(angle of elevation)} = \frac{b(a+c)}{ac}$$
$$\text{(vel. of projection)}^2 = \frac{g(a+c)}{2}\left\{\frac{ac}{b(a+c)} + \frac{b(a+c)}{ac}\right\}$$

N.B. Throughout the volume no account is taken of the resistance offered by the air to bodies moving in it. If the velocity is considerable this resistance becomes very great, and if taken into account seriously modifies the answers to question relating to falling bodies and projectiles; how seriously, may be judged from this, that Newton found that when glass globes, some about three-quarters of an inch in diameter, filled with mercury, others, about five inches in diameter, containing only air, were allowed to fall, the former took about four seconds, the latter about eight seconds, to describe 220 feet.*

---

* *Principia*, p. 352, 3rd ed.

CHAPTER III.

ON FORCE AND MOTION.

127. *Mass, density, momentum*.—If a body when placed in one pan of a perfectly just balance exactly counterpoises a second body when placed in the opposite pan, the bodies are said to contain equal quantities of *matter*. If the bodies were both lead they would contain equal quantities of lead, if both platinum they would contain equal quantities of platinum; if one were lead and the other platinum, we can say that they contain equal quantities of *matter*—the word matter being a general term, denoting *lead, platinum, wood, stone, water, air*, &c. When any two bodies contain equal quantities of matter they possess certain qualities in common, thus: —If they are successively placed in the same relative position to a third body, the mutual attractions between them severally and the third body will be equal; e.g. if the third body is the earth, this is the fact indicated by their counterpoising each other in a perfectly just balance. Again, if they both move with equal uniform velocities of translation, they will have equal kinetic energies, i.e. the same power of doing work. If they move with equal uniform velocities in equal circles they must be acted on by equal forces tending to the centres of those circles. Other physical properties might be named which they have in common. It is on account of all bodies possessing these and other properties in common that there is need for the use of the term *matter* in the science of mechanics.

The quantity of matter in a body is called its *mass*. Now, any substance which is an exact counterpoise to the standard pound contains as much matter as the standard pound. We may therefore take the standard pound as the unit of mass, and may speak of a body whose mass is three, four, or any number of *pounds*. If this is done, it must be borne in mind that we are using the word pound in a different sense from that in which it denotes a unit of force. That the senses are really different is evident on a little consideration, thus:—Suppose a perfectly just balance to be in equilibrium with a standard pound in one pan and an equal mass in the other, the equilibrium would be maintained if the force of gravity were altered by any amount, provided it continued to act equally on both bodies and along vertical lines. But if the force of gravity were altered, the pressure exerted by the standard pound on a horizontal plane supporting it would no longer be the same, and for this reason a particular place is mentioned in defining the unit of force, viz. the mutual sensible attraction in London between the earth and a pound of matter is a force of one pound (Art. 23), or, as it may be conveniently termed, a gravitation unit of force.*

The following definitions are connected with the definition of mass:—

(*a*) A body is any limited portion of matter.

(*b*) A particle is a body so small that for the purposes of a given investigation the distances between its parts may be neglected.

---

* The variations in the *weight* of a given body—the variations in the mutual sensible attraction between the earth and that body—at different points on the earth's surface, could be observed directly by means of a delicate spring.—Herschel's *Outlines of Astronomy*, Art. 234. (See *Ex*. 692.)

It may easily happen that in one investigation a body may be properly treated as a particle, and that in another the distances between the parts of a much smaller body must be attended to. Accordingly, when the word particle is used in the sense above defined the mass of a particle may be several pounds, or even tons.

(c) A body is said to be of *uniform density* when any equal volumes taken from any parts of the body contain equal quantities of matter.

(d) When a body has a uniform density, the quantity of matter in the unit of volume is called its *density*.

(e) The momentum of a moving body is a quantity which varies directly as its mass and velocity jointly.

If there are two bodies whose masses are in the ratio of M to M_1 and their velocities in the ratio of V to V_1, their momenta will be in the ratio of M V to $M_1 V_1$. If we agree to call the momentum of a body unity when its mass is one pound and its velocity one foot per second, the numerical value of the momentum of any body will be its mass in pounds multiplied by its velocity in feet per second.

It may be well to call the attention of the student to the fact that the word *weight* has two distinct meanings—first, 'quantity measured by the balance,' as when we say that the weight of a leg of mutton is ten pounds; in this sense weight is mass or quantity of matter; secondly, 'gravity or heaviness,' as when we say that a spring is pressed down by the weight of a body; in this sense weight is one kind of force. Practically speaking, the context will nearly always show in which of the two senses the word is used. In dynamics, where the distinction is of great importance, it is convenient, on account of the ambiguity of the word weight, to use the word mass where quantity of matter is intended.

128. *The absolute unit of force.*—The determination

of the numerical value of a force is effected by the principle that the force exerted on a body is proportional to the momentum it imparts to the body in a given time. Let P and P_1 be forces acting on bodies (A and B) whose masses are in the ratio M to M_1, and in any given time let them impart to these bodies velocities in the ratio of V to V_1; then the fundamental principle above stated enables us to draw the conclusion that

$$P : P_1 :: MV : M_1 V_1$$

Thus, if the masses of A and B are in the ratio of 3 to 4, and the velocities communicated to them in any equal times are in the ratio of 5 to 7, the force P must bear to the force P_1 the ratio of 15 to 28.

Now suppose that we agree to call that force unity which acting on a pound of matter for one second would communicate to it a velocity of one foot per second. Such a force is called an absolute unit of force, or a British absolute unit of force.* If a force (P) communicates in one second a velocity of f feet per second to a mass of m pounds we must have

$$P : 1 :: mf : 1 \times 1$$

or
$$P = mf$$

and this is the numerical value of the force (P) in absolute units. Thus if a force communicates in one second a velocity of 8 feet per second to a mass of 5 lbs., the force must be one of 40 absolute units. It will be observed that f is the acceleration of the velocity of m produced by P, and consequently the number of absolute units of force in P is the mass of the body in pounds multiqlied by the

---

* It would, of course, be possible to use any disunits of tance, mass, and time; the force which in one second communicates to a gramme of matter a velocity of one centimètre a second is an absolute unit of force and is called a C.G.S. unit, or a *dyne*. The British absolute unit is sometimes called a *poundal*.

THE ABSOLUTE UNIT OF FORCE. 269

accelerative effect of P on the body estimated in feet and seconds.

The relation between the absolute unit and the gravitation unit can be determined thus:—The sensible force of gravity on a pound of matter in London is, as we have seen, a gravitation unit; but if a body falls freely for a second in London it acquires a velocity of 32·1912 feet per second, so that the force producing this acceleration in a pound of matter must be one of 32·1912 absolute units, and, consequently, a gravitation unit of force equals 32·1912 absolute units. It will be observed that the force of gravity at the earth's surface on (about) half an ounce of matter is about an absolute unit.

Let g and g' denote the accelerative effect of gravity in London and in any other place; the force of gravity on a mass of m pounds at that place will be $m\,g'$ absolute units, and therefore $m\,g' \div g$ gravitation units. If the numbers registered in the last column of Table XV. are considered, it will now be seen that they admit of two interpretations. The entry for any one place gives the velocity in feet per second acquired in one second by a body falling freely at that place; it also gives in absolute units the sensible force of gravity exerted at that station on a mass of one pound.

129. *The unit of work.*—We have already seen (Art. 103) that when the point of application of a force equal to the unit of force is made to move through a distance equal to the unit of distance the work done is a unit of work, and when the unit of distance is a foot and the unit of force a pound (or gravitation unit) the unit of work is called a foot-pound. Of course the work done when a force equal to the absolute unit is exerted through a foot is also a unit of work, and we may call it an absolute unit of work.*

* It has been proposed to call this unit of work a *foot-poundal*. The work done when a C.G.S. unit, or dyne, is exerted through a centimètre is called an *erg*.

It is evident that a foot-pound equals 32·1912 absolute units of work.

Let a particle whose mass is M pounds be moving at any instant with a velocity of V feet a second, and suppose it to be brought to rest after describing a distance of h feet from that instant by a force of P absolute units acting in a direction opposite to that of its motion. If f be the accelerative effect of P on M we have the following relations:—

$$P = Mf$$

and $$2fh = V^2$$

therefore $$Ph = \tfrac{1}{2} M V^2$$

Now Ph is the number of absolute units of work done against P, and consequently $\tfrac{1}{2} M V^2$ is the kinetic energy of the body, or its capacity of doing work estimated in absolute units of work, when M is in pounds and V in feet per second. If the kinetic energy is required in foot-pounds its value in absolute units of work must be divided by 32·1912 (Art. 119).

It will of course follow from this that if the velocity of a particle is increased from V to v, a force or forces must have acted which have done on it $\tfrac{1}{2} M (v^2 - V^2)$ absolute units of work; and if its velocity is diminished from V to v it must have done against a force or forces $\tfrac{1}{2} M (V^2 - v^2)$ absolute units of work.

In the following examples, which illustrate the subject of Arts. 128, 129, the numerical value of g in London is taken strictly as 32·1912; in other parts of the book the appproximate value 32 is used except where the contrary is stated.*

---

* The pound has been used as the unit of mass throughout the remaining pages of the book. This accounts for the answers to several of the questions differing from those given in some of the previous editions.

THE ABSOLUTE UNIT OF FORCE. 271

Ex. 687.—A force estimated in British absolute units is 90; what would be the numerical value of the force if the unit of mass were 1 cwt. and the units of distance and time a yard and a minute? *Ans.* 964·3.

Ex. 688.—How many British absolute units would there be in an absolute unit of force if the kilogramme, mètre, and second were employed as the units of mass, distance, and time? A kilogramme equals 2·2055 lbs. and a mètre equals 39·37043 inches. *Ans.* 7·236.

Ex. 689.—Show that a British absolute unit or *poundal* equals 13825 dynes, and that a foot-poundal equals 421394 ergs; having given that a mètre equals 39·37043 inches and a gramme 15·43235 grains.

Ex. 690.—A mass of 500 lbs. is acted on by a force of 125 absolute units; what distance will it describe from rest in 8 seconds? *Ans.* 8 ft.

Ex. 691.—A cubic foot of cast iron is observed to increase its velocity by 5 ft. per second in every second of its motion; determine the force which produces this acceleration, assuming that a cubic foot of water weighs 1000 oz. *Ans.* (1) 2252·2 absolute units. (2) 69·96 grav. units.

Ex. 692.—A body whose mass is 10 lbs. moves from rest along a straight line under the action of a constant force; during the first second of its motion it describes a distance of 50 ft. What is the force which produces the motion? What is the mass of the body which that force would just support in London? What in Trinidad? (Table XV.)
Ans. (1) 1000 absolute units. (2) 31·06 lbs. (3) 31·16 lbs.

Ex. 693.—The accelerative effect of the moon's attraction on a particle situated on its surface is about 5·4.* A man can jump to a height of 5 ft. on the earth's surface; how high could he jump on the moon's surface?
Ans. 29·6 ft.

[In both cases the mass of the man's body is the same, and the force exerted by the muscles is the same, quite irrespectively of the force of gravity; consequently, in the act of jumping, the velocity with which he leaves the ground is the same in both cases.]

Ex. 694.—Suppose that near the surface of the moon a mass of 5 lbs. were placed on a plane, and it were observed that the plane descended with a uniform acceleration of 3·6 ft. per second; what pressure would the mass produce on the plane in absolute units? How much matter would a force equal to this pressure support in London? *Ans.* (1) 9. (2) 4·47 oz.

Ex. 695.—If near the moon's surface two bodies are connected by a fine thread passing over a smooth fixed cylinder, the former, containing 30 lbs. of matter, falls and pulls up the latter, which contains 20 lbs. of matter; find the tension of the thread (1) in absolute units, (2) in gravitation units.
Ans. (1) 129·6. (2) 4·02.

---

* Herschel's *Outlines of Astronomy* Art. 508.

Ex. 696.—A mass of 7 lbs. falls through 100 ft. near the surface of the moon; what kinetic energy does it acquire in the fall?
Ans. 3780 abs. units, or 117·4 ft.-pounds.

Ex. 697.—If equal forces (P) act on two unequal bodies for the same time, show that the bodies will acquire equal momenta.

Ex. 698.—Show that momenta of the bodies in the last example will be equal when P varies from instant to instant, provided the forces are the same at the same instant throughout their time of action.

[The results of the last two examples are of considerable importance; they are almost self-evident and therefore liable to be forgotten—for this reason the student's attention is particularly directed to them.]

Ex. 699.—When the powder in the bore of a cannon is ignited, the pressure on the ends of the bore and on the shot are at each instant equal: a shot weighing 6 lbs. is fired from a gun quite free to move and weighing 6 cwt.; the velocity with which the shot leaves the gun is 1000 ft. per second; what is the velocity of the gun's recoil? *Ans.* 8·93 ft. per sec.

Ex. 700.—Show that the kinetic energy of the gun is always small compared with that of the shot; and ascertain their values in foot-pounds in the case suggested in the last example.
Ans. 93,750 ft.-pds. in the shot and 837 in the gun.

Ex. 701.—If the trunnions of the gun in *Ex.* 699 are supported on two parallel smooth planes inclined at an angle of 30°, determine how far it will move along these planes. *Ans.* 2·5 ft.

Ex. 702.—The mercurial barometer (when placed on the sea-level) stands at any given height; suppose the force of gravity to undergo any change either of increase or decrease; explain why this, unaccompanied by any other change, would not alter the height of the barometer.

Ex. 703.—There are two mercurial barometers placed on the sea-level; the height of the mercury in the one is 30 in., and there is a perfect vacuum above it; the height of the mercury in the other is 28 in., and there is an imperfect vacuum of 6 in. above it; suppose the force of gravity to change from 32 to 24, at what height will the mercury now stand in the second barometer? *Ans.* $32 - 2\sqrt{5}$ in.

130. *Motion on an inclined plane.*—Let a particle whose mass is M lbs. slide on a smooth inclined plane. The force of gravity on the body, which is M g absolute units, may be resolved into two forces M g sin a along, and M g cos a at right angles to, the plane, where a denotes the inclination of the plane to the horizon. The latter of

MOTION ON AN INCLINED PLANE. 273

these forces is neutralised by the reaction of the plane, and consequently the motion is caused by the former force. If, therefore, f denotes the acceleration of the velocity of the body when moving down the plane, Mf is the force causing motion, and therefore

$$f = g \cos a$$

If the plane is rough there will be a friction $\mu M g \cos a$, where μ stands for the coefficient of friction between the body and the plane; consequently in this case the force causing motion is $Mg \sin a - \mu M g \cos a$; and as this must equal Mf we must have

$$f = g \sin a - \mu g \cos a$$

In like manner if the body moves up the plane the retardation (f) is given by the equation

$$f = g \sin a + \mu g \cos a$$

If the angle of friction is denoted by ϕ, the last two equations may be written

$$f \cos \phi = g \sin (a - \phi)$$

and $$f \cos \phi = g \sin (a + \phi)$$

The reader will not have failed to observe the distinction between the case of a body moving upward against gravity and that of a body sliding on a rough plane against friction. In the former case, when the upward velocity of the body is exhausted, it immediately begins to move downward. In the latter case the body simply comes to rest. Friction has no tendency to reverse the direction of the motion; it tends merely to stop the motion.

Ex. 704.—Find the velocity acquired by a body in descending a smooth inclined plane 50 ft. long and having an inclination of 23°; determine also the velocity that would be acquired if the limiting angle of resistance were 15°.   *Ans.* (1) 35·4 ft. per sec.   (2) 21·5 ft. per sec.

T

274   PRACTICAL MECHANICS.

Ex. 705.—There is a plane 50 ft. long and inclined to the horizon at an angle of 30°; the limiting angle of resistance between it and a given body is 15°; determine the velocity the body must have at the foot of the plane so as just to reach the top, and the time it will take to get there.
Ans. (1) 48·4 ft. per sec. (2) 2·07 sec.

Ex. 706.—A body just rests on a plane which is inclined at an angle of 30° to the horizon. Find the velocity acquired and the distance described from rest in 2 sec. by the body when the plane is inclined at an angle of 60° to the horizon.   *Ans.* (1) 36·9 ft. per sec. (2) 36·9 ft.

Ex. 707.—If a body slides down a gentle incline of 1 foot vertical to m horizontal, show that the acceleration very nearly equals $\left(\dfrac{1}{m} - \mu\right) g$. And if $m = 100$ show that the error equals about $\frac{1}{20000}$ part of the whole.

Ex. 708.—A train moving at the rate of 24 miles an hour comes to the top of an incline of 1 ft. in 350; the resistances are 8 lbs. per ton; the steam is cut off at the top of the incline, and the train comes to rest at its foot; determine—(1) the retardation of the train's velocity; (2) the length of the incline; (3) the time of motion.
Ans. (1) $\frac{4}{175}$. (2) 27,104 ft. (3) 1540 sec.

Ex. 709.—At the slide at Alpnach the first declivity has an inclination of 20° 30′ and is 500 ft. long; being kept continually wet the limiting angle of resistance is 14°; in how many seconds would a tree descend this first declivity were it not for the resistance of the air?   *Ans.* 14·3 sec.

Ex. 710.—A body slides from rest down a plane whose inclination is i and length L; it passes with the velocity acquired during the descent of the first plane to a second whose inclination i is less than the limiting angle of resistance ϕ; if l is the distance through which it slides before coming to rest show that

$$L = \frac{\sin(i-\phi)}{\sin(\phi-i)}$$

Ex. 711.—A train beginning to run down an incline with a given velocity, will go before stopping n times as far as it would do if it had begun to run up the incline with the same velocity; show that the tangent of the angle of inclination of the plane is $\mu (n-1) \div (n+1)$.

Ex. 712.—In the last example, if the friction is 8 lbs. a ton and the incline 1 in 420, show that it will run down the plane before stopping five times farther than up the plane; that the time which the train takes in coming to rest is also five times as great in the latter case as in the former.

Ex. 713.—Two planes equally inclined to the horizon ascend from the same point on opposite sides of the vertical line; a particle slides down the former and passes without change of velocity on to the seconds, up which it slides till its velocity is exhausted; it then slides down the second and up the first; if L and l are the lengths of the first plane describe in the descent and ascent respectively, show that

$$L \sin^2(\alpha - \phi) = l \sin^2(\alpha + \phi)$$

and that the time of the motion is

$$\frac{2t \sin \alpha \cos \phi}{\sin(\alpha + \phi)} \cdot \left\{ 1 + \sqrt{\frac{\sin(\alpha - \phi)}{\sin(\alpha + \phi)}} \right\}$$

where t is the time of descent of the first plane.

Ex. 714.—A body whose mass is M lbs. is pulled up an inclined plane by a force of Pg absolute units that acts parallel to the plane; show that the accelerative effect equals $\left(\dfrac{P}{M} - \dfrac{\sin(\alpha + \phi)}{\cos \phi} \right) g$, where α is the angle of inclination, and ϕ the limiting angle of resistance.

Ex. 715.—Let AC, CB be two planes sloping downward in contrary directions from the point C, and inclined to the horizon at angles A and B respectively; a mass P slides down CA and draws a mass Q up CB by means of a fine cord which passes over C and is tied to each body: if the limiting angle of resistance between the bodies and the planes is ϕ, show that

$$f = \frac{P \sin(A - \phi) - Q \sin(B + \phi)}{(P + Q) \cos \phi} g$$

Ex. 716.—In the last case if the inclines are equal and small, being 1 in m, show that

$$f = \left\{ \frac{1}{m} \cdot \frac{P - Q}{P + Q} - \mu \right\} g$$

Ex. 717.—If the resistances are 8 lbs. per ton and the incline 1 in 140, and a set of full trucks is required in their descent to pull up the incline an equal number of similar empty trucks, show that the contents of each truck should on the average be more than double the weight of the truck.

⁰*Ex.* 718.—If a circle be placed with its plane vertical, and through its highest point any chord be drawn, a body will descend along that chord (supposed to be smooth) in the same time as down the vertical diameter.

Ex. 719.—If through any point there is drawn a vertical line and any number of inclined planes on the same side of the line, and having a common limiting angle of resistance ϕ; then if bodies begin to slide from the point down these planes at the same instant, show that after any interval they will be found in the arc of the segment of a vertical circle cut off by the vertical line which subtends at the centre an angle equal to $\pi - 2\phi$.

131. If we are only required to find a relation between the velocity of the body and the distance it describes on

an inclined plane, we may reason on its kinetic energy; thus:—Suppose it is at any instant moving with a velocity V, and after it has moved up the plane a distance h let its velocity have changed to v; the change in its kinetic energy is $\frac{1}{2}\text{M V}^2 - \frac{1}{2}\text{M}v^2$, and this by *Ex.* 522 must equal the number of absolute units of work done against gravity ($\text{M}gh\sin a$) and the number done against friction ($\mu\text{M}gh\cos a$) during the motion. Therefore $\frac{1}{2}\text{M}(\text{V}^2-v^2)$ $=\text{M}gh(\sin a + \mu \cos a)$; and we may reason similarly in other cases.

Ex. 720.—A train moving at the rate of 15 miles an hour comes to the foot of an incline of 1 in 300, resistances 8 lbs. per ton; if the steam is cut off, how far will it go before stopping? *Ans.* 1095 ft.

Ex. 721.—A body slides down an inclined plane the height of which is 12 feet and length of base 20 feet; find how far will it slide along a horizontal plane at the bottom, supposing the coefficient of friction on both planes to be $\frac{1}{8}$, and that it passes from one plane to the other without loss of velocity? *Ans.* 52 ft.

Ex. 722.—A train weighing 90 tons comes to the foot of an incline of 1 in 160 with a velocity of 30 miles an hour; the resistances are 7 lbs. per ton, the length of the incline 2 miles; the train has at the top of the incline a velocity of 20 miles an hour; how many foot-pounds of work have been done by the steam in getting the train up the incline? and through how great a distance would the same expenditure of work have taken the train with a uniform velocity along a horizontal line?

Ans. (1) 16,570,400 ft.-pds. (2) 26,302 ft.

Ex. 723.—If a train begins to descend the incline in the last example with a velocity of 20 miles an hour, how far will it descend by its own weight before acquiring a velocity of 30 miles an hour? *Ans.* 5378 ft.

Ex. 724.—There are two points A and B on a railroad four miles apart on the same horizontal line; the railroad is in two equal inclines, one up and the other down, of 1 in 160; the train which weighs 50 tons and experiences resistances equal to 7 lbs. per ton, has a velocity of 30 miles an hour at A and B, and a velocity of 20 miles an hour at the top of the incline; the velocity being supposed to change uniformly from 30 to 20, and again from 20 to 30, and when the latter velocity is attained further acceleration is checked by putting on the break; determine—(1) the loss of work in foot-pounds; (2) the loss of time due to the inclines.

Ans. (1) 1,810,000 foot-pounds. (2) $72\frac{1}{2}$ sec.

Ex. 725.—A chest 6 ft. long and 2 ft. square stands on its end on the

NEWTON'S LAWS OF MOTION. 277

deck of a ship, one face being perpendicular to the direction of the motion; the ship is suddenly brought to rest; what must have been its velocity if the chest is just overthrown, it being supposed that all sliding is prevented?

Ans. 2·2 miles per hour.

[If v is the required velocity and M the mass of the chest in pounds, its kinetic energy in absolute units is $\frac{1}{2}Mv^2$; and to overthrow the chest requires $Mg(\sqrt{10}-3)$ absolute units of work.]

Ex. 726.—Show from the principles of the present article that the velocity acquired by the bodies in *Ex.* 715 while moving from rest over a length l of the planes is given by the formula

$$v^2 = 2gl \cdot \frac{P \sin(A-\phi) - Q \sin(B+\phi)}{(P+Q)\cos\phi}$$

Ex. 727.—There is an inclined plane of 1 in 90 along which a train weighing 80 tons is made to descend for a distance of 300 ft.; to the train is attached a rope which, after passing round a pulley at the top of the incline, is fastened by the other end to a lighter train weighing 16 tons; the rope is so long that the light train is at the foot of the incline when the heavy one is at the top; find—(1) the velocity with which the heavy train reaches the foot of the incline; (2) if the heavy train is disconnected from the light one at the foot of the incline, find the distance to which it will run before stopping on the horizontal plane, resistances on the incline being 7 lbs. per ton, on the level 8 lbs. per ton.

Ans. (1) 9·07 ft. per sec. (2) 359·7 ft.

132. *Newton's laws of motion.*—In the preceding pages we have taken for granted, as we have required them, several fundamental principles respecting the motion of bodies—such, for instance, as the equality of the mutual action and reaction of two bodies on each other; that a body has no inherent power of changing the velocity or direction of its motion, that such changes must be produced by action from without, i.e. by external or impressed force; that force is proportional to the momentum it produces; and, which is only a particular case of the last statement, that forces acting on the same or equal masses are in the same ratio as their accelerative effects. These fundamental principles were stated explicitly and with the utmost generality in the introductory part of the *Principia* * by Newton, under the name of axioms or laws of

* P. 13 (3rd edition).

motion. They are three in number and, with the illustrations he added to them, are as follows :—

1. Every body continues in its state of rest or of uniform motion in a straight line, except so far as it is compelled by impressed forces to change its state.

Projectiles continue in their state of motion, except so far as they are retarded by the resistance of the air, and urged downward by the force of gravity. The parts of a hoop by their cohesion continually draw one another back from their rectilinear motions; the hoop, however, does not cease to turn, except so far as it is retarded by the air. Moreover the greater bodies of planets and comets, which move in spaces offering less resistance than the air, preserve for a longer time their motions of translation and rotation.

2. Change of momentum is proportional to the impressed moving force, and takes place along the straight line in which that force is impressed.

If a force produce any momentum whatever, twice the force will produce twice the momentum, thrice the force will produce thrice the momentum, whether the forces are impressed at the same time and instantaneously, or gradually and successively. This momentum is always communicated in the direction of the force producing it; so that, when the body was previously moving, it is added to the body's momentum if the directions are the same, subtracted from it if the directions are opposite; it is added obliquely if the directions are inclined; and the momenta are compounded in accordance with their several directions.

3. Reaction is always contrary and equal to action; or the mutual actions of two bodies on each other are always equal and exerted in contrary directions.

If any body presses or draws another, it is just as much pressed or drawn by the second body. If any one presses a stone with his finger, the finger is also pressed by the stone. If a horse is drawing a stone tied to a rope, the horse is (so to speak) equally drawn back towards the stone; for the rope being stretched both ways will, in the same endeavour to slacken itself, urge the horse towards the stone, and the stone towards the horse; and it will impede the progress of the one as much as it advances the progress of the other. If a body strike on another body, and by its force change the momentum of the latter body in any way whatever, its own momentum will in turn undergo an equal change in a contrary direction from the force of the latter body, in consequence of the equality of the mutual pressure. By these actions equal changes are produced, not in the velocities, but in the momenta, i.e. in bodies otherwise free to move. The changes of velocities which also take place in opposite directions are reciprocally proportional to the masses; because the momenta are changed equally.

CHAPTER IV.

THE CONSTRAINED MOTION OF A PARTICLE.

Proposition 27.

The velocity acquired by a particle in sliding from one point to another on a smooth curve is the same as that acquired by a particle which falls freely through a distance equal to the vertical height of the higher above the lower point.

Let A and B be the two points, draw B B′ horizontal and A B′ vertical; let the particle leave A with a velocity V, and arrive at B with a velocity v; then, if M be the mass of the particle, the number of units of work accumulated in it while it moves from A to B will equal (Art. 129)

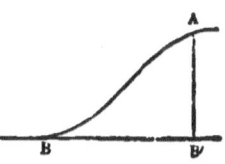

Fig. 165.

$$\tfrac{1}{2} M (v^2 - V^2)$$

Now, the only forces that have acted on the particle are gravity and the reaction of the curve; the work done by the former of these equals $M g \times A B'$, and the latter does no work, since its direction is always perpendicular to that in which its point of application is moving (Art. 103); therefore

$$\tfrac{1}{2} M (v^2 - V^2) = M g \times A B'$$

or
$$v^2 - V^2 = 2 g \times A B'$$

But this equation likewise gives the velocity (v) of a body supposed to leave A with a velocity V, and to fall freely to B′. Therefore, &c. Q. E. D.

Cor.—The above proposition is true, whether A B is a plane curve, e.g. a circle, or a curve of double curvature, e.g. the thread of a screw. It will be an instructive exercise for the reader to make out the kind of effect which friction would have on the velocity in both these cases: the actual calculation requires the Integral Calculus.

Proposition 28.

If a particle, whose mass is M, *move with a velocity* V *in a circle, whose radius is* r, *the force* (P) *tending to the centre necessary to keep the body moving in the circle is given by the formula* $P r = M V^2$.

(1) Let A B C D . . be any regular polygon inscribed in a circle whose centre is O. If A B is produced to H, so that B H and A B are equal, it is plain that H C is parallel to B O; if, then, K C is drawn parallel to B H, B H C K is a parallelogram whose diagonal B C is equal to the side B H. Now suppose a particle to move from A with a given velocity (V), and when it reaches B that another velocity (v) is impressed upon it in the direction B O, such that

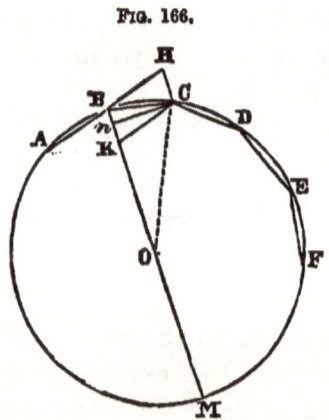

Fig. 166.

$$v : V :: B K : B H$$

At the point B the particle has simultaneously two velocities v and V, represented in magnitude and direction by B K and B H; consequently the resulting velocity is represented in magnitude and direction by B C; in other words, the consequence of the impressed velocity (v) is merely to change the direction of the motion, the velocity of the

MOTION IN A CIRCLE. 281

particle being still V, but the motion taking place along B C. If when the body reaches C a velocity equal to v is impressed on it along the line C O, the particle will describe the side C D with a velocity V. In a like manner it may be made to describe in succession all the sides of the polygon with the constant velocity V.

(2) This result is true whatever be the number of sides, and therefore when the number is very large; when this is the case the intervals between the successive impulses are very small. If, then, we suppose the number of sides to be indefinitely great we have the case in which the particle moves in a circle with a uniform velocity under the continuous action of a force directed towards the centre.

(3) To find the magnitude of the force (P) let its accelerative effect on M be f; then the velocity acquired by M in a time t will be ft. Now suppose the particle to describe one side of the polygon in the time t, and suppose the velocity ft or v to be impressed instantaneously, as in paragraph 1. Then we have B C $= Vt$ and B K $= vt$ $= ft^2$. Draw C n at right angles to B M; as B C equals C K, B n equals n K, and therefore B $n = \tfrac{1}{2} ft^2$. Now

$$MB : BC :: BC : Bn$$

or $$2r : Vt :: Vt : \tfrac{1}{2} ft^2$$

Therefore $$fr = V^2$$

This relation holds good independently of t, i.e. whether the sides of the polygon are long or short. It is therefore true in the limiting case, when the particle moves in a circle and the force acts continuously towards the centre. In this case, therefore, we have the force equal to Mf, i.e.

$$P = \frac{M V^2}{r}$$

N.B. If M is given in pounds, r in feet, and V in feet per second, the above formula will give P in absolute units. If we wish to obtain P in pounds or gravitation units we must divide its value in absolute units by 32·1912 or approximately by 32. The formula now proved is therefore identical with that given in Art. 121, and the student should refer to that article before going further. He will observe that when a body moves in a circle there must be some other guiding body to exert on it a force directed to the centre; this is the force P determined in the last proposition. The reaction of the moving body on the guiding body is the centrifugal force of the mass M. If, owing to the action of the other forces in addition to that tending to the centre, the velocity of the body varies, the force P (and therefore the centrifugal force of M) will undergo a corresponding variation, provided the body does not leave the circle. At any instant the value of P (and of the centrifugal force) will equal $MV^2 \div r$, where V is the velocity at that instant.

Ex. 728.—A locomotive engine weighing 9 tons passes round a curve 600 yards in radius at the rate of 30 miles an hour; what force tending towards the centre of the curve must be exerted to make it move in this curve? *Ans.* 677·6 lbs.

Ex. 729.—If this force is supplied by making the inner rail on a lower level than the outer, what ought to be the difference of the level if the distance between the rails is 4 ft. 9 in.? *Ans.* 1·92 in.

[The slope should be such that the resolved part of the weight along it shall equal the force determined in the last example.]

Ex. 730.—On the floor of a railway carriage are chalked two lines $x\,x'$, $y\,y'$, one perpendicular and the other parallel to the direction of the rails; the lines intersect in the point o; at a height of 4 ft. vertically over o is held a ball; the train moving at the rate of 30 miles per hour comes to a curve whose radius is 1000 ft. and centre in the prolongation of o x'; if the ball is dropped, where will it strike the floor of the carriage?

Ans. In o x at 2·9 in. from o.

Ex. 731.—A particle is tied to the end of a string whose length is l; it makes n revolutions per second; show that it will come into a position of

CENTRIFUGAL FORCE. 283

steady motion when the string makes an angle θ with the vertical given by the equation

$$\cos \theta = \frac{g}{4\pi^2 n^2 l}$$

[If T is the reaction of the fixed point on the body transmitted along the string, the vertical component of T must equal the weight of the body, and the horizontal component must be the force tending to the centre necessary to keep the body moving at the rate of n revolutions a second in a circle whose radius is $l \sin \theta$.]

Ex. 732.—A body weighing 12 lbs. is suspended by a cord 7 ft. long, and makes 80 revolutions per minute; determine the position of steady motion and the tension of the cord.

Ans. (1) 86° 15′ 55″ with the vertical. (2) 5895 abs. un.

Ex. 733.—A body weighing 20 lbs. is tied to the end of a string and suspended in a railway carriage the motion of which is perfectly steady; it comes to a curve 1000 feet in radius, round which it runs at the rate of 15 miles an hour; find the inclination of the string to the vertical, and the horizontal force that would have to be applied to the body to keep the string vertical. *Ans.* (1) 52′. (2) 0·3025 lbs.

Ex. 734.—If the earth were a perfect sphere and at rest, so that the accelerative effect of gravity at any point of its surface were g, show that the effect of its receiving its diurnal rotation will be to reduce the sensible accelerative effect of gravity to $g \left(1 - \frac{\cos^2 l}{289}\right)$ at a place whose latitude is l.

[See *Ex.* 639.]

Ex. 735.—Given that the accelerative effect of Jupiter's attraction at any point near its surface is 80 (in feet and seconds), that his radius is 11 times that of the earth, and that he makes one revolution in 10 hours, determine the ratio of the sensible force of gravity at his equator to that at his pole. *Ans.* 0·913 nearly.

Proposition 29.

To determine the time of a small oscillation of a simple pendulum (Art. 122).

Let S be the point of suspension of the particle, l the length of the pendulum. With centre S and radius equal to l describe a circle A C B H; let A C B be the arc of vibration, the middle point of which (C) will be at the lower extremity of the vertical diameter C H. Join A B cutting C H in D. Take P any point in A C, join P H, P C and draw P N at right angles to C H. The arc A C is supposed to be so small that the chord of A C may be used instead of the arc A C, and in

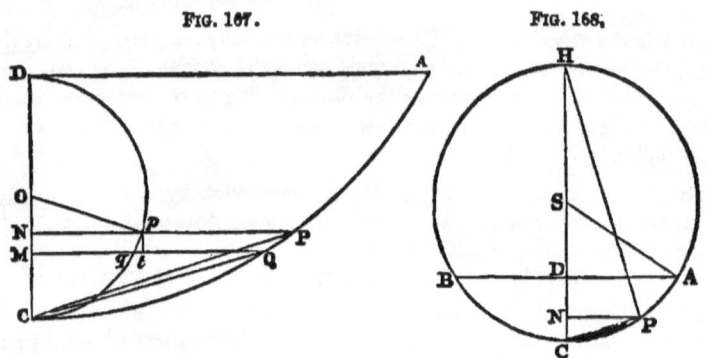

Fig. 167. Fig. 168.

like manner of any smaller arc and its chord. For convenience draw A C D on a larger scale; take O the middle point of C D: with centre O and radius O D describe the circle D p C cutting P N in p. Take Q a point very near to P; draw Q M at right angles to C D cutting D p C in q. Join O p, draw $p\,t$ at right angles to M Q.

Now, the velocity of the particle at P equals $\sqrt{2g\cdot\mathrm{DN}}$ (Prop. 27); therefore if δt is the time in which it passes over the arc P Q we have

$$\delta t = \frac{\mathrm{PQ}}{\sqrt{2g\cdot\mathrm{DN}}} \text{ ultimately (see App. Art. 3.),}$$

since in the limit the particle moves uniformly during δt. Also we have

$$\mathrm{PQ} = \text{arc C P} - \text{arc C Q} = \text{chd. C P} - \text{chd. C Q}$$

$$= \frac{\mathrm{CP}^2 - \mathrm{CQ}^2}{\mathrm{CP} + \mathrm{CQ}} = \frac{\mathrm{CP}^2 - \mathrm{CQ}^2}{2\,\mathrm{CP}} \text{ . ultimately.}$$

But by similar triangles H C : C P :: C P : C N

Therefore $\qquad \mathrm{CP}^2 = 2l\cdot\mathrm{CN}$

Similarly $\qquad \mathrm{CQ}^2 = 2l\cdot\mathrm{CM}$

TIME OF SMALL OSCILLATIONS. 285

Therefore $$PQ = \frac{l \cdot MN}{\sqrt{2l \cdot CN}}$$

and $$\delta t = \frac{l \cdot MN}{\sqrt{4gl \cdot DN \cdot CN}} = \sqrt{\frac{l}{4g}} \cdot \frac{MN}{Np}$$

Now, ultimately pq coincides with the tangent to the circle at p; therefore pqt is ultimately a right-angled triangle (App. Art. 2) whose sides pq and pt are severally perpendicular to op and pN; therefore by similar triangles

$$pq : pt :: op : Np$$

Therefore $$\frac{pq}{op} = \frac{pt}{Np} = \frac{MN}{Np}$$

Therefore $$\delta t = \sqrt{\frac{l}{4g}} \cdot \frac{pq}{op} \text{ ultimately,}$$

and the same being true of every interval of time while the body falls from A to C, the whole time, which is the limit of the sum of the intervals when their number becomes great, will equal the sum of the ultimate values

of δt, i.e. it will equal $\sqrt{\dfrac{l}{4g}} \cdot \dfrac{DpC}{op}$ or $\dfrac{\pi}{2}\sqrt{\dfrac{l}{g}}$

And the time of descending AC is plainly equal to the time of ascending CB. Therefore the time of one oscillation equals $\pi \sqrt{\dfrac{l}{g}}$. Q. E. D.

Ex. 736.—What is the length of a simple pendulum which at Greenwich oscillates in 1½ second? How much shorter is the simple pendulum which at Rawak (Table XV.) oscillates in the same time?

Ans. (1) 7·3387 ft. (2) 0·2824 in.

Ex. 737.—A pendulum whose length is L makes m oscillations in one day; its length changes, and it is now observed to make $m + n$ oscillations in one day; show that its length has been diminished by a part equal to $\dfrac{2n}{m}$L (nearly.)

[Since a mean solar day contains 86,400 seconds, we have

$$\frac{86{,}400}{m} = \pi\sqrt{\frac{L}{g}} \text{ and } \frac{86{,}400}{m+n} = \pi\sqrt{\frac{L-\delta L}{g}}$$

$$\therefore \frac{m+n}{m} = \sqrt{\frac{L}{L-\delta L}} \text{ whence } \delta L = \frac{2n}{m} L.]$$

Ex. 738.—A pendulum in a certain place makes in one day m oscillations; on transporting it to another place it is found to have the same length but to lose n oscillations a day; show that the force of gravity has been diminished by its $\frac{2n}{m}$th part.

Ex. 739.—Given the lengths of the seconds pendulums at Greenwich and Paris respectively (see Table XV.), find how many oscillations a day the Greenwich pendulum would make at Paris. *Ans.* 86,387.

Ex. 740.—Given that a pendulum oscillating seconds at the mouth of a coal pit gains 2·24 seconds per diem when removed to the bottom of the shaft; determine the decrease of the force of gravity. *Ans.* 0·0016.

Ex. 741.—A body leaves C (fig. 168) with such a velocity that A is the highest point it reaches, the arc A C being small. Let the arc A C be denoted by s_1 and the time of a double oscillation by T. Show that at a time t its distance s from C is given by the formula

$$s = s_1 \sin \frac{2\pi t}{T}$$

[From Prop. 29 it is plain that the time of describing C P bears to that of describing C A the same ratio that the arc C p bears to C p D or that the angle C O p bears to π.

Therefore $\qquad C O p = \dfrac{4\pi t}{T}$

therefore $\qquad C N = O p \left(1 - \cos\dfrac{4\pi t}{T}\right) = C D \sin^2 \dfrac{2\pi t}{T}$

but $\qquad 2 C N . C H = C P^2$ and $2 C D . C H = C A^2$

therefore $\qquad s = s_1 \sin \dfrac{2\pi t}{T}.]$

Ex. 742.—If V is the velocity at C, and v that at a time t, show that

$$v = V \cos \frac{2\pi t}{T}$$

Ex. 743.—In *Ex.* 741 and 742 obtain the value of s and v, when t is reckoned from the instant the body is at A.

[For t write $t + \frac{1}{4}$T.]

Ex. 744.—A body vibrates in a small circular arc, the velocity at the lowest point being V; when at its lowest point it has communicated to it a velocity V at right angles to its plane of vibration; show that its plane of vibration is turned through an angle of 45°, its arc of vibration is increased from s_1 to $s_1 \sqrt{2}$, and its time of oscillation is sensibly unchanged.

VIBRATIONS OF A ROD. 237

Ex. 745.—In the last case, if the velocity is communicated to the body when at A, show that it will now describe a horizontal circle round C, whose radius is s_1 and time of description $2\pi \sqrt{\dfrac{l}{g}}$.

Ex. 746.—If the angle A s c (fig. 168) is denoted by θ, the mass of the body by M, and the tension of the thread when the body is at c by T, show that $T = M g (3 - 2 \cos \theta)$.

Ex. 747.—In *Ex.* 741 show that the acceleration along the curve is $\dfrac{gs}{l}$.

Hence when a particle, whose mass is M, vibrates under the action of a force H s, where s is the distance of M from the middle point of the arc or line of vibration, show that the time of vibration equals $\pi \sqrt{\dfrac{M}{H}}$.

133. *Longitudinal vibrations of a rod.*—A rod suspended by one end whose length is L, area of section K, and modulus of elasticity E, has attached to the other end a particle whose weight is Q (which we will suppose so large that the weight of the rod can be neglected); if the rod is allowed to lengthen slowly, Q will descend through a small distance l equal to L Q ÷ K E and will continue at rest (see Art. 6 and *Ex.* 149); if, however, Q is allowed to descend at once, a certain number of units of work will be accumulated in it when it has descended through the distance l, so that it will continue to descend till the resistance to further elongation shall have neutralised them; a contraction will then ensue, and thus Q will vibrate in a vertical line about the point (A), at which in the former case it would have come to rest.

Ex. 748.—Show that when the particle, which contains Q lbs. of matter, is at a distance s from A it is moving under a force that varies as s, and that the time in which it proceeds from the highest to the lowest point is

$$\pi \sqrt{\dfrac{Q L}{g K E}}$$

Ex. 749.—In the last example suppose Q to be at a distance s below A; determine the number of units of work accumulated in it at that instant, and show that its velocity (v) is given by the equation

$$v^2 = \dfrac{g K E}{L Q}(l^2 - s^2)$$

[See *Ex.* 149. The student must bear in mind that the quantity $g \times s$ is the Modulus of Elasticity reckoned with reference to the whole cross-section of the rod and in absolute units. The same is true of the answer to *Ex.* 748.]

Ex. 750.—If a cylinder whose height is h and specific gravity s floats with its axis vertical in a fluid whose specific gravity is s_1, show that if it is depressed through *any* distance the time in which it will rise from its point of greatest depression to its greatest height is constant, and will be given by the formula

$$t = \pi \sqrt{\frac{h}{g} \frac{s}{s_1}}$$

CHAPTER V.

THE MOMENT OF INERTIA.

Def.—If we conceive a body to consist of a large number of particles, and multiply the *mass* of each by the square of its perpendicular distance from a given line or axis, the sum of all these products is the moment of inertia of the body with respect to that axis.

134. *Properties of the moment of inertia.*—It will appear hereafter that the moment of inertia is a quantity that enters into nearly every question in which the rotatory motion of a body is concerned; the present chapter will be devoted to proving some of its properties, and ascertaining its magnitude in certain particular cases. The first property we shall notice is one that follows immediately from the definition. Since the mass of a particle and the square of its perpendicular distance from a given axis are essentially positive, their product must be so too; consequently if we conceive any group of particles to consist of two or more subordinate groups, the sum of the moments of inertia of these separate groups with respect to a given axis will equal that of the whole group with respect to the same axis: hence if a body can be divided into a certain number of parts, and their respective moments of inertia are known with respect to a certain axis, that of the whole body, with respect to that axis, is found by adding them together.

Proposition 30.

If I is the moment of inertia of any body whose mass is M, about an axis passing through its centre of gravity,

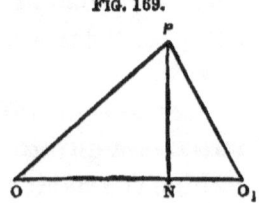

Fig. 169.

and I_1 the moment of inertia of the same body about a parallel axis situated at a perpendicular distance h from the former, then

$$I_1 = I + M h^2$$

Suppose the axes to be perpendicular to the plane of the paper, let the axis which passes through the centre of gravity meet that plane in O, and let the other axis meet it in O_1; let P be one of the particles of which the body is conceived to be made up, and let its mass be m_1; join P O, P O_1, and O O_1, and draw P N at right angles to O O_1; then (Eucl. II. 13)

$$O_1 P^2 = O P^2 + O O_1^2 - 2 O O_1 \cdot O N.$$

Let $O P = r_1$, $O_1 P = r_1'$, $O N = x_1$, and $O O_1 = h$, then

$$m_1 r_1'^2 = m_1 r_1^2 + m_1 h^2 - 2 m_1 h x_1 \qquad (1)$$

and the same algebraical formula will be true whatever be the position of P; hence if $m_2, r_2', r_2, x_2, m_3, r_3', r_3, x_3$, &c., ... are the magnitudes corresponding to other particles we shall have

$$m_2 r_2'^2 = m_2 r_2^2 + m_2 h^2 - 2 m_2 h x_2 \qquad (2)$$
$$m_3 r_3'^2 = m_3 r_3^2 + m_3 h^2 - 2 m_3 h x_3 \qquad (3)$$

and so on for every particle.

Now by the definition

$$m_1 r_1'^2 + m_2 r_2'^2 + m_3 r_3'^2 + \ldots = I_1$$
$$m_1 r_1^2 + m_2 r_2^2 + m_3 r_3^2 + \ldots = I$$

also $\qquad m_1 + m_2 + m_3 + \ldots = M$

and by the properties of the centre of gravity (Prop. 16)

$$m_1 x_1 + m_2 x_2 + m_3 x_3 \ldots = 0$$

since the weight of each particle is proportional to its mass.

MOMENT OF INERTIA. 291

Therefore by adding the equations (1), (2), (3), &c., we obtain

$$I_1 = I + Mh^2 \qquad \text{Q. E. D.}$$

Proposition 31.

If any number of particles lie in a plane, and if I_1 *and* I_2 *are respectively their moments of inertia about two rectangular axes in that plane, and if* I *is their moment of inertia about an axis perpendicular to the two others, and passing through their point of intersection, then*

FIG. 170.

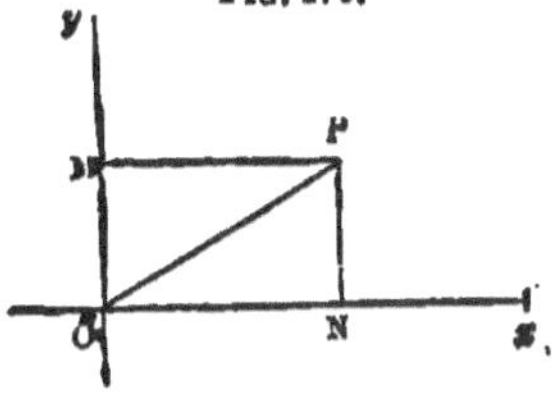

$$I = I_1 + I_2$$

For let Ox, Oy be the two axes, the third being perpendicular to the plane of the paper, and passing through O; let P be one of the particles whose mass is m; draw PM and PN at right angles to Oy and Ox, join OP, and let $PM = x$, $PN = y$, $OP = r$, then

$$r^2 = x^2 + y^2$$
$$\therefore mr^2 = mx^2 + my^2. \qquad (1)$$

Similarly, if other particles are taken, and the corresponding magnitudes are $m_1, r_1, x_1, y_1, m_2, r_2, x_2, y_2, \ldots$, we shall have

$$m_1 r_1^2 = m_1 x_1^2 + m_1 y_1^2 \qquad (2)$$
$$m_2 r_2^2 = m_2 x_2^2 + m_2 y_2^2 \qquad (3)$$

and so on, whatever be the number of particles. Now

$$mr^2 + m_1 r_1^2 + m_2 r_2^2 + \ldots = I$$
$$mx^2 + m_1 x_1^2 + m_2 x_2^2 + \ldots = I_1$$
$$my^2 + m_1 y_1^2 + m_2 y_2^2 + \ldots = I_2$$

Therefore, adding together (1), (2), (3), &c., we obtain

$$I = I_1 + I_2 \qquad \text{Q. E. D.}$$

Ex. 751.—If k is the area of the section of a thin rod, ρ the density of the material, and l its length, show that its moment of inertia about an axis passing through one end and perpendicular to it equals $\frac{1}{3}\rho k l^3$.

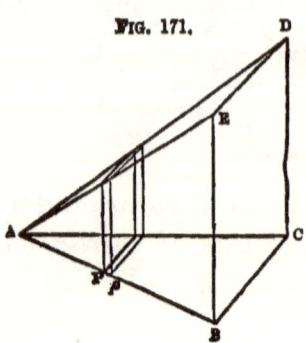

FIG. 171.

[If A B is the line, and a pyramid is constructed whose base B D is a square the side of which equals A B and its plane perpendicular to A B (compare the end of Art. 82); then if we consider a lamina contained by planes drawn parallel to the base through the extremities of any small portion Pp of A B, its volume will ultimately equal P$p \times$ A P^2; now, the moment of inertia of Pp equals mass of P$p \times$ A P^2, i.e. it equals $\rho k \times$ vol. of lamina; and hence the moment of inertia of the rod equals $\rho k \times$ volume of the pyramid.]

Ex. 752.—The moment of inertia of the rod in the last example about an axis perpendicular to its length and passing through its middle point equals $\frac{1}{12} \cdot \rho k l^3$.

[See Prop. 30.]

Ex. 753.—There is a rectangular lamina whose thickness is k, and sides a and b; show that with reference to an axis parallel to a and passing through the middle point of b the moment of inertia equals $\frac{1}{12} \cdot \rho k a b^3$.

[The lamina can be divided into a number of lines whose length is b; the moments of inertia of these can be added together by Art. 134.]

Ex. 754.—If in the last example the axis is perpendicular to the plane and passes through the centre of gravity, show that the moment of inertia of the lamina equals $\frac{1}{12} \cdot \rho k a b (a^2 + b^2)$.

[See Prop. 31.]

Ex. 755.—There is a rectangular parallelopiped whose edges are a, b, c; an axis is drawn through the centre of gravity, and parallel to the edge c; show that the moment of inertia about that axis equals $\frac{1}{12} \rho a b c (a^2 + b^2)$.

[See Art. 134.]

Ex. 756.—There is a right prism whose base is a right-angled triangle, the sides containing the right angle of which are a and b, the height of the prism is c. Show that if an axis be drawn through the centres of gravity of the ends the moment of inertia about that axis equals $\frac{1}{36} \cdot \rho a b c \times (a^2 + b^2)$.

MOMENT OF INERTIA.

[By Art. 134 and *Ex.* 755 the amount of inertia about a parallel axis joining the middle points of the hypothenuses of the ends is $\frac{1}{24} \cdot \rho\, a\, b\, c$ $\times (a^2 + b^2)$; the result is then obtained by Prop. 30.]

Ex. 757.—There is a right prism whose height is c and base an isosceles triangle, the base of which is a and height b; if an axis be drawn passing through the centres of gravity of the ends its moment of inertia about that axis equals $\frac{1}{12} \cdot \rho\, a\, b\, c \left(\frac{a^2}{4} + \frac{b^2}{3}\right)$.

[This prism can be divided into two resembling that in the last example.]

Ex. 758.—There is a right prism whose mass is M and base a regular polygon the radius of whose inscribed circle is r, and length of one side a; show that its moment of inertia about its geometrical axis is $\frac{1}{2} \cdot M \times \left(\frac{a^2}{12} + r^2\right)$.

[This prism can be divided into prisms like that in the last example.]

Ex. 759.—If there is a cylinder whose height is h and radius of base r, show that its moment of inertia about its geometrical axis equals $\frac{\pi}{2} \rho\, h\, r^4$.

[If the cylinder reduces to a circular lamina whose thickness is h, the same formula is of course true.]

Ex. 760.—There is a thin circular lamina whose radius is r and thickness k; show that the moment of inertia about a diameter equals $\frac{\pi}{4} \rho\, k\, r^4$.

[See Prop. 31.]

Ex. 761.—There is a drum the length of which is a, the mean radius of the end r, and the thickness t; show that its moment of inertia about its axis very nearly equals $2\pi\rho\, a\, t\, r^3$; and that if t equals $\frac{r}{n}$, the error in the above determination of the moment of inertia is the $\frac{1}{4n^2}$th part of that quantity.

Ex. 762.—There is a cylinder the length of which is h and the radius of whose base is r; show that its moment of inertia about a diameter of one end equals $\pi \rho\, h\, r^2 \left(\frac{h^2}{3} + \frac{r^2}{4}\right)$.

[If we consider a lamina contained between two planes parallel to the end and at distances x and $x + \delta x$, it appears from *Ex.* 760 and Prop. 30 that the moment of inertia of the lamina equals $\frac{1}{4} \cdot \pi \rho\, r^4 \delta x + \pi \rho\, r^2 x^2 \delta x$; whence the required moment of inertia equals the mass of a line the mass of each unit of whose length is $\frac{1}{4} \cdot \pi \rho\, r^4$, together with the moment of inertia about one end of a line the mass of each unit of whose length is $\pi \rho\, r^2$.]

Ex. 763.—Determine the moment of inertia of a cylinder about a generating line.

Ex. 764.—There is a cone the height of which is h and radius of base r; show—(1) that its moment of inertia about its axis equals $\frac{1}{10}\pi\rho h r^4$; (2) that its moment of inertia about an axis drawn through the vertex and perpendicular to the axis of the cone equals $\frac{1}{5}\pi\rho h r^2\left(h^2 + \frac{r^2}{4}\right)$.

Ex. 765.—Show that the moment of inertia of a sphere about any diameter equals $\frac{8}{15}\pi\rho r^5$.

[The results in the last two examples cannot be easily obtained without the aid of the integral calculus.]

Ex. 766.—In the mass of iron described in *Ex.* 12 let an axis be drawn passing through the end of the longer rectangular piece and bisecting those sides of the end which are 6 in. long; determine the moment of inertia of the mass with respect to that axis.*   *Ans.* 73,900.

Ex. 767.—There is a cast-iron cone 16 in. high, radius of base 8 in.; determine its moment of inertia—(1) about an axis through its centre of gravity and parallel to its base; (2) about a parallel axis distant 4 ft. from the former.   *Ans.* (1) 37·27.   (2) 4509.

Ex. 768.—Determine the moment of inertia about a vertical edge of the oak door described in *Ex.* 17.   *Ans.* 460.

Ex. 769.—There is a cube of oak whose edge is 8 in. long; through the middle of it at right angles to one of its faces passes a cylinder of oak 4 ft. long and 3 in. in diameter; the centres of gravity of the two figures coincide; determine the moment of inertia of the whole about an axis passing through the common centre of gravity and perpendicular to the axis of the cylinder and also to a face of the cube.   *Ans.* 16·54.

Ex. 770.—Determine the moment of inertia of the hollow leaden cylinder described in *Ex.* 15 about a diameter of its mean section.   *Ans.* ·6462.

Ex. 771.—If a cylinder like that in the last example is fitted to each arm of the figure described in *Ex.* 769, determine the moment of inertia of the whole about the specified axis—(1) when the ends of the leaden cylinders coincide with those of the arms; (2) when the other ends of the cylinders are in contact with the cube.   *Ans.* (1) 200·8.   (2) 28·16.

Ex. 772.—Determine the moment of inertia, about the axis, of a grindstone 4 ft. in diameter and 8 in. thick.   *Ans.* 2244.

Ex. 773.—There is a cast-iron fly-wheel consisting of a rim, four spokes at right angles to each other, and an axle; the external and internal radii of the rim are 4 and $3\frac{1}{2}$ ft. respectively, and its thickness 8 in.; the sections

---

* In this, as in all examples of moments of inertia, mass is reckoned in pounds and distance in feet.

RADIUS OF GYRATION. 295

of the spokes are each 4 square in., the axle 12 in. in diameter and 18 in. long; determine the moment of inertia of the whole about the geometrical axis of the axle; and also determine the error if the spokes and axle were neglected and the moment of inertia of the rim calculated by *Ex.* 761.

Ans. (1) 50,750. (2) 1000.

Ex. 774.—If the moment of inertia of any body with reference to an axis coinciding with a given line of particles be represented by I, and if the body be uniformly expanded by heat, so that the linear dimensions before expansion are to those after in the ratio of $1 : 1+a$, show that the moment of inertia with reference to the axis coinciding with the same line of particles becomes $(1+a)^2 $I.

135. *The radius of gyration.*—It is evident from the definition of the moment of inertia of a body with respect to a given axis, that there will be, with respect to that axis, a line of a certain determinate length k, such that

$$\text{I} = \text{M}k^2$$

where I is the moment of inertia, and M the mass of the body; the line k is called the radius of gyration with respect to that axis, and may be defined to be that distance from the axis at which the whole mass of the body may be supposed to be collected without producing any change in the moment of inertia. Thus, it is evident that in *Ex.* 751, 754, and 759 the values of the radius of gyration are respectively $\dfrac{l}{\sqrt{3}}$, $\sqrt{\dfrac{a^2+b^2}{12}}$, and $\dfrac{r}{\sqrt{2}}$. Moreover, if k be the radius of gyration of a body with reference to an axis passing through the centre of gravity, and k_1 its radius of gyration with reference to an axis parallel to the former, and at a perpendicular distance from it equal to h, then it is evident from Prop. 30 that

$$k_1^2 = k^2 + h^2$$

It is to be observed that the moment of inertia is essentially a mechanical magnitude, while the radius of gyration is simply a line; now, suppose k to be the radius of gyration of any lamina of uniform density, the area of one face of which is A, it is not unusual to speak of that *area* as having a moment of inertia; when this is done it means that

$$I = Ak^2.$$

For instance, in the case of a rectangular area whose sides are b and c, the moment of inertia, with respect to an axis bisecting the two parallel edges c, is

$$\frac{1}{12} b^3 c$$

as is plain from *Ex.* 753. It is in this sense that the term Moment of Inertia is used in Prop. 21. Strictly speaking, an *area* has a moment of inertia no more than it has weight.

CHAPTER VI.

D'ALEMBERT'S PRINCIPLE.

136. *Account of problem to be solved.*—The manner in which the dimensions of a body influence its motion may be illustrated as follows :—If we suppose a rigid bar to be suspended by one end and to oscillate, the velocities with which the different points are, at any instant, moving stand to one another in a fixed relation; thus the free end moves twice as fast as the middle point; moreover, with one exception, each point has a different velocity from what it would have if it were detached from the rest, and swung freely at the same distance from the centre of suspension; this difference must depend upon the cohesive forces which bind the parts of the bar together. The consideration of this simple case points out the two chief additional conceptions required for the investigation of the motion of a body whose form has to be taken into account.

(1) A means must be obtained for comparing the velocities of different points of a rigid body revolving round an axis, which is done by introducing the conception of *Angular Velocity*.

(2) A principle is required by means of which we can avoid the consideration of the cohesive forces which hold together the parts of the body: this is generally called D'Alembert's Principle.

137. *Angular velocity.*—If a rigid body revolves round an axis, it is plain that the perpendiculars let fall from each

point of the body on the axis will, in any given time, describe equal angles; hence arises the following

Def.—If a body revolves uniformly round an axis, the angle (estimated in *circular measure*) described in one second by the perpendicular let fall from any point on the axis of rotation is called the *angular velocity* of the body.

If the velocity is variable, it is measured at any instant by the angle that would be so described if, from that instant, the velocity continued uniform for one second.

In the following pages ω and Ω are used to denote angular velocity.

Ex. 775.—A body makes 30 uniform revolutions in one minute; what is its angular velocity?   *Ans.* π.

Ex. 776.—A point moves at the rate of 12 ft. per second in a circle whose radius is 15 ft.; what is its angular velocity?   *Ans.* $\frac{4}{5}$.

Ex. 777.—Determine the angular velocity of the earth round its axis.

Ans. $\dfrac{\pi}{43{,}082}$.

[See *Ex.* 569.]

Ex. 778.—If a body has an angular velocity 2·5, how many revolutions will it make per hour?   *Ans.* 1432·4.

Ex. 779.—If a body has a uniform angular velocity ω, show that the accelerative effect of the centrifugal force of a point in it, situated at a distance r from the axis, is $r\omega^2$.

138. *Impressed forces.*—All forces acting on a body which do not arise out of the mutual cohesion of its parts, are called the *impressed forces* that act on the body.

Thus, when a cricket-ball is thrown in vacuo, the impressed force is gravity; if the ball were pierced by a spindle and caused to revolve round it, the impressed forces would be gravity and the reaction of the points of support of the spindle; and so on in other cases.

139. *Effective forces.*—It must be remembered that when a solid body is in motion each point in it moves

EFFECTIVE FORCES. 299

along a determinate line, straight or curved according to circumstances. As this fact should be distinctly conceived by the student, it may be mentioned by way of illustration that, when a cart moves along a perfectly even road, each point on the circumference of one of its wheels describes a cycloid, the centre of the wheel describes a straight line, while any point in one of the spokes describes a curve called a trochoid. A similar, though much more complicated, kind of motion belongs to the different points of a cricket-ball; when in the act of being thrown it receives a rotatory motion. The only fact, however, that we are concerned with here is that, whatever be the motion of the body, each point in it will describe a *determinate* path.

Let m be the mass of a particle of a moving body, and suppose that particle to describe the path H K; at M let it be moving with a velocity v, and at M' with a velocity v', having described the small space between M M' in the short time t. Let it now be inquired what forces acting on an isolated particle would make it move as the particle actually does when forming part of the moving body. The points M M' may be considered to be on the circumference of the circle of curvature at the point M, whose radius r can be determined from the nature of the curve H K; hence at M the isolated particle must be acted on by a normal force equal to $\dfrac{m v^2}{r}$, and the change of velocity must be produced by a tangential force $m \cdot \dfrac{v'-v}{t}$, the time t being supposed indefinitely small. If, then, at M we suppose the particle to become isolated, its motion retaining the same velocity and direction, it will continue to move as it actually does during the next short time, if acted on by the resultant of

FIG. 172.

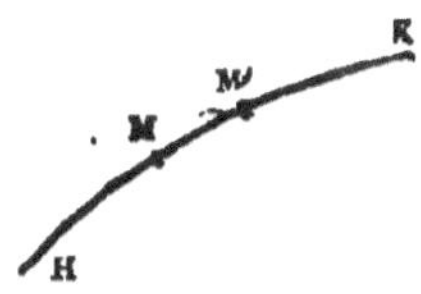

the forces $m \cdot \dfrac{v^2}{r}$, and $m \cdot \dfrac{v'-v}{t}$; this resultant is called the effective force on the particle at M. Hence we may define it in general terms as follows:—

Def.—If the velocity and direction of the motion of a particle, forming part of a rigid body, undergo a certain change in an indefinitely short time beginning at a given instant; then if we suppose the particle to be at that instant disconnected from the body, and to be acted on by a force which produces in that indefinitely short time the same change in the velocity and direction, the force is called the effective force on the particle at that instant.

140. *Effective forces in the case of rotatory motion.*—Suppose a particle whose mass is m to be situated at a distance r from the axis of rotation of a body, of which the particle forms a part; let ω be the angular velocity of the body at a given instant, and at the end of a short time t let the angular velocity become ω', then at the given instant the effective force will consist of two components $m\,r\,\omega^2$ along r towards the centre, and $m \cdot \dfrac{r(\omega'-\omega)}{t}$ in the direction of the tangent; if the angular velocity is uniform, the second component is zero, and the effective force is $m\,r\,\omega^2$ acting along r towards the centre. A force equal and opposite to the effective force is plainly the centrifugal force of the particle.

141. *D'Alembert's principle.*—Let it now be inquired what are the forces that act on any particle M of the moving body A B; it will be remarked that they can only be of two kinds, (1) the impressed force P transmitted to it, (2) the resultant Q of the cohesive forces which bind it to the rest of the body. These two forces must have at any given instant a determinate resultant R, and this must be the effective force on M at that instant, since if M were isolated for a short time, and were acted on by R,

its motion would experience the same change in velocity and direction that it actually experiences. Now, if a force equal and opposite to R were to act on M at the instant under consideration, it would be in equilibrium with P and Q; and the same is true of every other particle of the body; consequently if we suppose that to each particle of the body a force is applied equal and opposite to the effective force on that particle, these forces will be in equilibrium with the impressed and cohesive forces, and we shall have three systems of forces constituting a system in equilibrium, viz. (1) a system of impressed forces, (2) a system of cohesive forces, (3) a system of effective forces applied to the particles in the opposite direction to that in which they must act to produce the actual motion of the particles. Now, D'Alembert's principle asserts that the cohesive forces are separately in equilibrium, and infers the conclusion that *if forces equal and opposite to the effective forces at any instant were at that instant applied to each particle of the body, they would be in equilibrium with the impressed forces.*

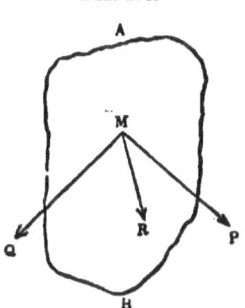

FIG. 173.

Proposition 32.

A body, whose mass equals M, *is symmetrical with reference to a plane containing a certain axis and the centre of gravity, the distance of which from the axis is denoted by* $\bar{x}$; *if the body revolves round the axis with a uniform angular velocity* ω, *the resultant of the effective forces equals* $M \bar{x} \omega^2$.

Let O A be the axis, and B C the revolving body, the plane of the paper being the plane of symmetry, we may suppose it divided into a number of laminæ, such as D E, by planes perpendicular to A O; then if we find the effective force of each lamina, their resultant will be the required force.

(1) Let G_1 be the centre of gravity of D E, and $G_1 N_1$ its perpendicular distance (x_1) from the axis A O; in its plane draw the axes $N_1 y$, $N_1 z$, and take P, any small portion of it, and suppose the mass of P to be m, and its co-ordinates to be y and z; also let the angle $P N_1 y$ be denoted by θ. Now (Art. 140) since P describes a circle round N with a uniform velocity, its effective force is $m \omega^2 . P N_1$, which can be resolved into two components, viz. $m \omega^2 y$ parallel to $N_1 y$, and $m \omega^2 z$ parallel to $N_1 z$. In the same manner, if m_1, y_1, z_1, m_2, y_2, z_2, ... are the corresponding values for other elements of the lamina, we shall have forces $m_1 \omega^2 y_1$, $m_2 \omega^2 y_2$ .... parallel to $N_1 y$, and $m_1 \omega^2 z_1$, $m_2 \omega^2 z_2$ .... parallel to $N_1 z$. Hence (Prop. 16) the effective forces on the lamina are equivalent to the two

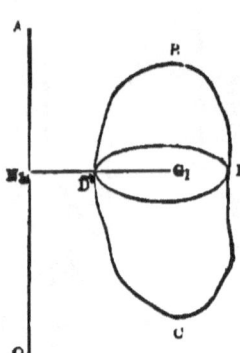

Fig. 174.

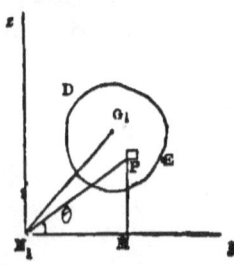

Fig. 175.

$$\omega^2 \{my + m_1 y_1 + m_2 y_2 + \ldots \} = \omega^2 M_1 \bar{y} \text{ parallel to } N_1 y$$
$$\text{and } \omega^2 \{mz + m_1 z_1 + m_2 z_2 + \ldots \} = \omega^2 M_1 \bar{z} \text{ parallel to } N_1 z$$

where $\bar{y}, \bar{z}$ are the co-ordinates of G_1, and M_1 is the mass of the lamina D E; if we compound these two forces, we shall obtain $\omega^2 . M_1 x_1$ as their resultant acting along $G_1 N_1$.

(2) Let the masses of the several laminæ into which B C is divided be respectively M_1, M_2, M_3, ... and let the respective distances of their centres of gravity from A O

be $x_1, x_2, x_3, \ldots$ then their effective forces are severally $\omega^2 \text{M}_1 x_1, \omega^2 \text{M}_2 x_2, \omega^2 \text{M}_3 x_3, \ldots$ Now, it follows from the symmetry of the figure, that all these centres of gravity are in the same plane, viz. the plane of the paper; the effective forces are therefore parallel, and their resultant will equal their sum, viz.

$$\omega^2 \left(\text{M}_1 x_1 + \text{M}_2 x_2 + \text{M}_3 x_3 + \ldots \right)$$

which equals $\omega^2 \text{M} \bar{x}$.   Q. E. D.

Cor.—The point at which the direction of the resultant of the effective forces cuts the axis is determined thus:— Take any point o on the axis, and let o N_1 be denoted by z_1, and let $z_2, z_3, \ldots$ correspond to the other laminæ, then if $\bar{z}$ is the distance of the required point from o, we have (Prop. 16)

$$\omega^2 \text{M} \bar{x} \bar{z} = \omega^2 \left\{ \text{M}_1 x_1 z_1 + \text{M}_2 x_2 z_2 + \text{M}_3 x_3 z_3 + \ldots \right\}$$

Now in general the right-hand side of this equation cannot* be obtained except by means of the integral calculus: one important exception, however, may be mentioned, viz. when the body is symmetrical with reference to a plane perpendicular to the axis of rotation as well as with reference to a plane containing the axis of rotation and the centre of gravity; in this case it is evident that if we take o at the point where this plane cuts the axis, the right-hand side of the above equation will equal zero, i.e. the force P must act along the intersection of two planes of symmetry, so that the direction of the resultant of the effective forces must pass through the centre of gravity of the body. Examples of this case are supplied by a sphere revolving round any axis, a cylinder

---

* The right-hand side of the equation is commonly written $\omega^2 \Sigma\, m\, z\, x$; and it may be added that $\Sigma\, m\, z\, x$ is one of the three quantities $\Sigma\, m\, x\, y$, $\Sigma\, m\, y\, z$, $\Sigma\, m\, z\, x$, that occur in systematic treatises on the dynamics of a solid body.—See Poisson, *Mécanique*, vol. ii. c. 2.

revolving round an axis either parallel or perpendicular to its geometrical axis, and a cone about an axis perpendicular to its geometrical axis.

N.B. The resultant of the effective forces is plainly equal in magnitude to the centrifugal force of a particle of the same mass as the body, placed at the centre of gravity and revolving with the same velocity about the same axis.

Ex. 780.—A thin rod whose length is l is fastened by one end to a spindle, to which it is inclined at an angle a and round which it revolves; show that the direction of the resultant of the effective forces cuts the spindle at a distance from that end equal to $\frac{2}{3}l \cos a$.

Ex. 781.—A cone of cast iron 1 ft. high, the radius of whose base is 6 in., revolves 30 times a minute round an axis parallel to its geometrical axis, and passing through a point in the circumference of the base; find the centrifugal force, or the resultant of the effective forces.

Ans. 582 abs. un.

Ex. 782.—A cylinder of cast iron 3 ft. high, whose base is 6 in. in diameter, revolves 100 times a minute with its axis vertical round a parallel axis at a distance of $1\frac{1}{2}$ ft.; find the centrifugal force. *Ans.* 43644 abs. un.

Ex. 783.—A wrought-iron rod 10 ft. long and section 1 in. in radius is made to revolve 60 times in a minute round an axis perpendicular to its length and passing through one extremity; find the centrifugal force.

Ans. 20962 abs. un.

Ex. 784.—Two balls of cast iron, one 10 in. and the other 6 in. in diameter, have their centres joined by a horizontal rod 3 ft. long; they are made to revolve 100 times a minute about a vertical spindle, whose distance from the centre of the heavier ball is 1 ft.; find the pressure due to centrifugal force of the spindle. *Ans.* 8502 abs. un.

Ex. 785.—Two rods in all respects equal are made to revolve about a vertical spindle; they are always in the same vertical plane, but on different sides of the spindle, and are quite free to move round the top of the spindle in that plane; if the spindle makes n revolutions per second, determine the position of steady motion. $\quad$ *Ans.* $\cos a = \dfrac{3g}{8\pi^2 n^2 l}$.

Ex. 786.—A shaft of cast iron whose section is 8 in. by 4 in. and whose length is 4 ft., revolves in a horizontal plane round a vertical axis of wrought iron 6 in. in diameter whose centre is 4 in. from the end of the shaft; if it makes 200 revolutions per minute, determine the number of units of work expended on the friction of the axle caused by the centrifugal force, the axle being well greased ($\mu = 0.075$).

Ans. 6897000 abs. un. of work per min.

142. *Pressure on a fixed axis of rotation.*—The student must be on his guard against supposing that $M\bar{x}\omega^2$ is the whole of the pressure on the fixed axis; though it is frequently the most important part of it. The complete investigation of that pressure lies beyond the scope of the present work; to prevent misapprehension, however, it may be well to add one or two of the results of the investigation.

(1) The body being supposed symmetrical, as in Prop. 32, and it being further supposed that no external force, such as gravity, acts upon the body, the only impressed force will be the reaction of the axis, which (Art. 141) will therefore equal $M\bar{x}\omega^2$.

(2) If in the last case the axis were vertical, and the body acted on by gravity, the horizontal pressure is still $M\bar{x}\omega^2$; but there is also a vertical pressure acting along the axis equal to the weight of the body.

(3) If in the last case (2) the body were not symmetrical with reference to a plane containing the axis and the centre of gravity, there will in general be the following pressures:—(a) a pressure equal to the weight of the body acting along the axis; (b) in the plane containing the axis and the centre of gravity a pressure equal to $M\bar{x}\omega^2$ acting perpendicularly to the axis through a certain point whose position depends on the form of the body; (c) in a plane containing the axis and perpendicular to the former plane, a pair of equal parallel pressures acting towards contrary parts constituting a couple (Art. 54), whose moment depends on the angular velocity and the form of the body.

In most other cases the pressures on the axis vary from instant to instant, and are of a much more complicated character than those mentioned above.

CHAPTER VII.

ON THE KINETIC ENERGY OF, OR THE WORK ACCUMULATED IN, A BODY THAT ROTATES ON A FIXED AXIS.

143. *The work accumulated in a moving body.*—If all the forces that act on a body are considered, viz. both those which tend to accelerate and those which tend to retard its motion, it will be evident that the number of units of work accumulated in a given interval is the excess of the number of units done by the former over those done by the latter; in other words, it is the (algebraical) sum of the units of work done by the impressed forces: let this be denoted by the letter U. Now, it will be remembered that the effective forces at any instant applied in opposite directions would be in equilibrium with the impressed forces (Art. 141), and consequently (Art. 104) the sum of the units of work done by the impressed forces will equal the sum of the units of work done by the effective forces; in fact, as the body is rigid, its particles undergo no relative displacement, so that no work is done by or against the forces of cohesion (Art. 103). Let now the different particles of which the body is made up be considered, let their masses be severally m_1, m_2, m_3, . . . and at the beginning of the given interval let their velocities be severally V_1, V_2, V_3, ... and at the end of it v_1, v_2, v_3, ... then (Art. 129) if they had moved separately the number of units of work accumulated in them respectively would have been $\frac{1}{2} m_1 (v_1{}^2 - V_1{}^2)$, $\frac{1}{2} m_2 (v_2{}^2 - V_2{}^2)$, $\frac{1}{2} m_3 (v_3{}^2 - V_3{}^2)$ . . . . . Now, these must be the works done by the effective forces, and therefore

KINETIC ENERGY OF ROTATION. 307

$$U = \tfrac{1}{2} m_1 (v_1^2 - v_1^2) + \tfrac{1}{2} m_2 (v_2^2 - v_2^2) + \tfrac{1}{2} m_3 (v_3^2 - v_3^2) + \ldots$$

i.e. the work done by the impressed forces in the interval equals the change in the kinetic energy of the body.

Proposition 33.

If a body moves round a fixed axis, and in a given interval of time its angular velocity is changed from Ω to ω, the algebraical sum of the number of units of work done upon it during that interval equals $\tfrac{1}{2} I_1 (\omega^2 - \Omega^2)$, where I_1 is the moment of inertia of the body with reference to the axis.

For conceive the body to be made up of particles whose respective masses are $m_1, m_2, m_3, \ldots$ and whose perpendicular distances from the axis are $r_1, r_2, r_3, \ldots$ then, using the notation of the last article, we have

$$V_1 = r_1 \Omega, \quad V_2 = r_2 \Omega, \quad V_3 = r_3 \Omega \ldots$$
and $\quad v_1 = r_1 \omega, \quad v_2 = r_2 \omega, \quad v_3 = r_3 \omega \ldots$

therefore the number of units of work done upon it during the interval equals

$$\tfrac{1}{2} m_1 r_1^2 (\omega^2 - \Omega^2) + \tfrac{1}{2} m_2 r_2^2 (\omega^2 - \Omega^2) + \tfrac{1}{2} m_3 r_3^2 (\omega^2 - \Omega^2) + \ldots$$

which equals $\tfrac{1}{2} (\omega^2 - \Omega^2)(m_1 r_1^2 + m_2 r_2^2 + m_3 r_3^2 + \ldots)$;

now $m_1 r_1^2 + m_2 r_2^2 + m_3 r_3^2 + \ldots$ is the moment of inertia of the body (I_1) with respect to the axis of rotation; consequently the number of units of work done upon the body equals $\tfrac{1}{2} (\omega^2 - \Omega^2) I_1$, or $\tfrac{1}{2} (\omega^2 - \Omega^2) M k_1^2$, where k_1 is the radius of gyration with reference to the axis of rotation.

Q. E. D.

Cor. 1. If the units of distance, time, and mass, are a foot, a second, and a pound, the formula $\tfrac{1}{2} (\omega^2 - \Omega^2) M k_1^2$ gives the change in the kinetic energy of the body in absolute units of work. It can be reduced to foot-pounds by dividing by 32·1918 or approximately by 32. The angular velocities ω and Ω are expressed in circular measure.

Cor. 2. If we suppose the axis round which the body revolves to be a geometrical line, and the body to move from a state of rest under the action of gravity, the angular velocity (ω) acquired while the centre of gravity falls through a vertical height h can be thus determined:—The kinetic energy of the body is $\frac{1}{2} \text{M} k_1^2 \omega^2$, and this must equal the work done by the impressed forces, viz. the reactions of the bearings and the weight of the body; now the former forces do no work as their points of application do not move (Art. 101), and the work done by the latter is $\text{M} g h$; therefore

$$\tfrac{1}{2} \text{M} k_1 \omega^2 = \text{M} g h$$

or

$$\omega^2 = \frac{2 g h}{k_1^2}$$

Ex. 787.—A rod 3 ft. long turns round one of its ends from a position in which it makes an angle of 45° with the horizon; find the angular velocity it has when in a horizontal position. *Ans.* 4·757.

[See *Ex.* 751.]

Ex. 788.—In the last example determine—(1) the velocity in feet per second with which the end of the rod moves, and (2) the number of degrees through which the rod would move in one second if it continued to move uniformly with the angular velocity acquired.

Ans. (1) 14·271. (2) 272° 33'.

Ex. 789.—A cone turns round a horizontal spindle, passing through its vertex at right angles to its axis: what angular velocity will it acquire in falling from its highest to its lowest position? *Ans.* $\omega^2 = \dfrac{20hg}{4h^2 + r^2}$.

[See *Ex.* 764.]

Ex. 790.—In the last example, if the cone is of brass, and is 4 ft. high and its base 1 ft. in radius, what pressure will be produced on the axis by its centrifugal force when in its lowest position? and how many times greater than the weight is this pressure?

Ans. (1) 8,116 g abs. un. (2) $3\tfrac{9}{13}$ times.

Ex. 791.—If the mass of cast iron described in *Ex.* 12 move round the axis assigned in *Ex.* 766, determine—(1) the angular velocity it acquires in falling from an inclination of 30° to a horizontal position, and (2) the number of foot-pounds of work accumulated in it.

Ans. (1) 2·21. (2) 5,638 ft.-pds.

KINETIC ENERGY OF ROTATION.

Ex. 792.—A cone of cast iron 16 in. high, the radius of whose base is 8 in., is fastened to the end of a shaft 4 ft. long at right angles to its axis, and whose end coincides with its centre of gravity; the whole moves about a horizontal axis at right angles to the shaft and passing through its extremity; the centre of gravity of the cone descends through a vertical height of 2 ft.: find the angular velocity acquired.   *Ans.* 2·817.
[See *Ex.* 767.]

O Ex. 793.—If the oak door described in *Ex.* 17 is pushed open by a force of 5 lbs. acting at every instant perpendicularly to its face, and at a distance of two feet from the inner edge of the door, determine the angular velocity required in moving through an angle of 90°.
Ans. 1·477.

[The number of foot-pounds of work done on the door is 5π, and therefore the number of absolute units of work is $5\pi g$, so that $\omega^2 \mathrm{I} = 10 g \pi$; where g is 32·1912, or approximately 32. See Cor. 1, Prop. 33, and *Ex.* 768.]

Ex. 794.—A pulley whose moment of inertia is $\mathrm{M} k^2$ and radius r turns freely round a horizontal axis, a fine thread is wrapped round it to the end of which a mass of M_1 lbs. is tied; the weight of the string and the passive resistances being neglected, show that if ω is the angular velocity of the pulley when M_1 has descended through h feet, then

$$\omega^2 = \frac{2 \mathrm{M}_1 g h}{\mathrm{M}_1 r^2 + \mathrm{M} k^2}.$$

[The number of absolute units of work done by gravity is $\mathrm{M}_1 g h$. The kinetic energy of M is $\frac{1}{2} \mathrm{M} k^2 \omega^2$, and, as the velocity of M_1 is $r\omega$, its kinetic energy is $\frac{1}{2} \mathrm{M}_1 r^2 \omega^2$.]

Ex. 795.—A cylinder with its axis vertical turns round a fine spindle coinciding with its axis; a thread is wrapped round the cylinder, and then passes horizontally over a pulley capable of revolving round a horizontal axis; to the end of the thread is tied a mass M_1; if $m k^2$, $\mathrm{M} \mathrm{K}^2$ are the moments of inertia of the pulley and cylinder, and r and R their radii, and ω the angular velocity of the cylinder after M_1 has fallen through a height h, show that if the passive resistances are neglected

$$\omega^2 = \frac{2 g h}{\mathrm{R}^2} \times \frac{\mathrm{M}_1}{\mathrm{M}_1 + \mathrm{M} \frac{\mathrm{K}^2}{\mathrm{R}^2} + m \frac{k^2}{r^2}}$$

144. *Smeaton's machine.*—For the purpose of testing the truth of the formula for the angular velocity, and consequently of the principles from which that formula is deduced, a machine was invented by Smeaton, which may be described as follows:—A B is a cylinder capable of revolving round a very fine and smooth vertical spindle

coinciding with its axis; it is crossed at right angles by an arm C D, whose axis is bisected by that of A B, on which are two masses of lead of a hollow cylindrical form, and capable of being shifted backward and forward on their respective arms. The whole is set in motion by a weight w attached to the end of a string, which, after passing horizontally over a small pulley P, is wrapped round the cylinder A B.

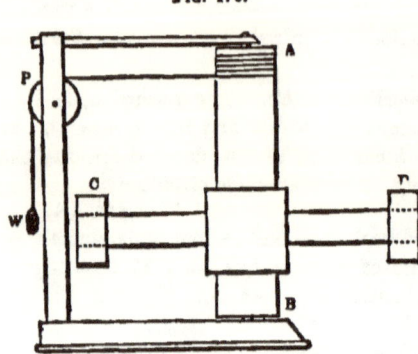

Fig. 176.

Ex. 796.—In Smeaton's machine, given the following dimensions, A B is 3 ft. 8 in. long and 6 in. in diameter, C D is 4 ft. long and 3 in. in diameter, they are joined by a centre, in shape a cube 8 in. along the edge, all of oak, the masses of lead are 6 in. in external diameter and 3 in. long; the string is long enough to cause the machine to make 15 turns before it is unwound; determine the angular velocity communicated to the machine by a weight of 20 lbs.—(1) when the leaden cylinders are placed at the ends of the arms; (2) when they touch the faces of the cube—the inertia of the pulley, the weight of the string, and the passive resistances being neglected.

Ans. (1) 12·04. (2) 29·16.

[Employing the results obtained in *Ex.* 771, it is easily shown that the moment of inertia of the revolving piece is 208·14 in the first case, and 35·46 in the second case.]

Ex. 797.—In the first case of the last example determine approximately the error in the angular velocity that results from omitting the inertia of the pulley, supposing it to be of brass, and to be 2 in. in radius and ½ an inch thick. *Ans.* 0·0023.

Ex. 798.—There is a pulley whose radius is r, and radius of axle ρ, the limiting angle of resistance between the axle and its bearings is ϕ; a rope (whose weight is to be neglected) is wrapped round the pulley, and carries at its end a body whose mass is P; given M the mass of the pulley and $M k^2$ its moment of inertia; determine the angular velocity acquired by the pulley when P has fallen through h feet.

[If the force of gravity on a mass P' at the end of the rope could just overcome the friction of the axle we should have

$$P' g (r - \rho \sin \phi) = M g \rho \sin \phi.$$

This is, of course, the equation established in Ex. 411, when the rigidity of the rope is put out of account, and Q is equal to zero; the forces, however, are expressed in absolute units. The force therefore which produces motion equals the force of gravity on a mass $P - P'$ i.e. it equals $(P - P')g$, or

$$\left(P - \frac{M \rho \sin \phi}{r - \rho \sin \phi}\right)g.$$

Hence the work done by the impressed forces is

$$\left(P - \frac{M \rho \sin \phi}{r - \rho \sin \phi}\right)gh$$

in absolute units; and this must equal the kinetic energy of the system viz. $\frac{1}{2}(M k^2 + P r^2)\omega^2$. Therefore

$$\omega^2 = \frac{2gh(Pr - (P+M)\rho \sin \phi)}{(M k^2 + P r^2)(r - \rho \sin \phi)}.]$$

Ex. 799.—A cylinder turns round an axle whose radius is ρ; it starts with an angular velocity ω: if g is the force of gravity at the station, show that it will be brought to rest by friction after n turns, where

$$n = \frac{r^2 \omega^2}{8 \pi \rho g \mu}.$$

Ex. 800.—The grindstone described in Ex. 16 turns on a bearing of cast iron; it makes 15 turns per minute: determine the number of turns it will make when left to itself, the axle being well greased ($\mu = 0.075$, see p. 159). *Ans.* 1·94.

[The moment of inertia may be taken as equal to that found in Ex. 772, and the mass to that found in Ex. 16.]

Ex. 801.—Round the wheel described in Ex. 773 is wound a rope 30 ft. long, to the end of which is attached a weight of 250 lbs.; the coefficient of friction between the axle and its bearing is 0·075; the weight is allowed to run down: determine the number of revolutions made by the wheel after the rope has run out, supposing that the rope does not slide on the surface of the wheel during any part of the motion. *Ans.* 6·35 times.

145. *Atwood's machine* was invented for the purpose of determining the accelerative effect of gravity; for practical purposes this can be far more accurately done by means of observations of the pendulum; it however, presents a case of terrestrial motion which admits of very accurate observation, and thus supplies a means of testing the truth of the fundamental principles of dynamics. The annexed figure represents an elevation of this machine, which can be sufficiently described as follows:—A and B

are boxes containing equal weights, and connected by a thread A C B passing over a pulley C, which is supported either on friction wheels or by the points of screws, one of which is seen at D. The box A is made to descend either by a flat weight placed on it or by a bar E, which is intercepted by the ring F, through which the box passes and continues to descend till it strikes the stage G; the distance passed over is measured by a scale on H I, and the time by a pendulum K, which may be kept in motion by a clock escapement with a weight: the machine is levelled by the screws L M.* The weight E produces a certain velocity while moving over a given distance, viz. till E comes to F; the velocity acquired is then determined by observing the time in which A moves from F to G; for when E is removed, the boxes A and B will of course move uniformly with the velocity acquired.

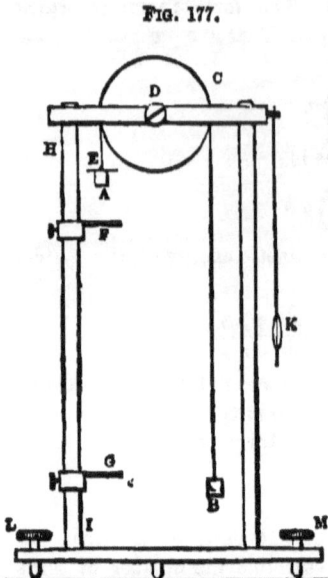

FIG. 177.

Ex. 802.—In Atwood's machine if w is the mass of A or B, and P the mass of the bar, and if mk^2 is the moment of inertia of the pulley and r its radius, then v, the velocity acquired by each of the boxes while P moves through a distance h, is given by the formula

$$v^2 = \frac{2gh\,\text{P}r^2}{(2\text{w}+\text{P})r^2 + mk^2}$$

[The weight of the thread and the passive resistances are neglected; consequently in comparing this result with experiment great care must be taken to suspend the axis of the pulley so that it turns without friction; and a very fine strong thread should be employed.]

---

* Young's *Lectures*, p. 758.

Ex. 803.—If in Atwood's machine the pulley were a solid cylinder of cast iron 2 ft. in diameter and 3 in. thick, the equal weights 28 lbs. each, the bar 2 lbs., what velocity will the weights have acquired when the preponderating weight has fallen through 15 ft.? *Ans.* 2·858 ft. per sec.

[It may be observed that in the ordinary form of Atwood's machine the wheels are light brass wheels—not at all resembling that described in the example.]

146. *The flywheel.*—When a steam-engine is employed as a prime mover, it is desirable that the angular velocity communicated to the principal shaft should be as nearly as possible uniform; now it commonly happens that the driving pressure is variable, or else acts at a variable distance (as in the case of a crank); it may also happen that the work to be done by the shaft is intermittent; for instance, it may be required to lift a tilt hammer. Now, if a sufficiently large flywheel is made to turn with the shaft, there will be accumulated in it a number of units of work very much greater than that done by a single turn of the crank, or than the number expended on a single lift of the hammer, and consequently the variations produced in the angular velocity will be very small—the diminution of these variations being the end to be attained by the flywheel. In the examples that follow, it is supposed that the mass of the wheel (M) is distributed uniformly along the circumference of the circle described by the mean radius (r). The moment of inertia of the wheel is therefore M r^2. A more accurate determination of the moment of inertia could be obtained as in *Ex.* 773.

Ex. 804.—An engine of 35 horse-power makes 20 revolutions (i.e. up and down strokes) per minute, the flywheel is 20 ft. in diameter, and weighs 20 tons; determine the number of foot-pounds of work accumulated in it; and if the work done during half a revolution were lost, determine what part of the angular velocity would be lost by the flywheel.

Ans. (1) 307,000 ft.-pds. (2) $\frac{1}{21}$.

Ex. 805.—If the engine in the last example were employed to lift a tilt hammer weighing 4,000 lbs., the centre of gravity of which is raised 3 ft.

at each stroke, and if this were done once merely by the work accumulated in the flywheel, what part of its angular velocity would it lose?

$$Ans.\ \frac{1}{51}.$$

Ex. 806.—If the axis of the flywheel in *Ex.* 804 were 6 in. in diameter, and were of wrought iron turning on cast iron well greased ($\mu = 0.075$), determine approximately the fractional part of the 35 horse-powers expended on turning the flywheel for one minute.

$$Ans.\ \frac{1}{11}.$$

Ex. 807.—If the flywheel in *Ex.* 804 were divided into two pieces along a diameter, and if each piece were connected with the axle by a spoke at right angles to that diameter, determine the tension of each spoke arising from centrifugal force; if the velocity of the wheel were liable to be raised to 40 turns per minute, what ought to be the section of a wrought-iron spoke which would bear this tension *with safety*?

Ans. (1) 625,500 abs. un. (2) 11·5 sq. in.

[See *Ex.* 328 and Art. 9.]

147. *M. Morin's experiments on friction.*—A full account of M. Morin's experiments will be found in his 'Notions Fondamentales,' already frequently referred to; it would be inconsistent with the plan of the present work to enter into the details of the methods he employed; it may, however, be stated that the arrangement adopted was in principle the same as that described in *Ex.* 611; to which it must be added that the rope supporting P was of considerable thickness, and passed over a pulley on the edge of the table. Now, it will be remarked that in *Ex.* 611 and 616, it is implicitly assumed that the tension of the horizontal portion of the rope is equal to the tension of the vertical portion; but as in the present case the rope is thick, the axle of the pulley rough, and work is expended in overcoming the inertia of the pulley, this assumption is untrue; the formulæ actually employed will be seen in the following questions; the student will probably find little difficulty in investigating them. The notation adopted is as follows:—P denotes the weight producing motion, T the tension of the horizontal portion of

MORIN'S EXPERIMENTS ON FRICTION. 315

the rope; w the weight of the pulley, I its moment of inertia, r its radius, r_1 the radius of its axle, μ the coefficient of friction between the axle and its bearing, a the coefficient of the rigidity of the rope, so that $(1+a)\,\text{T}$ is the force to be overcome by P in its descent, f the acceleration of P's motion, g the accelerative effect of gravity. The acceleration produced by the weight of the rope is neglected. The mode of determining f will be understood from the next question.

Ex. 808.—If a drum revolves in such a manner that a point on its circumference receives a uniform acceleration f, and if a sheet of paper is wrapped on it, and a pencil with its point resting on the paper is made to move in a direction parallel to the axis with a uniform velocity of v feet per second, show that the curve described on the paper will be a portion of a parabola, and that if c is the semi-latus rectum measured in feet, we shall have $f = \dfrac{\text{v}^2}{c}$.

[In the experiments the parabolic curve was unmistakably obtained, whence immediately follows the important law that friction is independent of velocity.]

Ex. 809.—In M. Morin's experiments show that the pressure between the axis of the pulley and its bearings is given by the formula

$$\sqrt{\left(\text{P}+w-\text{P}\dfrac{f}{g}\right)^2+\text{T}^2} \quad \text{or} \quad 0\cdot 96\,\text{P}\left(1-\dfrac{f}{g}\right)+0\cdot 96w+0\cdot 4\,\text{T}.^*$$

Ex. 810.—The second formula in the last example being employed, show that T is given by the formula

$$\text{T}\left(1+a+0\cdot 4\dfrac{\mu r_1}{r}\right) = \text{P}\left(1-\dfrac{f}{g}\right)\left(1-0\cdot 96\dfrac{\mu r_1}{r}\right)-0\cdot 96\dfrac{\mu w r_1}{r}-\dfrac{\text{I}f}{r^2}.$$

---

* The theorem that $\sqrt{a^2+b^2} = 0\cdot 96a+0\cdot 4b$ where $a > b$, with an error not exceeding $\frac{1}{25}$th part of the true value, is due to M. Poncelet; it may be proved as follows:—Let $a = r \sin\theta$, $b = r\cos\theta$ $\therefore r = \sqrt{a^2+b^2}$, and θ must have some value between $45°$ and $90°$. Now, if $r' = 0\cdot 96a+0\cdot 4b$ we have $r' = r\,(0\cdot 96 \sin\theta + 0\cdot 4 \cos\theta)$; but $0\cdot 4 = 0\cdot 96 \tan 22° \, 30'$, therefore $r' = r \times 0\cdot 96 \dfrac{\sin(\theta + 22°\,30')}{\cos 22°\,30'}$. Then, as θ increases from $45°$ up to $67°\,30'$, r' will increase from $0\cdot 96\,r$ to $1\cdot 04\,r$, and as θ increases from $67°\,30'$ up to $90°$, r' decreases from $1\cdot 04r$ to $0\cdot 96r$, and consequently r' never differs from r by more than $\dfrac{r}{25}$.

Ex. 811.—A body whose weight is W is caused to slide on a rough horizontal plane by a force T; after moving through s ft. it acquires a velocity v: show that the coefficient of friction (μ) is given by the equation

$$\mu = \frac{T}{W} - \frac{v^2}{2gs}.$$

148. *Compound pendulums*.—The terms centre of suspension and centre of oscillation have already been explained (Art. 123); their properties are proved in the following propositions.

Proposition 34.

If k_1 is the radius of gyration of a body with reference to its axis of suspension, and h the distance of the centre of gravity below the centre of suspension, then $\dfrac{k_1^2}{h}$ is the distance of the centre of oscillation from the latter point.

Let A B be the body (whose mass is M) oscillating about an axis passing through s at right angles to the plane of the paper, which also contains the centre of gravity G; join S G, draw the vertical line S C, let G_1 be the position of G at the commencement of the motion, draw $G_1 M_1$ and G M at right angles to S C, and denote G_1 S C and G S C by θ_1 and θ respectively. Now, when S G_1 falls to S G the centre of gravity descends through a vertical height M_1 M or h (cos θ − cos θ_1). If then g denotes the accelerative effect of gravity at the station, and ω the angular velocity acquired, we have (Prop. 33, Cor. 2)

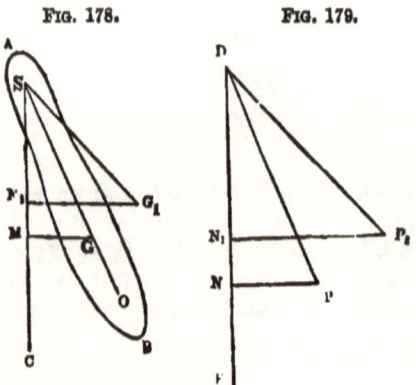

Fig. 178.   Fig. 179.

COMPOUND PENDULUMS. 317

$$\omega^2 = \frac{2gh}{k_1^2}(\cos\theta - \cos\theta_1).$$

Let D P be a simple pendulum oscillating about D, draw the vertical line D E, and let P_1 be the position from which P begins to move; draw PN and $P_1 N_1$ at right angles to D E, and let D P be denoted by l, and let P_1 D E equal θ_1, and P D E equal θ; then if v is the velocity acquired by the point in falling from P_1 to P, we have

$$v^2 = 2g \times N N_1 = 2gl(\cos\theta - \cos\theta_1)$$

and therefore, if ω' is the angular velocity of P, we have

$$\omega'^2 = \frac{2g}{l}(\cos\theta - \cos\theta_1).$$

Now, if l equals $\frac{k_1^2}{h}$, ω' will equal ω for all values of θ, and since A B and D P are moving at each instant with the same angular velocity, their oscillations will be performed in the same time, and therefore $\frac{k_1^2}{h}$ is the length of the simple pendulum oscillating in the same time as A B; hence, if in S G produced a point O be taken, such that S O equals $\frac{k_1^2}{h}$, that point will be the centre of oscillation.

Cor.—The time of a small oscillation of B A will equal $\frac{\pi k_1}{\sqrt{gh}}$ by Prop. 29.

Proposition 35.

The centres of oscillation and suspension are reciprocal.

Let A B be the body, G its centre of gravity, S a centre of suspension through which the axis of rotation passes at right angles to the plane of the paper, and O the corresponding centre of oscillation, it is to be proved that these points are reciprocal, i.e. if o is made the centre of suspension, s will be the corresponding centre of oscillation. Let k be the radius of gyration round a parallel axis through the centre of gravity, let s G, G o be respectively denoted by h and x,

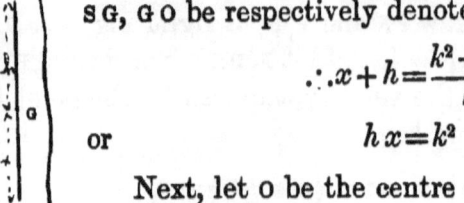

Fig. 180.

$$\therefore x + h = \frac{k^2 + h^2}{h}$$

or
$$h x = k^2$$

Next, let o be the centre of suspension, and y the distance from G to the corresponding centre of oscillation, then

$$y + x = \frac{k^2 + x^2}{x}$$

or
$$y x = k^2$$

and therefore $y = h$, or s is the centre of oscillation.

Q. E. D.

Ex. 812.—A thin rod of steel 10 ft. long oscillates about an axis passing through one end of it: determine the time of a small oscillation; the number of oscillations it makes in a day; and the number it will lose in a day if the temperature is increased by 15° F.

Ans. (1) 1·434 sec.   (2) 60,254.   (3) 3.

Ex. 813.—A pendulum oscillates about an axis passing through its end; it consists of a steel rod 60 in. long, with a rectangular section $\frac{1}{2}$ by $\frac{1}{4}$ of an inch; on this rod is a steel cylinder 2 in. in diameter and 4 in. long; when the ends of the rod and cylinder are set square, determine the time of a small oscillation. *Ans.* 1·174.

Ex. 814.—Determine the radius of gyration with reference to the axis of suspension of a body that makes 73 oscillations in 2 minutes, the distance of the centre of gravity from the axis being 3 ft. 2 in.   *Ans.* 5·267 ft.

o *Ex.* 815.—Determine the distance between the centres of suspension and oscillation of a body that oscillates in $2\frac{1}{2}$ sec.   *Ans.* 20·264 ft.

Ex. 816.—If $\frac{k_1^2}{h}$ is the length of a simple pendulum corresponding to an oscillating rod, show that if it expands uniformly in the proportion of $1 + a : 1$ that the length of the simple pendulum becomes $(1 + a) \frac{k_1^2}{h}$.

Ex. 817.—Miaran determined the length of the seconds pendulum at Paris to be 39·128 inches; he employed a ball of lead 0·533 inch in diameter, suspended by an exceedingly fine fibre whose weight could be neglected;* supposing the measurements made with perfect accuracy, upon the supposition that the distance from the point of suspension to the centre of the ball is the length of the pendulum, show that the error is less than the 0·001 of an inch.

Ex. 818.—A pendulum consists of a brass sphere 4 in. in diameter suspended by a steel wire $\frac{1}{10}$ of an inch in diameter; the centre of the sphere is 40 in. below the point of support: † determine the number of oscillations it will make in a day, and what number would be obtained on the supposition that the centre of oscillation coincides with the centre of the sphere ($g = 32$). *Ans.* (1) 85,766. (2) 85,212.

Ex. 819.—If a sphere whose radius is r is suspended successively from two points by a very fine thread, and if the distances of the centre of the sphere from the points of suspension are respectively h and h', and if l and l' are the distances of the corresponding centres of oscillation from the points of suspension, show that

$$l - l' = (h - h')\left(1 - \frac{2r^2}{5hh'}\right)$$

Ex. 820.—If t and t' are the times of a small oscillation of the pendulum in the last example corresponding respectively to l and l', show that the accelerative effect of gravity is given by the equation

$$g = \frac{\pi^2(h - h')}{t^2 - t'^2}\left(1 - \frac{2r^2}{5hh'}\right)$$

149. *M. Bessel's determination of the accelerative effect of gravity.*—The last two examples contain the principle of the method by which M. Bessel determined the accelerative effect of gravity at Königsberg.‡ The pendulum was first allowed to swing from a point of support at a distance h above the centre of the sphere, and the number of oscillations made in a given time was noted, by which t was determined with great accuracy; the wire was then grasped firmly at a point lower down, so that the oscillations were now performed about a point distant h' from the centre of the sphere, and t' noted as before; now $h - h'$ being the distance between two fixed

* Airy, *Figure of the Earth*, p. 224. † *Ibid.* p. 225. ‡ *Ibid.* p. 223.

points can be very accurately determined; the lengths h and h' cannot be determined without some liability to error, but as they only appear in the small term $\dfrac{2r^2}{5hh'}$, that error will hardly affect the determination of g, which can by this method be ascertained with extreme accuracy.

Ex. 821.—In the last example let r, h, and h' be respectively reckoned 1, 50, and 40 inches, so that $h-h'$ is exactly 10 in., but it is doubtful whether the separate values of h and h' are not as much as $\frac{1}{15}$th of an inch longer than the values assigned: determine the possible error in the value of g.

Ans. $\dfrac{g}{1,115,000}$.

150. *Captain Kater's method of determining the accelerative effect of gravity.*—This method depends on the reciprocity of the centres of oscillation and suspension; the pendulum has two axes (or 'knife edges,' as they are called, though they are really wedges of very hard steel), by either of which it can be suspended; now, if the time of oscillation about either axis be the same, the distance between the edges (l) will be the length of the simple pendulum; the distance, being that between two fixed points, admits of very accurate measurement, and then g is obtained by the formula

$$g = \frac{\pi^2}{t^2} \cdot l$$

The difficulty of giving the edges their exact position is overcome as follows:—On the pendulum rod is placed a weight that can be moved up or down by screws; the edges are fixed as nearly as possible in the right position; and then by moving the weight up or down, the values of k_1 and h can be changed until $k_1^2 \div h$ equals the distance between the edges, i.e. until the number of oscillations made in a given time about either edge is the same.

CHAPTER VIII.

ON IMPACT.

151. *Impulsive action.*—Suppose a sphere A to overtake a sphere B, their centres moving in the same line; it is a matter of common observation that they will strike, and then separate, A moving after impact with a less, and B with a greater, velocity than before; the problem we are to solve is this:—Given the masses of the bodies and their velocities at the instant before impact, to determine the velocities they will have at the instant after impact.

Now, it will be observed that though the bodies are in contact during a very short time, yet that time is really finite, and the pressure which the one exerts on the other must increase from zero at the instant of contact, till it attains a very considerable magnitude, and must then decrease down to zero at the instant of separation. Moreover, it appears from *Ex.* 698, that if A exerts at each instant against B a pressure equal to that which B exerts against A—in other words, if the action and reaction are equal and opposite forces—then the momentum lost by A must equal that gained by B, and the total amount of momentum in A and B before impact must equal the total amount after impact. Now, that this is a fact was ascertained by numerous experiments made by Newton,* and this we shall take as our fundamental principle, viz. *that the momentum lost during the impact by one body equals that gained by the other.* To prevent misunderstanding,

* Introduction to the *Principia.*

Y

it may be added that the sum of the momenta of the two bodies means their *algebraical* sum.

152. *The mean pressure exerted during impact.*—The following example is intended to illustrate the fact that during impact there is really called into play a very large force which is exerted during a very short time.

Ex. 822.—A hard mass weighing 50 lbs. falls from a height of 6 ft. on to a plane surface which at the instant of greatest compression has yielded to the extent of $\frac{1}{20}$th of an inch—the mass itself being supposed to be entirely uncompressed; determine the mean mutual pressure, and the duration of compression supposing it produced by the *mean* pressure.*

Ans. (1) 72,000 g abs. un. (2) 0·000425 sec.

[The pressure by acting through $\frac{1}{20}$th of an inch brings the mass to rest.]

153. *Impact of inelastic bodies.*—When A overtakes B, it is plain that so long as A moves faster than B, the two surfaces of contact will be compressed, and the compression will continue to increase until A and B are moving with the same velocity; if the mutual action then ceases, the bodies are said to be inelastic.

Now, let the masses be denoted by A and B respectively, let R be the momentum lost by the one and gained by the other during impact, and let their velocities before impact be V and U, and their common velocity after impact be v; then we obtain from the fundamental principle (Art. 151)

$$A v = A V - R \quad \ldots \ldots \ldots (1)$$
$$B v = B U + R \quad \ldots \ldots \ldots (2)$$

whence
$$R = \frac{A B (V - U)}{A + B} \quad \ldots \ldots (3)$$

and
$$v = \frac{A V + B U}{A + B} \quad \ldots \ldots (4)$$

In working examples the student is recommended to proceed from the general principle, or, in other words, to

* Poncelet, *Introd. à la Méc. ind*, p. 166.

IMPACT OF INELASTIC BODIES. 323

form and then solve the equations (1) and (2), and not to substitute particular values in (3) and (4). If A meet B, one of the velocities must be reckoned negative, and the bodies will move after impact in the direction of that velocity if v be negative.

Ex. 823.—If A weighing 2 lbs. and moving with a velocity of 20 ft. per second overtakes B weighing 5 lbs. and moving with a velocity of 5 ft. per second, determine the common velocity after impact. *Ans.* $9\frac{2}{7}$ ft. per sec.

Ex. 824.—In the last example if the bodies had met, determine the common velocity after impact. *Ans.* $2\frac{1}{7}$ ft. per sec. in A's direction.

Ex. 825.—In Art. 153 show that the number of absolute units of work lost during impact equals $\dfrac{A B (V-U)^2}{2 (A+B)}$.

Ex. 826.—If a shot weighing P lbs. is fired with a velocity v into a mass of wood weighing Q lbs. which is quite free to move, show that the velocity with which the wood begins to move is $\dfrac{P V}{P+Q}$; and state why this case must be one of inelastic impact.

Ex. 827.—If in the last example $Q = n P$, show that, in consequence of the impact, n units of work are lost in every $n + 1$.

154. *Impact of elastic bodies.*—It commonly happens that the mutual action does not entirely cease with the compression, but when that ends the bodies begin to recover their shapes, and thereby continue to press on each other till the impact terminates. Now, let R be the momentum lost by the one body and gained by the other during compression, and R' that lost and gained during expansion; then the whole momentum lost by the one body and gained by the other will equal R + R'. But it is found by experiment that for the same substances R bears to R' a fixed ratio 1 : λ;* therefore $R' = \lambda R$, and $R + R' = (1 + \lambda) R$; where λ is a constant quantity depending on the materials of the impinging bodies called the *coefficient of restitution*; it is often called the coefficient of elasticity, but must on no account be mistaken for the modulus

* This follows from Newton's experiments already referred to.

of elasticity (Art. 6). In the two extreme cases of inelasticity and perfect elasticity, λ equals 0 and 1 respectively; in other cases λ is a proper fraction. We have already seen that if a body whose mass is A, moving with a velocity V, overtakes another whose mass is B, moving with a velocity U, then the momentum lost by the one and gained by the other at the end of compression equals $\dfrac{AB(V-U)}{A+B}$. Hence the total momentum gained and lost will equal $(1+\lambda) \times \dfrac{AB(V-U)}{A+B}$. And therefore if v and u are their respective velocities after impact, we shall have

$$Av = AV - (1+\lambda)R$$
$$Bu = BU + (1+\lambda)R$$

or
$$v = V - \dfrac{(1+\lambda)B(V-U)}{A+B}$$

and
$$u = U + \dfrac{(1+\lambda)A(V-U)}{A+B}$$

It may be added that the remarks made in Art. 153, relative to the working of examples, are applicable to the case of elastic bodies.

Ex. 828.—Show that v and u are given by the following formulæ—

$$v = \dfrac{AV + BU}{A+B} - \dfrac{\lambda B(V-U)}{A+B}$$

$$u = \dfrac{AV + BU}{A+B} + \dfrac{\lambda A(V-U)}{A+B}.$$

Ex. 829.—Determine the velocities after impact of a ball (A) weighing 20 lbs. which, moving with a velocity of 100 ft. per second, overtakes a ball (B) weighing 50 lbs. and moving with a velocity of 40 ft. per second, their coefficient of restitution being ½.

Ans. A's velocity 35⅝. B's velocity 65⅝.

Ex. 830.—In the last case suppose the heavier body (B) to be at rest: determine the velocities after impact.

Ans. A rebounds with a velocity 7⅙; B moves forward with a velocity 42⅚.

IMPACT OF ELASTIC BODIES.   325

Ex. 831.—Obtain the velocities after impact in *Ex.* 829, upon the supposition that the bodies meet.

Ans. A rebounds with a velocity 50, and B with a velocity 20.

Ex. 832.—If there are two perfectly elastic balls A and B of equal masses, and A moving with a velocity v impinges on B at rest, show that A is brought to rest and B takes the velocity v. If there is a number of equal and perfectly elastic balls, B, C, D, E, placed in a line, what would be the result of A striking B, the direction of the impact coinciding with the line?

Ex. 833.—If a ball whose weight is A moving with a velocity V meets a ball whose weight is B moving with a velocity U, show that in the case of perfect elasticity the velocities of rebound are given by the following construction:—Draw any line A B, divide it in G in the inverse ratio of the weights of A and B, and in C in the ratio of their velocities ; on the other side of G measure off G D equal to G C, then A's velocitiy of rebound : B's velocity of rebound :: A D : B D.*

Ex. 834.—Two balls weighing respectively 12 and 8 lbs. are suspended by threads in such a manner that their centres are 4 ft. below the points of support ; when at rest the line joining their centres is horizontal ; if the smaller one is raised so as to fall through a quadrant, determine the angle described by the other after impact, if the coefficient of restitution equals $\frac{2}{3}$.

Ans. 56° 14'.

Ex. 835.—If A and B are the weights of two perfectly elastic balls, if V and U are their velocities before impact and v and u their velocities after impact, show that

$$A V^2 + B U^2 = A v^2 + B u^2$$

Ex. 836.—If a ball impinges perpendicularly on a fixed plane with a velocity V, show that the velocity of rebound equals λ V.

[It must be remembered that at the end of compression the velocity is entirely destroyed, consequently $0 = A V - R$; hence, if v is the velocity at the end of the impact $A v = A V - (1 + \lambda) R$, whence $v = -\lambda V$.]

Ex. 837.—If bodies are dropped from equal heights on to a fixed horizontal plane, show that their coefficients of restitution are in the same ratio as the square roots of the heights to which they rebound.

O*Ex.* 838.—A ball is dropped from a height h: show that the whole distance it describes before coming to rest equals

$$h \frac{1 + \lambda^2}{1 - \lambda^2}$$

---

* It was in this form that the problem of impact was originally solved by Sir C. Wren (vide *Montucla*, vol. ii. p. 411).

Ex. 839.—A ball (A) is thrown upward with a velocity of 160 ft. per second; when it has reached a height of 300 ft. it is met by an equal ball (B) which has fallen from a height of 100 ft.; determine the time after the instant of impact in which each will reach the ground, assuming that λ equals unity. *Ans.* A after $2\frac{1}{2}$ sec. B after $7\frac{1}{2}$ sec.

155. *Oblique impact of smooth bodies.*—Suppose a smooth ball A, moving with a velocity V, to impinge obliquely on a smooth ball B, moving with a velocity U; draw the line of centres, and resolve V into component velocities V_1 and V_2, the former along, the latter at right angles to, the line of centres; in like manner resolve U into U_1 and U_2; now V_2 and U_2 will remain unchanged by the impact, but V_1 and U_1 will be changed into v_1 and u_1 exactly as if the bodies had impinged directly with the velocities V_1 and U_1: hence, by compounding v_1 and V_2 and also u_1 and U_2, we obtain the required velocities after impact. The general formulæ commonly given for these velocities are of very little value, as any particular example is much more easily worked by proceeding from first principles: the following example will sufficiently exhibit the method of treating these cases.

Ex. 840.—Let A and B be two perfectly elastic balls which at the instant of impact are moving along the lines P A and Q B, the line of centres C D being in the same plane as P A and Q B; A weighs 10 lbs., moves with a velocity of 16 ft. per second, and the angle P A C contains 30°; B weighs 15 lbs., moves with a velocity of 8 ft. per second, and the angle Q B D contains 60°: determine the velocities after impact and their directions.

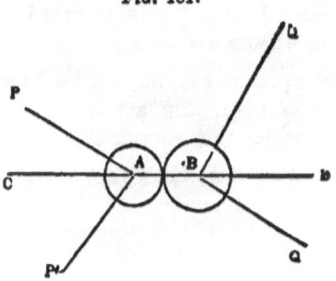

FIG. 181.

(*a*) Before impact A's velocity at right angles to C D is 8, and B's $4\sqrt{3}$; they are unchanged by the impact.

(*b*) Before impact A's velocity along C D is $8\sqrt{3}$ and B's velocity is -4; they are changed by impact into $-\frac{8}{5}(3+\sqrt{3})$ and $\frac{4}{5}(-1+8\sqrt{3})$ respectively.

OBLIQUE IMPACT. 327

(c) Hence A's velocity after impact equals $\frac{8}{5}(37+6\sqrt{3})^{\frac{1}{2}}$, and B's velocity $\frac{4}{5}(268-16\sqrt{3})^{\frac{1}{2}}$; i.e. A's velocity equals 11·02 ft. per second, and B's equals 12·4 ft. per second.

(d) The directions of the motion of P and Q after impact are respectively A P' and B Q' where $\tan$ P' A C equals $\dfrac{5}{3+\sqrt{3}}$, and $\tan$ Q' B D equals $\dfrac{5\sqrt{3}}{8\sqrt{3}-1}$

i.e. P' A C equals 46° 35' and Q' B D equals 33° 58'.

By this means the motion of A and B at the instant after impact is completely determined.

Ex. 841.—If a ball A moving in a direction making an angle of 30° with the line of centres overtakes B moving along the line of centres, determine the velocities after impact, if A weighs 12 lbs. and its velocity is 12 ft. per second, and B weighs 30 lbs. and its velocity 4 feet per second, and the coefficient of restitution equals ½.

Ans. (1) A's vel. 7, C A P' (fig. 179) = 120° 34'. (2) B's vel. 7.

Ex. 842.—Referring to the last two examples, how does it appear that the impact has produced no change in the total momentum of the two bodies?

Ex. 843.—A body whose coefficient of restitution is ⅓ impinges with a velocity of 30 ft. per second on a fixed plane in a direction making an angle of 27° with the perpendicular; determine the magnitude and direction of the velocity after impact. *Ans.* (1) 19·1 ft. (2) 45° 32'.

Ex. 844.—If in the last example the body had been inelastic, how would it begin to move after impact?

Ex. 845.—If in *Ex.* 843 the angle of impact is α and the angle of rebound β, and the coefficient of restitution λ, show that

$$\tan \beta = \frac{\tan \alpha}{\lambda}$$

Ex. 846.—Give a geometrical construction by which to determine the direction in which a billiard ball must begin to move so that after one rebound it may strike another ball whose position is given, (1) if the coefficient of restitution equals unity, (2) if the coefficient of restitution equals λ.

Ex. 847.—Extend the construction in the last example to the case in which the ball makes two rebounds from cushions at right angles to each other.

Remark.—If the surfaces of the impinging bodies are rough, the effect of the tangential impact will generally be to produce a rotatory motion, as well as to modify the previous motion of the bodies: the complete solution of this case lies beyond the scope of the present work. The same remark applies to the case in which the motion of one or both bodies sustains a resistance appreciable in comparison with the mean impulsive pressure.

156. *Application of D'Alembert's principle to the case of impulsive action.*—It will be remarked that a case of impulsive action does not differ essentially from any other case of motion produced by force; the difference in the mode of treating these cases arises solely from our inability to determine the force exerted at each instant of the duration of the impact; it follows, therefore, that at each instant during the collision the effective forces applied in the opposite directions would be in equilibrium with the impressed forces; and consequently the momenta produced by the effective forces so applied, and those actually produced by the impressed forces, will satisfy the conditions of the equilibrium of forces. We shall apply this principle to determine the angular velocity communicated by a blow to a body capable of revolving round a fixed axis, and the impulse produced on the axis by that blow.

Proposition 36.

A body capable of turning round a given axis, and symmetrical with reference to the plane passing through the centre of gravity at right angles to the axis, is struck by a blow of given magnitude along a line lying in that plane, to determine the angular velocity communicated to the body, and the impulse on the axis.

Let the plane of the paper be the plane of symmetry, and let the axis of rotation pass through o: let R be the magnitude of the blow which is delivered along the line R N; draw o y at right angles, and o x parallel to R N; let M be the mass of the body, Mk_1^2 its moment of inertia with reference to the given axis, $\bar{x}, \bar{y}$ the co-ordinates of its centre of gravity, and let O N equal a; let X and Y be the impulsive reactions of the axis in the directions of Ox and Oy respectively; and let ω be the angular velocity of the body communicated by the blow. Consider any particle P whose co-ordinates are x_1, y_1, whose distance from O is r_1 and mass m_1 and let the angle x O P equal θ_1, also suppose a similar notation to be employed for the other particles composing the body. Now, the velocity of P is $r_1\omega$ in a direction perpendicular to O P or is equivalent to velocities $\omega r_1 \sin \theta$, or ωy_1 parallel to O X and $-\omega r_1 \cos \theta$, or $-\omega x_1$ parallel to O y, and therefore the momentum communicated to P is equivalent to the two $m_1 y_1 \omega$ parallel to O x, and $-m_1 x_1 \omega$ parallel to O y; the expressions for all the other particles being precisely similar. Now, these

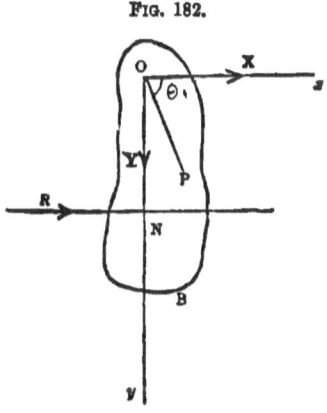

FIG. 182.

are the momenta that would be communicated by the effective forces, the impressed forces being R, Y, and X; also it will be observed that the moment of P's momentum round O is $m_1 r_1^2 \omega$; consequently (Prop. 15)—

$$R + X = m_1 y_1 \omega + m_2 y_2 \omega + m_3 y_3 \omega + \ldots$$
$$= \omega (m_1 y_1 + m_2 y_2 + m_3 y_3 + \ldots)$$
$$-Y = m_1 x_1 \omega + m_2 x_2 \omega + m_3 x_3 \omega + \ldots$$
$$= \omega (m_1 x_1 + m_2 x_2 + m_3 x_3 + \ldots)$$
$$a R = m_1 r_1^2 \omega + m_2 r_2^2 \omega + m_3 r_3^2 \omega + \ldots$$
$$= \omega (m_1 r^2 + m_2 r_2^2 + m_3 r_3^2 + \ldots)$$

Or by Prop. 16 and Art. 134—

$$R + X = M \bar{y} \omega$$
$$-Y = M \bar{x} \omega$$
$$a R = M k_1^2 \omega$$

or
$$\omega = \frac{R}{M} \cdot \frac{a}{k_1^2} \quad \ldots \ldots (1)$$

$$-Y = \frac{R\,a\,\bar{x}}{k_1^2} \quad \ldots \ldots (2)$$

$$-X = R - \frac{R\,a\,\bar{y}}{k_1^2} \quad \ldots \ldots (3)$$

The first of these equations gives the angular velocity communicated to the body; the second and third equations give the components of the reaction of the axis, which is of course equal and opposite to the blow sustained by the axis.

N.B.—It will be an instructive exercise for the student to ascertain for what positions of the centre of gravity the reactions of the axis will be as indicated in the figure: it will commonly happen, as he will find, that the reactions will in reality act in the contrary directions to those indicated.

Ex. 848.—A uniform rod 12 ft. long and weighing 10 lbs. is suspended at one end; it receives at the other, in a direction perpendicular to its length, a blow whose momentum is 32: determine—(1) the angular velocity with which it begins to move; (2) the impulsive pressure on the axis; and (3) find how many times this impulse exceeds the blow given by a weight of $\frac{1}{4}$ of a pound which has fallen through a height of 4 ft.
Ans. (1) 0·8. (2) 16. (3) 4 times.

Ex. 849.—A beam of oak 10 ft. long and 1 ft. square is suspended by an axis perpendicular to one face and passing through the axis of the beam, at a distance of 1 ft. from the end; it is struck at a point 8 ft. below the axis by a bullet weighing 1 lb. and moving with a velocity of 1000 ft. per second: determine—(1) the impulse on the axis; (2) the angular velocity communicated to the beam; (3) the angle through which the beam will revolve.
Ans. (1) 310. (2) 0·56. (3) 14° 5′.

Ex. 850.—A hammer's head weighs 10 lbs. and makes 60 strokes per minute on an anvil: if the time of ascending equals that of descending, and the blow is entirely due to the velocity it acquires in falling, compare that blow with the impulse on the axis in the last example.
Ans. One half.

THE CENTRE OF PERCUSSION. 331

Ex. 851.—Determine the impulse on the axis if the mass of cast iron in *Ex.* 791 strikes an anvil after falling through the 30°, the blow on the anvil being supposed to be given by the extreme edge of the cube.

Ans. 3150.

[It will be observed that in this case the impulse on the axis is greater than that which would be produced by a shot weighing 3 lbs. and moving at the rate of 1000 ft. per second; it is obvious that a succession of such impulses would tear to pieces the masonry on which the axis of such a hammer is supported; and accordingly it becomes a point of great practical importance to suspend a tilt hammer in such a manner that there shall be no impulse on the axis. The following article explains the principle on which this is done.]

157. *The centre of percussion.*—Referring to the equations (2) and (3) of Prop. 35, we see that if the blow is delivered in such a manner that $\bar{x}$ equals zero, and k^2 equals $a\bar{y}$, then X and Y equal zero separately, and there is no impulsive pressure on the axis of suspension; hence if O be the centre of suspension, G the centre of gravity of the body, and a point O_1 be taken in O G produced so that

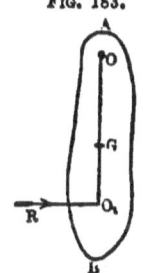

FIG. 183.

$$O O_1 = \frac{k_1^2}{O G}$$

then if the body be struck by a blow whose direction passes through O_1 at right angles to O O_1, there will be no impulsive pressure on the axis, and the point O_1 is therefore called the centre of percussion; it evidently coincides with the centre of oscillation with respect to the centre of suspension O. It must be remembered that the body is supposed to be symmetrical with regard to the plane of the paper, as specified in the enunciation of Prop. 35.

158. *Axis of spontaneous rotation.*—Since the body in the last article when struck begins to rotate round the axis through O without any constraint, it follows that if the body were entirely free, it would begin to move round that axis, which is therefore called the axis of spontaneous

rotation. If it is given that a body is struck by a blow R along a given line, the axis of spontaneous rotation is determined as follows:—Consider the plane passing through G the centre of gravity and the direction of the blow; through G draw a line at right angles to this plane, and let k be the radius of gyration of the body with respect to it: through the centre of gravity draw a line at right angles to the direction of the blow and cutting it in o_1, and on the other side of the centre of gravity take in the line a point O such that

$$o\,G \cdot G\,o_1 = k^2$$

then an axis through O perpendicular to the given plane is the axis of spontaneous rotation, provided the body is symmetrical with reference to that plane.

It will be observed that if the axis of spontaneous rotation is to pass through the centre of gravity, we must have in equations (2) and (3) of Prop. 35, both $\bar{x}=0$ and $\bar{y}=0$, and therefore $R=0$; but from equation (1) ω having a finite value $a\,R$ must also have a finite value; or in other words the body must be struck by an impulsive couple whose moment is $a\,R$, and whose plane passes through the centre of gravity of the body; it will then begin to revolve with an angular velocity $\dfrac{a\,R}{M\,k^2}$ round an axis at right angles to the plane of the couple, and passing through the centre of gravity.

Ex. 852.—A hammer turns round a given axis, the weight of the head is w, and its radius of gyration is k with respect to an axis parallel to the given axis and passing through its centre of gravity; the weight of the handle is w_1, its radius of gyration with respect to the axis is k_1, and the distance of its centre of gravity from the axis a. If the head of the hammer is so placed that its centre of gravity is at the same distance (x) from the axis as the centre of percussion of whole hammer, then

$$x = \frac{w_1\,k_1^2 + w\,k^2}{w_1\,a}$$

BALLISTIC PENDULUM. 333

Ex. 853.—If the head of the hammer in *Ex.* 851 is shifted so as to fulfil the conditions of the last example, determine the distance of its centre of gravity from the axis of rotation. *Ans.* 5·35 ft.

Ex. 854.—A sledge hammer A B is movable round an axis through A; it is 6 ft. long and weighs 4 cwt., it is held in a horizontal position by a weight of 3 cwt. attached to the end of a string which after passing over a small pulley is fastened to B (the parts of the string being vertical); the hammer when allowed to fall into a vertical position makes 50 oscillations per minute round A: determine—(1) the centre of percussion, and (2) the radius of gyration about an axis parallel to the axis of suspension and passing through its centre of gravity. *Ans.* (1) 4·67 ft. (2) 0·87 ft.

Ex. 855.—A cylindrical bolt of cast iron 4 in. in diameter and 8 in. long is struck simultaneously by two equal blows in contrary directions, each at right angles to an extremity of a diameter of its mean section; in consequence the bolt rotates 250 times in a second: determine the magnitude of each blow, and compare it with that which the bolt itself would give if moving with a velocity of 1000 ft. per second.

$$Ans. \ (1) \ 1715. \quad (2) \ \frac{\pi}{48}.$$

159. *Robin's ballistic pendulum.*—This machine is employed to ascertain the velocity with which a shot leaves the mouth of a cannon. The principle on which it is constructed will be most easily understood by describing it in its original form; at present the gun itself is suspended and the recoil observed; but at first it was constructed as follows :—A large mass of wood is carefully suspended so as to turn freely round a knife edge (Art. 150); the shot is fired into this mass, which is backed with iron plates to prevent the ball passing through or shivering it, so that the shot stays in it, and by the blow causes it to revolve through a certain angle (θ), the magnitude of which can be ascertained by a riband attached to a point of the pendulum which is pulled through a spring sufficiently strong to keep the riband straight while the mass moves up, and also to prevent any of it returning when the mass moves back; it is evident that the length of the riband gives the chord of the arc described by the point to which it is fastened, and thus θ is observed; the weight W of the

pendulum includes that of the shot w; the distance h of the centre of gravity of W from the knife edge is determined in the manner suggested by *Ex.* 854. The radius of gyration is inferred from n, the number of small oscillations made in a minute; the distance (a) below the point of support of the point in which the shot strikes the pendulum is measured; and it is (of course) endeavoured that this point should as nearly as possible coincide with the centre of percussion. From these data the velocity V of the shot can be found.

Ex. 856.—In the ballistic pendulum show that

$$V = \frac{120\, g\, h\, W}{\pi\, n\, a\, w} \sin \frac{\theta}{2}$$

APPENDIX.

ON LIMITS.

THROUGHOUT the present work particular geometrical limits have been used instead of the formulæ of the differential and integral calculus—at least, this has been done as far as possible: if the reader has not been accustomed to reason on limits, he may perhaps find a difficulty in understanding the propositions in which they occur : should this be so, the following remarks may prove useful.

1. *Definition of a limit.*—Let there be any variable magnitude X, and let there be a fixed magnitude A ; also suppose that X in the course of its successive changes continually approaches A, but never becomes equal to it, though the difference between the two magnitudes can be made less than any assigned magnitude, however small ; A is then said to be the limit of X. Thus, suppose that X denotes the area of a polygon of n sides inscribed in a circle whose area is A ; if we continually increase the number of sides, X will continually approach A ; also if we assign any magnitude, say one square inch, a polygon with a certain number of sides can be found, whose area will differ from A by less than one square inch ; in like manner if $\frac{1}{10}$, $\frac{1}{100}$, &c., of a square inch had been assigned ; therefore the area of a circle is the limit of the area of the inscribed polygon.

The simplest form which the reasoning on limits can assume is the following :—Suppose it can be proved that two variable quantities X and Y remain equal throughout their variations, and suppose that X continually approaches a limit A, while Y approaches B, then it follows that A must equal B. Thus it can be proved that the area of the regular polygon inscribed in a circle equals the rectangle between the semi-perimeter and the perpendicular let fall from the centre on one side; now the limit of the former is the area of the circle, and of the latter the rectangle between the semi-circumference and semi-diameter, and therefore the area of the circle *equals* that rectangle; not, the reader will observe, nearly equals it, but actually equals it. Prop. 1 supplies a good example of the same form of reasoning.

2. *On ultimate ratios.*—Suppose there are two variable magnitudes x and y whose separate limits are zero; what, it may be asked, is the limit of their ratio $\frac{x}{y}$? The value of this limit depends upon circumstances, and in different cases may have values differing to any extent whatever. Suppose x denotes the sine of an arc, and y the length of that arc, when x continually diminishes y continually diminishes, and their separate limits are zero; it is capable of proof that in this case the limit of $\frac{x}{y}$ is unity; but if x denotes the base and y the hypotenuse of a right-angled triangle, whose dimensions continually diminish in such a manner that the angle (A) between x and y continues unchanged, then although the separate limits of x and y are zero, the limit of $\frac{x}{y}$ is cos A; in the former case x is frequently said to be ultimately equal to y; in the latter, x ultimately equals y cos A.

As this point is of great importance, we will illustrate it by the following case :—Let A P a be a semicircle; take P any point in

Fig. 184.

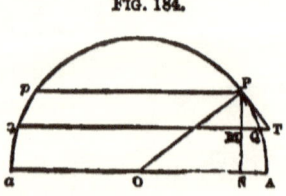

its circumference, join P with the centre O, and draw P N at right angles to A O; take Q a point between A and P, draw Q M q and

APPENDIX. 337

Pp parallel to Aa; let PT be a tangent to the circle at P, and produce MQ to meet PT in T. Now, suppose Q to move along the circumference up to P, then it is plain that the limiting values of PM, PQ, PT, MQ, MT, and QT are separately zero, while Pp is the limiting value of qM, qQ, and qT. Under these circumstances it is commonly stated that PMQ *is ultimately a triangle similar to* OPN; this means that the limit of $\frac{MQ}{PM}$ equals $\frac{PN}{ON}$, from whence it will of course follow that the limit of $\frac{PM}{PQ}$ equals $\frac{ON}{OP}$, and that of $\frac{MQ}{PQ}$ equals $\frac{PN}{OP}$. Now it will be remarked that $\frac{MT}{PM}$ equals $\frac{PN}{ON}$ under all circumstances, and therefore in the limit; so that what we have to prove will be done if we can show that the limit of $\frac{MQ}{PM}$ equals that of $\frac{MT}{PM}$, i.e. equals that of $\frac{MQ}{PM} + \frac{QT}{PM}$, or, in other words, we have to show that the limit of $\frac{QT}{PM}$ is zero. But QT . Tq = PT^2 (Eucl. 36—III.)

$$\therefore \frac{QT}{PM} = \frac{PT}{Tq} \cdot \frac{PT}{PM} = \frac{PT}{Tq}. \text{ Sec. AOP.}$$

Now the limit of PT is zero, while that of Tq is Pp, consequently in the limit the right-hand side of this equation equals zero, and therefore the limit of $\frac{QT}{PM} = 0$. The reader is requested to remark particularly, that not only does QT vanish in the limit, for so also do QM and PQ, but that in the limit *it vanishes in comparison with them*. Hence, if we are reasoning upon the relations that exist between the limits of the ratios of the sides of PQM, we may substitute for them those of PTM, or *vice versâ*, the two being ultimately equal. This is done in Prop. 29.

3. *Quantities of the second and higher orders.*—Suppose there are quantities x, y, z, &c.; such that $y = m\,x^2$, $z = n\,x^3$, &c., then y, z, &c., are said to be of the second, third, &c., orders, x being of the first order. If we have quantities, x_1, x_2, &c., which are severally equal to $p\,x, q\,x$, &c., p, q, &c., being finite quan-

Z

tities, x_1, x_2, &c., are said to be of the first order. Thus in fig. 184, PM, PT, MT, are of the first order, while QT being equal to $PT^2 \div Tq$ is of the second order.

Now, suppose we have an equation of the following kind :—

$$P x + P_1 x_1 + Q y = 0 \qquad (1)$$

and suppose we wish to obtain the relation existing between P, P_1, and Q (which are finite quantities) when x becomes indefinitely small. The equation is plainly equivalent to

$$P + p P_1 + Q m x = 0$$

which when x is indefinitely small becomes

$$P + p P_1 = 0 \qquad (2)$$

It is in many cases convenient to keep the ultimate values of the quantities x and x_1 in the equation instead of the limit of their ratio (p). In this case (1) must be written

$$P x + P_1 x_1 = 0 \qquad (3)$$

It is plain that (3) is equivalent to (2), and thus we obtain the rule that when an equation consists of the sum of quantities of the first, and of higher orders, it is reduced to its ultimate form by striking out all terms but those of the first order. The following are some of the cases in which this mode of reasoning has been employed :—

(a) In the 'equation of virtual velocities' (p. 202), if any one of the quantities (e.g. p_3) is of the second order the term in which it appears is struck out of the equation.

(b) In the lemma on p. 205, let AX be denoted by a, and AOY by θ, so that $\cos AOY$ equals $1 - \tfrac{1}{2}\theta^2 + \ldots$, then we have

$$X m - A n = AX - AX \cos AOX = \tfrac{1}{2} a \theta^2$$

consequently

$$X m - A n = 0$$

unless they are of the second order.

(c) In Prop. 29, it is assumed that the particle describes the arc PQ with a velocity that is ultimately uniform. This can be proved as follows :—

APPENDIX. 339

Let v denote the velocity of the particle at P, s the arc P Q, f the acceleration of the velocity of the particle when at P. Then (Prop. 24)

$$s = v \cdot \delta t + \tfrac{1}{2} f \cdot \delta t^2$$

the ultimate form of this equation is

$$s = v \cdot \delta t$$

or
$$\delta t = \frac{PQ}{\sqrt{2g \cdot PN}} \text{ ultimately.}$$

www.ingramcontent.com/pod-product-compliance
Lightning Source LLC
Chambersburg PA
CBHW030744250426
43672CB00028B/388